SCHAUM'S OUTLINE OF

THEORY AND PROBLEMS

OF

BOOLEAN ALGEBRA

and

SWITCHING CIRCUITS

•

BY

ELLIOTT MENDELSON, Ph.D.

Professor of Mathematics
Queens College
City University of New York

•

SCHAUM'S OUTLINE SERIES
McGRAW-HILL BOOK COMPANY
New York, St. Louis, San Francisco, London, Sydney, Toronto, Mexico, and Panama

ISBN 07-041460-2

6 7 8 9 10 11 12 13 14 15 SH SH 8 7 6 5 4 3

To Julia, Hilary, and Peter

Preface

This book is devoted to two separate, but related, topics: (1) the synthesis and simplification of switching and logic circuits, and (2) the theory of Boolean algebras.

Those people whose primary interest is in switching and logic circuits can read Chapter 4 immediately after a quick perusal of Chapter 1. We have confined our treatment of switching and logic circuits to combinational circuits, i.e. circuits in which the outputs at a given time depend only on the present value of the inputs and not upon the previous values of the inputs. The extensive theory of sequential circuits, in which the outputs depend also upon the history of the inputs, may be pursued by the reader in *Introduction to Switching Theory and Logical Design* by F. J. Hill and G. R. Peterson (reference 34, page 202), *Introduction to Switching and Automata Theory* by M. A. Harrison (ref. 33, page 202), and other textbooks on switching theory.

The treatment of Boolean algebras is somewhat deeper than in most elementary texts. It can serve as an introduction to graduate-level books such as *Boolean Algebras* by R. Sikorski (ref. 148, page 207) and *Lectures on Boolean Algebras* by P. R. Halmos (ref. 116, page 207).

There is no prerequisite for the reading of this book. Each chapter begins with clear statements of pertinent definitions, principles and theorems together with illustrative and other descriptive material. This is followed by graded sets of solved and supplementary problems. The solved problems serve to illustrate and amplify the theory, bring into sharp focus those fine points without which the student continually feels himself on unsafe ground, and provide the repetition of basic principles so vital to effective learning. A few problems which involve modern algebra or point-set topology are clearly labeled. The supplementary problems serve as a complete review of the material of each chapter. Difficult problems are identified by a superscript ^{D} following the problem number.

The extensive bibliography at the end of the book is divided into two parts, the first on Switching Circuits, Logic Circuits and Minimization, and the second on Boolean Algebras and Related Topics. It was designed for browsing. We have listed many articles and books not explicitly referred to in the body of the text in order to give the reader the opportunity to delve further into the literature on his own.

Queens College
July 1970

ELLIOTT MENDELSON

CONTENTS

CONTENTS

Chapter 1

The Algebra of Logic

1.1 TRUTH-FUNCTIONAL OPERATIONS

There are many ways of operating on propositions to form new propositions. We shall limit ourselves to those operations on propositions which are most relevant to mathematics and science, namely, to truth-functional operations. An operation is said to be *truth-functional* if the truth value (truth or falsity) of the resulting proposition is determined by the truth values of the propositions from which it is constructed. The investigation of truth-functional operations is called the *propositional calculus,* or, in old-fashioned terminology, the *algebra of logic,* although its subject matter forms only a small and atypically simple branch of modern mathematical logic.

Negation

Negation is the simplest common example of a truth-functional operation. If **A** is a proposition, then its denial, not-**A**, is true when **A** is false and false when **A** is true. We shall use a special sign $\neg$ to stand for negation. Thus, $\neg$**A** is the proposition which asserts the denial of **A**. The relation between the truth values of $\neg$**A** and **A** can be made explicit by a diagram called a truth table.

A	¬A
T	F
F	T

Fig. 1-1

In this truth table, the column under **A** gives the two possible truth values T (truth) and F (falsity) of **A**. Each entry in the column under $\neg$**A** gives the truth value of $\neg$**A** corresponding to the truth value of **A** in the same row.

Conjunction

Another truth-functional operation about which little discussion is necessary is *conjunction.* We shall use **A** & **B** to stand for the conjunction (**A** and **B**). The truth table for & is

A	B	A & B
T	T	T
F	T	F
T	F	F
F	F	F

Fig. 1-2

There are four possible assignments of truth values to **A** and **B**. Hence there are four rows in the truth table. The only row in which **A** & **B** has the value T is the first row, where each of **A** and **B** is true.

Disjunction

The use of the word "or" in English is ambiguous. Sometimes, "**A** or **B**"† means that at least one of **A** and **B** is true, but that both **A** and **B** may be true. This is the *inclusive* usage of "or". Thus to explain someone's success one might say "he is very smart or he is very lucky", and this clearly does not exclude the possibility that he is both smart and lucky. The inclusive usage of "or" is often rendered in legal documents by the expression "and/or".

Sometimes the word "or" is used in an *exclusive* sense. For example, "Either I will go skating this afternoon or I will stay at home to study this afternoon" clearly means that I will not both go skating and stay home to study this afternoon. Whether the exclusive usage is intended by the speaker or is merely inferred by the listener is often difficult to determine from the sentence itself.

In any case, the ambiguity in usage of the word "or" is something that we cannot allow in a language intended for scientific applications. It is necessary to employ distinct symbols for the different meanings of "or", and it turns out to be more convenient to introduce a special symbol for the inclusive usage, since this occurs more frequently in mathematical assertions.††

"**A** ∨ **B**" shall stand for "**A** or **B** or both". Thus in its truth table (Fig. 1-3) the only case where **A** ∨ **B** is false is the case where both **A** and **B** are false. The expression **A** ∨ **B** will be called a *disjunction* (of **A** and **B**).

A	**B**	**A** ∨ **B**
T	T	T
F	T	T
T	F	T
F	F	F

Fig. 1-3

Conditionals

In mathematics, expressions of the form "If **A** then **B**" occur so often that it is necessary to understand the corresponding truth-functional operation. It is obvious that, when **A** is T and **B** is F, "If **A** then **B**" must be F. But in natural languages (like English) there is no established usage in the other cases (when **A** is F, or when both **A** and **B** are T). In fact when the meanings of **A** and **B** are not related (such as in "If the price of milk is 25¢ per quart, then high tide is at 8:00 P.M. today"), the expression "If **A** then **B**" is not regarded as having any meaning at all.

†Strictly speaking, we should employ quotation marks whenever we are talking about an expression rather than using it. However, this would sometimes get the reader lost in a sea of quotation marks, and we adopt instead the practice of omitting quotation marks whenever misunderstanding is improbable.

††In some natural languages, there are different words for the inclusive and exclusive "or". For example, in Latin, "vel" is used in the inclusive sense, while "aut" is used in the exclusive sense.

Thus if we wish to regard "If **A** then **B**" as truth-functional (i.e. the truth value must be determined by those of **A** and **B**), we shall have to go beyond ordinary usage. To this end we first introduce → as a symbol for the new operation. Thus we shall write "**A** → **B**" instead of "If **A** then **B**". **A** → **B** is called a *conditional* with *antecedent* **A** and *consequent* **B**. The truth table for → contains so far only one entry, in the third row.

A	B	A → B
T	T	
F	T	
T	F	F
F	F	

Fig. 1-4

As a guideline for deciding how to fill in the rest of the truth table, we can turn to "If (**C** & **D**) then **C**", which seems to be a proposition which should always be true. When **C** is T and **D** is F, (**C** & **D**) is F. Thus the second line of our truth table should be T (since (**C** & **D**) is F, **C** is T, and (If (**C** & **D**) then **C**) is T). Likewise when **C** is F and **D** is F, (**C** & **D**) is F. Hence the fourth line should be T. Finally, when **C** is T and **D** is T, (**C** & **D**) is T, and the first line should be T. We arrive at the following truth table:

A	B	A → B
T	T	T
F	T	T
T	F	F
F	F	T

Fig. 1-5

A → **B** is F when and only when **A** is T and **B** is F.

To make the meaning of **A** → **B** somewhat clearer, notice that **A** → **B** and (¬**A**) ∨ **B** always have the same truth value. (Just consider each of the four possible assignments of truth values to **A** and **B**.) Thus the intuitive meaning of **A** → **B** is "not-**A** or **B**". This is precisely the meaning which is given to "If **A** then **B**" in contemporary mathematics.

A proposition **A** → **B** is T whenever **A** is F, irrespective of the truth value of **B**. Notice also that **A** → **B** is automatically T whenever **B** is T, without regard to the truth value of **A**. In these two cases, one sometimes says that **A** → **B** is *trivially true* by virtue of the falsity of **A** or the truth of **B**.

Example 1.1.

The propositions $2+2=5 \rightarrow 1 \neq 1$ and $2+2=5 \rightarrow 1=1$ are both trivially true, since $2+2=5$ is false.

Biconditionals

At this time we shall introduce a special symbol for just one more truth-functional operation: **A** if and only if **B**. Let **A** ↔ **B** stand for "**A** if and only if **B**", where we understand the latter expression to mean that **A** and **B** have the same truth value (i.e. if **A** is T, so is **B**, and vice versa). This gives rise to the truth table:

A	**B**	**A ↔ B**
T	T	T
F	T	F
T	F	F
F	F	T

Fig. 1-6

A proposition of the form **A** ↔ **B** is called a *biconditional*. Notice that **A** ↔ **B** always takes the same truth value as (**A** → **B**) & (**B** → **A**); this is reflected in the mathematical practice of deriving a biconditional **A** ↔ **B** by proving **A** → **B** and **B** → **A** separately.

1.2 CONNECTIVES

Up to this point, we have selected five truth-functional operations and introduced special symbols for them: ¬, &, ∨, →, ↔. Of course if we limit ourselves only to two variables, then there are $2^4 = 16$ different truth-functional operations. With two variables, a truth table has four rows:

A	*B*	
T	T	–
F	T	–
T	F	–
F	F	–

Fig. 1-7

A truth-functional operation can have either T or F in each row. Hence there are $2 \cdot 2 \cdot 2 \cdot 2$ possible binary truth-functional operations.

Corresponding to any truth-functional operation (i.e. to any truth table) we can introduce a special symbol, called a *connective*, to indicate that operation. Thus the symbols ¬, &, ∨, →, ↔ are connectives. These five connectives will suffice for all practical purposes.

Example 1.2.

The operation corresponding to the exclusive usage of "or" could be designated by a connective +, having as its truth table:

A	**B**	**A + B**
T	T	F
F	T	T
T	F	T
F	F	F

Fig. 1-8

1.3 STATEMENT FORMS

To study the properties of truth-functional operations we introduce the following notions.

By a *statement form* (in the connectives ¬, &, ∨, →, ↔) we mean any expression built up from the *statement letters* $A, B, C, \ldots, A_1, B_1, C_1, \ldots$ by a finite number of applications of the connectives ¬, &, ∨, →, ↔. More precisely, an expression is a statement form if it can shown to be one by means of the following two rules:

(1) All statement letters (with or without positive integral subscripts) are statement forms.

(2) If $\mathbf{A}$ and $\mathbf{B}$ are statement forms, so are $(\neg \mathbf{A})$, $(\mathbf{A} \& \mathbf{B})$, $(\mathbf{A} \vee \mathbf{B})$, $(\mathbf{A} \rightarrow \mathbf{B})$, and $(\mathbf{A} \leftrightarrow \mathbf{B})$.†

Example 1.3.

Examples of statement forms:

(i) $(A \rightarrow (B \vee (C \& (\neg A))))$; (ii) $(\neg(A \leftrightarrow (\neg B_2)))$; (iii) $((\neg(\neg A_2)) \rightarrow (A_2 \rightarrow A_1))$.

Clearly we can talk about statement forms in any given set of connectives (instead of just $\neg$, &, $\vee$, $\rightarrow$, $\leftrightarrow$) by using the given connectives in clause (2) of the definition.

1.4 PARENTHESES

The need for parentheses in writing statement forms seems obvious. An expression such as $A \vee B \& C$ might mean either $((A \vee B) \& C)$ or $(A \vee (B \& C))$, and these two statement forms are not, in any sense, equivalent.

While parentheses are necessary, there are many cases in which some parentheses may be conveniently and unambiguously omitted. For that purpose, we adopt the following *conventions for omission of parentheses.*

(1) Every statement form other than a statement letter has an outer pair of parentheses. We may omit this outer pair without any danger of ambiguity. Thus instead of $((A \vee B) \& (\neg C))$, we write $(A \vee B) \& (\neg C)$.

(2) We omit the pair of parentheses around a denial $(\neg \mathbf{A})$. Thus instead of $(\neg A) \vee C$, we write $\neg A \vee C$. This cannot be confused with $\neg(A \vee C)$, since the parentheses will not be dropped from the latter. As another example consider $(A \& B) \vee (\neg(\neg(\neg B)))$. This becomes $(A \& B) \vee \neg\neg\neg B$.

(3) For any binary connective, we adopt the principle of *association to the left.* For example, $\mathbf{A} \& \mathbf{B} \& \mathbf{C}$ will stand for $(\mathbf{A} \& \mathbf{B}) \& \mathbf{C}$, and $\mathbf{A} \rightarrow \mathbf{B} \rightarrow \mathbf{C}$ will stand for $(\mathbf{A} \rightarrow \mathbf{B}) \rightarrow \mathbf{C}$.

Example 1.4.

Applying (1)-(3) above, the statement forms in the column on the left below are reduced to the equivalent expressions on the right.

$((\neg(\neg(A \& C))) \vee (\neg A))$	$\neg\neg(A \& C) \vee \neg A$
$((A \vee (\neg B)) \& (C \& (\neg A)))$	$(A \vee \neg B) \& (C \& \neg A)$
$(((A \vee (\neg B)) \& C) \& (\neg A))$	$(A \vee \neg B) \& C \& \neg A$
$((\neg A) \rightarrow (B \rightarrow (\neg(A \vee C))))$	$\neg A \rightarrow (B \rightarrow \neg(A \vee C))$

More far-reaching conventions for omitting parentheses are presented in Appendix A. In addition, Appendix B contains a method of rewriting statement forms so that no parentheses are required at all.

---

†An even more rigorous definition is: $\mathbf{B}$ is a statement form if and only if there is a finite sequence $\mathbf{A}_1, \ldots, \mathbf{A}_n$ such that

(1) $\mathbf{A}_n$ is $\mathbf{B}$;

(2) if $1 \leq i \leq n$, then either $\mathbf{A}_i$ is a statement letter or there exist $j, k < i$ such that $\mathbf{A}_i$ is $(\neg \mathbf{A}_j)$ or $\mathbf{A}_i$ is $(\mathbf{A}_j \& \mathbf{A}_k)$ or $\mathbf{A}_i$ is $(\mathbf{A}_j \vee \mathbf{A}_k)$ or $\mathbf{A}_i$ is $(\mathbf{A}_j \rightarrow \mathbf{A}_k)$ or $\mathbf{A}_i$ is $(\mathbf{A}_j \leftrightarrow \mathbf{A}_k)$.

1.5 TRUTH TABLES

Every statement form **A** defines a truth-function: for every assignment of truth values to the statement letters in **A**, we can calculate the corresponding truth value of **A** itself. This calculation can be exhibited by means of a *truth table*.

Example 1.5.

The statement form $(\neg A \vee B) \leftrightarrow A$ has the truth table

A	B	$\neg A$	$\neg A \vee B$	$(\neg A \vee B) \leftrightarrow A$
T	T	F	T	T
F	T	T	T	F
T	F	F	F	F
F	F	T	T	F

Fig. 1-9

Each row corresponds to an assignment of truth values to the statement letters. The columns give the corresponding truth values for the statement forms occurring in the step-by-step construction of the given statement form.

Example 1.6.

The statement form $(A \vee (B \& C)) \to B$ has the truth table

A	B	C	$B \& C$	$A \vee (B \& C)$	$(A \vee (B \& C)) \to B$
T	T	T	T	T	T
F	T	T	T	T	T
T	F	T	F	T	F
F	F	T	F	F	T
T	T	F	F	T	T
F	T	F	F	F	T
T	F	F	F	T	F
F	F	F	F	F	T

Fig. 1-10

When there are three statement letters, notice that the truth table has eight rows. In general, when there are n statement letters, there are 2^n rows in the truth table, since there are two possibilities, T or F, for each statement letter.

Abbreviated Truth Tables

By the *principal connective* of a statement form (other than a statement letter), we mean the last connective used in the construction of the statement form. For example, $(A \vee B) \to C$ has $\to$ as its principal connective, $A \vee (B \to C)$ has $\vee$ as its principal connective, and $\neg(A \vee B)$ has $\neg$ as its principal connective.

There is a way of abbreviating truth tables so as to make the computations shorter. We just write down the given statement form once, and, instead of devoting a separate column to each statement form forming a part of the given statement form, we write the truth value of every such part under the principal connective of that part.

Example 1.7.

Abbreviated truth table for $(\neg A \vee B) \leftrightarrow A$. We begin with Fig. 1-11. Notice that each occurrence of a statement letter requires a repetition of the truth assignment for that letter.

($\neg$	A	$\vee$	B)	$\leftrightarrow$	A
	T		T		T
	F		T		F
	T		F		T
	F		F		F

Fig. 1-11

Then the negation is handled:

($\neg$	A	$\vee$	B)	$\leftrightarrow$	A
F	T		T		T
T	F		T		F
F	T		F		T
T	F		F		F

followed by the disjunction

($\neg$	A	$\vee$	B)	$\leftrightarrow$	A
F	T	T	T		T
T	F	T	T		F
F	T	F	F		T
T	F	T	F		F

and, finally, the biconditional

($\neg$	A	$\vee$	B)	$\leftrightarrow$	A
F	T	T	T	T	T
T	F	T	T	F	F
F	T	F	F	F	T
T	F	T	F	F	F

Of course our use of four separate diagrams was only for the sake of illustration. In practice all the work can be carried out in one diagram.

1.6 TAUTOLOGIES AND CONTRADICTIONS

A statement form **A** is said to be a *tautology* if it takes the value T for all assignments of truth values to its statement letters. Clearly **A** is a tautology if and only if the column under **A** in its truth table contains only T's.

Example 1.8. $A \to A$ is a tautology.

A	$A \to A$
T	T
F	T

Fig. 1-12

Example 1.9. $A \vee \neg A$ is a tautology.

A	$\neg A$	$A \vee \neg A$
T	F	T
F	T	T

Fig. 1-13

Example 1.10. $(A \vee B) \leftrightarrow (B \vee A)$ is a tautology.

A	B	$A \vee B$	$B \vee A$	$(A \vee B) \leftrightarrow (B \vee A)$
T	T	T	T	T
F	T	T	T	T
T	F	T	T	T
F	F	F	F	T

Fig. 1-14

Example 1.11. $\underbrace{[A \,\&\, (B \vee C)] \leftrightarrow [(A \,\&\, B) \vee (A \,\&\, C)]}_{\mathbf{A}}$ is a tautology.†

A	B	C	$B \vee C$	$A \,\&\, (B \vee C)$	$A \,\&\, B$	$A \,\&\, C$	$(A \,\&\, B) \vee (A \,\&\, C)$	**A**
T	T	T	T	T	T	T	T	T
F	T	T	T	F	F	F	F	T
T	F	T	T	T	F	T	T	T
F	F	T	T	F	F	F	F	T
T	T	F	T	T	T	F	T	T
F	T	F	T	F	F	F	F	T
T	F	F	F	F	F	F	F	T
F	F	F	F	F	F	F	F	T

Fig. 1-15

Theorem 1.1. If **K** is a tautology, and statement forms **A**, **B**, **C**, ... are substituted for the statement letters $A, B, C, \ldots$ of **K** (the same statement form replacing all occurrences of a statement letter), then the resulting statement form $\mathbf{K}^\#$ is a tautology.

Example 1.12.

$(A \vee B) \leftrightarrow (B \vee A)$ is a tautology. Replace A by $(B \vee C)$ and simultaneously replace B by A. The new statement form $[(B \vee C) \vee A] \leftrightarrow [A \vee (B \vee C)]$ is a tautology.

Proof of Theorem 1.1. **K** determines a truth-function $f(A, B, C, \ldots)$ which always takes the value T no matter what the truth values of $A, B, C, \ldots$ may be. Let $g_1, g_2, g_3, \ldots$ be the truth-functions determined by **A**, **B**, **C**, ... Then the truth-function determined by $\mathbf{K}^\#$ must have the form $f^\# = f(g_1(\ldots), g_2(\ldots), g_3(\ldots), \ldots)$, and, since f always takes the value T, $f^\#$ also always takes the value T. ▶

A *contradiction* is a statement form which always takes the value F. Hence **A** is a contradiction if and only if $\neg$**A** is a tautology, and **A** is a tautology if and only if $\neg$**A** is a contradiction.

Example 1.13. $A \,\&\, \neg A$ is a contradiction.

A	$\neg A$	$A \,\&\, \neg A$
T	F	F
F	T	F

Fig. 1-16

†In writing this statement form, we have replaced some parentheses by brackets to improve legibility. For the same purpose, we also shall use braces.

Example 1.14. $A \leftrightarrow \neg A$ is a contradiction.

A	$\neg A$	$A \leftrightarrow \neg A$
T	F	F
F	T	F

Fig. 1-17

Example 1.15. $(A \vee B)$ & $\neg A$ & $\neg B$ is a contradiction.

A	B	$A \vee B$	$\neg A$	$\neg B$	$(A \vee B)$ & $\neg A$	$(A \vee B)$ & $\neg A$ & $\neg B$
T	T	T	F	F	F	F
F	T	T	T	F	T	F
T	F	T	F	T	F	F
F	F	F	T	T	F	F

Fig. 1-18

1.7 LOGICAL IMPLICATION AND EQUIVALENCE

We say that a statement form **A** *logically implies* a statement form **B** if and only if every assignment of truth values making **A** true also makes **B** true.

Example 1.16. **A** logically implies **A**.

Example 1.17. **A** logically implies **A** $\vee$ **B**. For, whenever **A** is true, **A** $\vee$ **B** also must be true.

Example 1.18. **A** & **B** logically implies **A**.

Theorem 1.2. **A** logically implies **B** if and only if **A** $\rightarrow$ **B** is a tautology.

Proof. **A** logically implies **B** if and only if, whenever **A** is true, **B** must also be true. Therefore **A** logically implies **B** if and only if it is never the case that **A** is true and **B** is false. But the latter assertion means that **A** $\rightarrow$ **B** is never false, i.e. that **A** $\rightarrow$ **B** is a tautology. ▶

Since we can effectively determine by a truth table whether a given statement form is a tautology, Theorem 1.2 provides us with an effective procedure for checking whether **A** logically implies **B**.

Example 1.19. Show that $(A \rightarrow B) \rightarrow A$ logically implies A.

Proof. Fig. 1-19 shows that $((A \rightarrow B) \rightarrow A) \rightarrow A$ is a tautology.

A	B	$A \rightarrow B$	$(A \rightarrow B) \rightarrow A$	$((A \rightarrow B) \rightarrow A) \rightarrow A$
T	T	T	T	T
F	T	T	F	T
T	F	F	T	T
F	F	T	F	T

Fig. 1-19

Statement forms **A** and **B** are called *logically equivalent* if and only if **A** and **B** always take the same truth value for any truth assignment to the statement letters. Clearly this means that **A** and **B** have the same entries in the last column of their truth tables.

Example 1.20. $A \leftrightarrow B$ is logically equivalent to $(A \rightarrow B)\ \&\ (B \rightarrow A)$.

A	B	$A \leftrightarrow B$	$A \rightarrow B$	$B \rightarrow A$	$(A \rightarrow B)\ \&\ (B \rightarrow A)$
T	T	T	T	T	T
F	T	F	T	F	F
T	F	F	F	T	F
F	F	T	T	T	T

Fig. 1-20

Theorem 1.3. $\mathbf{A}$ and $\mathbf{B}$ are logically equivalent if and only if $\mathbf{A} \leftrightarrow \mathbf{B}$ is a tautology.

Proof. $\mathbf{A} \leftrightarrow \mathbf{B}$ is T when and only when $\mathbf{A}$ and $\mathbf{B}$ have the same truth value. Hence $\mathbf{A} \leftrightarrow \mathbf{B}$ is a tautology (i.e. always takes the value T) if and only if $\mathbf{A}$ and $\mathbf{B}$ always have the same truth value (i.e. are logically equivalent). ▶

Example 1.21. $A \rightarrow (B \rightarrow C)$ is logically equivalent to $(A\ \&\ B) \rightarrow C$.

Proof. $[A \rightarrow (B \rightarrow C)] \leftrightarrow [(A\ \&\ B) \rightarrow C]$ is a tautology as shown in Fig. 1-21.

A	B	C	$B \rightarrow C$	$A \rightarrow (B \rightarrow C)$	$A\ \&\ B$	$(A\ \&\ B) \rightarrow C$	$[A \rightarrow (B \rightarrow C)] \leftrightarrow [(A\ \&\ B) \rightarrow C]$
T	T	T	T	T	T	T	T
F	T	T	T	T	F	T	T
T	F	T	T	T	F	T	T
F	F	T	T	T	F	T	T
T	T	F	F	F	T	F	T
F	T	F	F	T	F	T	T
T	F	F	T	T	F	T	T
F	F	F	T	T	F	T	T

Fig. 1-21

Corollary 1.4. If $\mathbf{A}$ and $\mathbf{B}$ are logically equivalent and we replace statement letters in $\mathbf{A}$ and $\mathbf{B}$ by statement forms (all occurrences of the same statement letter being replaced in both $\mathbf{A}$ and $\mathbf{B}$ by the same statement form), then the resulting statement forms are also logically equivalent.

Proof. This is a direct consequence of Theorems 1.3 and 1.1. ▶

Example 1.22.

$A \rightarrow (B \rightarrow C)$ and $(A\ \&\ B) \rightarrow C$ are logically equivalent. Hence so are $(C \vee A) \rightarrow (B \rightarrow (A \vee B))$ and $((C \vee A)\ \&\ B) \rightarrow (A \vee B)$ (and, in general, so are $\mathbf{A} \rightarrow (\mathbf{B} \rightarrow \mathbf{C})$ and $(\mathbf{A}\ \&\ \mathbf{B}) \rightarrow \mathbf{C}$ for any statement forms $\mathbf{A}, \mathbf{B}, \mathbf{C}$).

Theorem 1.5 (Replacement). If $\mathbf{B}$ and $\mathbf{C}$ are logically equivalent and if, within a statement form $\mathbf{A}$, we replace one or more occurrences of $\mathbf{B}$ by $\mathbf{C}$, then the resulting statement form $\mathbf{A}^{\%}$ is logically equivalent to $\mathbf{A}$.

Proof. In the calculation of the truth values of $\mathbf{A}$ and $\mathbf{A}^{\%}$, the distinction between $\mathbf{B}$ and $\mathbf{C}$ is unimportant, since $\mathbf{B}$ and $\mathbf{C}$ always take the same truth value. ▶

Example 1.23.

Let $\mathbf{A}$ be $(A \vee B) \rightarrow C$. Since $A \vee B$ is logically equivalent to $B \vee A$, $\mathbf{A}$ is logically equivalent to $(B \vee A) \rightarrow C$.

The following examples of logically equivalent pairs of statement forms will be extremely useful in the rest of this book, for the purpose of finding, for a given statement form, logically equivalent statement forms which are simpler or have a particularly revealing structure. We leave verification of their logical equivalence as an exercise.

Example 1.24. $\neg\neg \mathbf{A}$ and $\mathbf{A}$ (Law of Double Negation)

Example 1.25. (Idempotence)
(*a*) $\mathbf{A} \,\&\, \mathbf{A}$ and $\mathbf{A}$
(*b*) $\mathbf{A} \vee \mathbf{A}$ and $\mathbf{A}$

Example 1.26. (Commutativity)
(*a*) $\mathbf{A} \,\&\, \mathbf{B}$ and $\mathbf{B} \,\&\, \mathbf{A}$
(*b*) $\mathbf{A} \vee \mathbf{B}$ and $\mathbf{B} \vee \mathbf{A}$

Example 1.27. (Associativity)
(*a*) $(\mathbf{A} \,\&\, \mathbf{B}) \,\&\, \mathbf{C}$ and $\mathbf{A} \,\&\, (\mathbf{B} \,\&\, \mathbf{C})$
(*b*) $(\mathbf{A} \vee \mathbf{B}) \vee \mathbf{C}$ and $\mathbf{A} \vee (\mathbf{B} \vee \mathbf{C})$

As a result of the associative laws, we can leave out parentheses in conjunctions or disjunctions, if we do not distinguish between logically equivalent statement forms. For example, $A \vee B \vee C \vee D$ stands for $((A \vee B) \vee C) \vee D$, but the statement forms $(A \vee (B \vee C)) \vee D$, $A \vee ((B \vee C) \vee D)$, $(A \vee B) \vee (C \vee D)$ and $A \vee (B \vee (C \vee D))$ are logically equivalent to it.

Terminology: In $\mathbf{A}_1 \vee \mathbf{A}_2 \vee \cdots \vee \mathbf{A}_n$, the statement forms $\mathbf{A}_i$ are called *disjuncts,* while in $\mathbf{A}_1 \,\&\, \mathbf{A}_2 \,\&\, \ldots \,\&\, \mathbf{A}_n$ the statement forms $\mathbf{A}_i$ are called *conjuncts*.

Example 1.28. De Morgan's Laws.
(*a*) $\neg(\mathbf{A} \vee \mathbf{B})$ and $\neg\mathbf{A} \,\&\, \neg\mathbf{B}$
(*b*) $\neg(\mathbf{A} \,\&\, \mathbf{B})$ and $\neg\mathbf{A} \vee \neg\mathbf{B}$

Example 1.29. Distributive Laws (or Factoring-out Laws).
(*a*) $\mathbf{A} \,\&\, (\mathbf{B} \vee \mathbf{C})$ and $(\mathbf{A} \,\&\, \mathbf{B}) \vee (\mathbf{A} \,\&\, \mathbf{C})$
(*b*) $\mathbf{A} \vee (\mathbf{B} \,\&\, \mathbf{C})$ and $(\mathbf{A} \vee \mathbf{B}) \,\&\, (\mathbf{A} \vee \mathbf{C})$

Notice that there is a distributive law in arithmetic: $a \cdot (b+c) = (a \cdot b) + (a \cdot c)$; but the other distributive law, $a + (b \cdot c) = (a+b) \cdot (a+c)$ is false. (Take $a = b = c = 1$.)

Example 1.30. Absorption Laws.
(I) (*a*) $\mathbf{A} \vee (\mathbf{A} \,\&\, \mathbf{B})$ and $\mathbf{A}$
(*b*) $\mathbf{A} \,\&\, (\mathbf{A} \vee \mathbf{B})$ and $\mathbf{A}$

(II) (*a*) $(\mathbf{A} \,\&\, \mathbf{B}) \vee \neg\mathbf{B}$ and $\mathbf{A} \vee \neg\mathbf{B}$
(*b*) $(\mathbf{A} \vee \mathbf{B}) \,\&\, \neg\mathbf{B}$ and $\mathbf{A} \,\&\, \neg\mathbf{B}$

(III) If $\mathbf{T}$ is a tautology and $\mathbf{F}$ is a contradiction,
(*a*) $(\mathbf{T} \,\&\, \mathbf{A})$ and $\mathbf{A}$ (*c*) $(\mathbf{F} \,\&\, \mathbf{A})$ and $\mathbf{F}$
(*b*) $(\mathbf{T} \vee \mathbf{A})$ and $\mathbf{T}$ (*d*) $(\mathbf{F} \vee \mathbf{A})$ and $\mathbf{A}$

We shall often have occasion to use the logical equivalence between $(\mathbf{A} \,\&\, \neg\mathbf{B}) \vee \mathbf{B}$ and $\mathbf{A} \vee \mathbf{B}$, and between $(\mathbf{A} \vee \neg\mathbf{B}) \,\&\, \mathbf{B}$ and $\mathbf{A} \,\&\, \mathbf{B}$. We shall justify this by reference to Example 1.30(II), since it amounts to substituting $\neg\mathbf{B}$ for $\mathbf{B}$ in Example 1.30(II) and then using Example 1.24.

Example 1.31. $\mathbf{A} \rightarrow \mathbf{B}$ and $\neg\mathbf{B} \rightarrow \neg\mathbf{A}$ (Contrapositive)

Example 1.32. Elimination of conditionals.
(*a*) $\mathbf{A} \rightarrow \mathbf{B}$ and $\neg\mathbf{A} \vee \mathbf{B}$
(*b*) $\mathbf{A} \rightarrow \mathbf{B}$ and $\neg(\mathbf{A} \,\&\, \neg\mathbf{B})$

Example 1.33. Elimination of biconditionals.

(*a*) $\mathbf{A} \leftrightarrow \mathbf{B}$ and $(\mathbf{A} \,\&\, \mathbf{B}) \vee (\neg\mathbf{A} \,\&\, \neg\mathbf{B})$

(*b*) $\mathbf{A} \leftrightarrow \mathbf{B}$ and $(\neg\mathbf{A} \vee \mathbf{B}) \,\&\, (\neg\mathbf{B} \vee \mathbf{A})$

Examples 1.32 and 1.33 enable us to transform any given statement form into a logically equivalent statement form which contains neither $\rightarrow$ nor $\leftrightarrow$.

1.8 DISJUNCTIVE NORMAL FORM

By *literals* we mean the statement letters $A, B, C, \ldots$ and the denials of statement letters $\neg A, \neg B, \neg C, \ldots$. By a *fundamental conjunction* we mean either (i) a literal or (ii) a conjunction of two or more literals no two of which involve the same statement letter. For instance, A_2, $\neg B$, $A \,\&\, B$, $\neg A_1 \,\&\, A \,\&\, C$ are fundamental conjunctions, while $\neg\neg A$, $A \,\&\, B \,\&\, A$, $B \,\&\, A \,\&\, C \,\&\, \neg B$ are not fundamental conjunctions.

One fundamental conjunction $\mathbf{A}$ is said to be *included* in another $\mathbf{B}$ if all the literals of $\mathbf{A}$ are also literals of $\mathbf{B}$†. For example, $A \,\&\, B$ is included in $A \,\&\, B$, $B \,\&\, (\neg C)$ is included in $(\neg C) \,\&\, B$, B is included in $A \,\&\, B$, and $\neg C \,\&\, A$ is included in $A \,\&\, B \,\&\, \neg C$, while B is not included in $A \,\&\, \neg B$.

A statement form $\mathbf{A}$ is said to be in *disjunctive normal form* (dnf) if either (i) $\mathbf{A}$ is a fundamental conjunction, or (ii) $\mathbf{A}$ is a disjunction of two or more fundamental conjunctions, of which none is included in another.

Example 1.34. The following statement forms are in dnf.

(*a*) B

(*b*) $\neg C \vee C$

(*c*) $A \vee (\neg B \,\&\, C)$

(*d*) $(A \,\&\, \neg B) \vee (\neg A \,\&\, \neg B \,\&\, C)$

(*e*) $(B \,\&\, \neg A) \vee (\neg A \,\&\, \neg B \,\&\, D) \vee A \vee (B \,\&\, C \,\&\, \neg D)$.

Example 1.35. The following statement forms are not in dnf.

(*a*) $C \,\&\, \neg C$

(*b*) $(C \vee D) \,\&\, A$

(*c*) $(C \,\&\, A \,\&\, \neg B) \vee (\neg C \,\&\, A) \vee (C \,\&\, \neg B)$.

Replacing statement forms by logically equivalent ones, we can transform a statement form into one in disjunctive normal form.

Example 1.36. $(A \vee B) \,\&\, (A \vee C \vee \neg B)$

$A \vee (B \,\&\, (C \vee \neg B))$	(Distributive Laws, Example 1.29)
$A \vee (B \,\&\, C)$	(Absorption Law, Example 1.30(II*b*))

Example 1.37. $\neg(A \vee C) \vee (A \rightarrow B)$

$\neg(A \vee C) \vee (\neg A \vee B)$	(Example 1.32(*a*))
$(\neg A \,\&\, \neg C) \vee \neg A \vee B$	(Example 1.28(*a*), De Morgan's Laws)
$\neg A \vee B$	(Example 1.30(I*a*))

†More precisely, if all literals of $\mathbf{A}$ which do not occur within another literal of $\mathbf{A}$ are also literals of $\mathbf{B}$ which do not occur within another literal of $\mathbf{B}$. Thus $B \,\&\, A$ is not included in $C \,\&\, A \,\&\, \neg B$, and $\neg B \,\&\, A$ is not included in $B \,\&\, A$.

Example 1.38. $(A \,\&\, \neg B) \leftrightarrow (B \vee A)$

$[(A \,\&\, \neg B) \,\&\, (B \vee A)] \vee [\neg(A \,\&\, \neg B) \,\&\, \neg(B \vee A)]$ Example 1.33(a)

↓ Examples 1.27(a), 1.30(IIb) ↓ Examples 1.28(b), 1.24 ↓ Example 1.33(a)

$[A \,\&\, \neg B \,\&\, A] \vee [(\neg A \vee B) \,\&\, (\neg B \,\&\, \neg A)]$

↓ Example 1.25 ↓ Example 1.30(IIb)

$[A \,\&\, \neg B] \vee [(\neg A \,\&\, \neg B) \,\&\, \neg A]$

↓ Example 1.25(a)

$[A \,\&\, \neg B] \vee [\neg A \,\&\, \neg B]$

This is in disjunctive normal form. However, it is logically equivalent to

$(A \vee \neg A) \,\&\, \neg B$ (Example 1.29(a))

$\neg B$ (Example 1.30(IIIa))

This example shows that there are two logically equivalent statement forms, both of which are in disjunctive normal form.

The fact illustrated in Examples 1.36-1.38 is codified in the following proposition.

Theorem 1.6. Every statement form which is not a contradiction is logically equivalent to a statement form in disjunctive normal form.

Proof. By Examples 1.32 and 1.33 we may find a logically equivalent statement form in the connectives $\neg$, &, $\vee$, and then, by De Morgan's Laws (Example 1.28) we can move the negation signs inward so that negation signs apply only to statement letters. Thus we may confine our attention to statement forms built up from literals by means of & and $\vee$. The proof proceeds by induction on the number n of the connectives & and $\vee$ in the given statement form $\mathbf{A}$. If $n = 0$, $\mathbf{A}$ is a literal, and every literal is already in dnf. Assume now that $\mathbf{A}$ contains k of the connectives & and $\vee$, and that the theorem is true for all natural numbers $n < k$.

Case 1: $\mathbf{A}$ is $\mathbf{B} \vee \mathbf{C}$. By inductive hypothesis, $\mathbf{B}$ and $\mathbf{C}$ are logically equivalent to statement forms $\mathbf{B}^\#$ and $\mathbf{C}^\#$, respectively, in dnf. Hence $\mathbf{A}$ is logically equivalent to $\mathbf{B}^\# \vee \mathbf{C}^\#$. Now if any disjuncts $\mathbf{D}_1$ of $\mathbf{B}^\#$ or of $\mathbf{C}^\#$ are included in any other disjuncts $\mathbf{D}_2$ of $\mathbf{B}^\#$ or of $\mathbf{C}^\#$, then we drop the disjuncts $\mathbf{D}_2$ (by Example 1.30(Ia)). The resulting statement form is in dnf and is logically equivalent to $\mathbf{A}$.

Case 2: $\mathbf{A}$ is $\mathbf{B} \,\&\, \mathbf{C}$. By inductive hypothesis, $\mathbf{B}$ and $\mathbf{C}$ are logically equivalent to statement forms $\mathbf{B}^\#$ and $\mathbf{C}^\#$, respectively, in dnf. Hence $\mathbf{A}$ is logically equivalent to $\mathbf{B}^\# \,\&\, \mathbf{C}^\#$. Let us assume that $\mathbf{B}^\#$ is $(\mathbf{B}_1 \vee \cdots \vee \mathbf{B}_r)$ and $\mathbf{C}^\#$ is $(\mathbf{C}_1 \vee \cdots \vee \mathbf{C}_s)$, where the $\mathbf{B}_i$'s and $\mathbf{C}_j$'s are fundamental conjunctions, and $r \geq 1$, $s \geq 1$. Then $\mathbf{B}^\# \,\&\, \mathbf{C}^\#$ is

$$(\mathbf{B}_1 \vee \cdots \vee \mathbf{B}_r) \;\&\; (\mathbf{C}_1 \vee \cdots \vee \mathbf{C}_s)$$

which by a Distributive Law (Example 1.29(b)) is logically equivalent to

$$[(\mathbf{B}_1 \vee \cdots \vee \mathbf{B}_r) \;\&\; \mathbf{C}_1] \vee \cdots \vee [(\mathbf{B}_1 \vee \cdots \vee \mathbf{B}_r) \;\&\; \mathbf{C}_s]$$

and, again by a Distributive Law (Example 1.29(a)), each $(\mathbf{B}_1 \vee \cdots \vee \mathbf{B}_r) \,\&\, \mathbf{C}_j$ is logically equivalent to $(\mathbf{B}_1 \,\&\, \mathbf{C}_j) \vee \cdots \vee (\mathbf{B}_r \,\&\, \mathbf{C}_j)$. Thus we obtain the disjunction of all $\mathbf{B}_i \,\&\, \mathbf{C}_j$, where $1 \leq i \leq r$, $1 \leq j \leq s$. Each $\mathbf{B}_i \,\&\, \mathbf{C}_j$ is a conjunction of literals. We can omit repeated literals in $\mathbf{B}_i \,\&\, \mathbf{C}_j$ (by Example 1.25(a)), and, if both a statement letter and its denial occur as conjuncts in $\mathbf{B}_i \,\&\, \mathbf{C}_j$, then the latter is a contradiction and can be dropped (by Example 1.30(IIId)). (Not all the $\mathbf{B}_i \,\&\, \mathbf{C}_j$ will be dropped, since, in that case, $\mathbf{A}$ would be logically equivalent to a disjunction of contradictions and hence, would be a contradiction itself.) The resulting disjunction is in dnf. ▶

Remark (1) on Theorem 1.6. A statement form in dnf cannot be a contradiction. For, if $\mathbf{L}_1 \& \ldots \& \mathbf{L}_k$ is one of its disjunctions, where each $\mathbf{L}_i$ is a literal, then we assign to the statement letter appearing in $\mathbf{L}_i$ the value T if $\mathbf{L}_i$ is the statement letter itself, and the value F if $\mathbf{L}_i$ is the denial of the statement letter. This assignment of truth values makes each $\mathbf{L}_i$ true and hence $\mathbf{L}_1 \& \ldots \& \mathbf{L}_k$ true, and therefore the whole disjunction must be true (since one of its disjuncts is true). Thus the disjunction cannot be a contradiction.

Example 1.39.

In $(A \& \neg B \& C) \vee (\neg A \& \neg B \& C)$, if we make A true, B false and C true, then the first disjunct $A \& \neg B \& C$ is true (and, alternatively, if we make A false, B false and C true, then the second disjunct is true).

Remark (2) on Theorem 1.6. From the proof it is clear that the logically equivalent statement form in dnf may be chosen so that its statement letters already occur in the given statement form, i.e. no new statement letters are introduced.

There is a special type of dnf which will be very useful. A statement form $\mathbf{A}$ in dnf is said to be in *full disjunctive normal form* (with respect to the statement letters $\mathbf{S}_1, \ldots, \mathbf{S}_k$) if

(i) any statement letter in $\mathbf{A}$ is one of the letters $\mathbf{S}_1, \ldots, \mathbf{S}_k$, and

(ii) each disjunct in $\mathbf{A}$ contains all the letters $\mathbf{S}_1, \ldots, \mathbf{S}_k$.

Example 1.40.

The statement forms $(A \& B \& \neg C) \vee (\neg A \& B \& C) \vee (A \& \neg B \& \neg C)$ and $\neg B \& A \& \neg C$ are in full disjunctive normal form (with respect to A, B, C). However, $(A \& B) \vee (\neg A \& B \& C)$ and $\neg A \vee (A \& \neg B \& \neg C)$ are not in full disjunctive normal form with respect to A, B, C.

Example 1.41.

The statement form $\neg B$ is in full dnf with respect to B, but not with respect to A and B. The statement form $(A \& \neg B) \vee (A \& B)$ is in full dnf with respect to A and B, but not with respect to any other collection of letters.

Theorem 1.7. Every non-contradictory statement form $\mathbf{A}$ containing $\mathbf{S}_1, \ldots, \mathbf{S}_k$ as its statement letters is logically equivalent to a statement form in full dnf (with respect to $\mathbf{S}_1, \ldots, \mathbf{S}_k$).

Proof. $\mathbf{A}$ is known by Theorem 1.6 to be logically equivalent to a statement form $\mathbf{B}$ in dnf, and the statement letters of $\mathbf{B}$ already occur in $\mathbf{A}$. Now assume that some statement letter $\mathbf{S}_i$ is missing from a disjunct $\mathbf{D}_j$ of $\mathbf{B}$. However, $\mathbf{D}_j$ is logically equivalent to $\mathbf{D}_j \& (\mathbf{S}_i \vee \neg \mathbf{S}_i)$ (by Example 1.30(III*a*)), which in turn is logically equivalent to $(\mathbf{D}_j \& \mathbf{S}_i) \vee (\mathbf{D}_j \& \neg \mathbf{S}_i)$ (by Example 1.29(*a*)). Hence we replace $\mathbf{D}_j$ by $(\mathbf{D}_j \& \mathbf{S}_i) \vee (\mathbf{D}_j \& \neg \mathbf{S}_i)$. In this way we can introduce the letters $\mathbf{S}_1, \ldots, \mathbf{S}_k$ into any of the disjuncts from which they are missing. The final result is in full dnf with respect to $\mathbf{S}_1, \ldots, \mathbf{S}_k$. ▶

Example 1.42.

$(A \& \neg B) \vee B \vee (\neg A \& \neg B \& \neg C)$ is in dnf, but not in full dnf with respect to A, B, C. We obtain a logically equivalent full dnf as follows:

$$(A \& \neg B \& C) \vee (A \& \neg B \& \neg C) \vee B \vee (\neg A \& \neg B \& \neg C)$$

$$(A \& \neg B \& C) \vee (A \& \neg B \& \neg C) \vee (B \& A) \vee (B \& \neg A) \vee (\neg A \& \neg B \& \neg C)$$

$$(A \& \neg B \& C) \vee (A \& \neg B \& \neg C) \vee (B \& A \& C) \vee (B \& A \& \neg C) \vee (B \& \neg A \& C)$$
$$\vee (B \& \neg A \& \neg C) \vee (\neg A \& \neg B \& \neg C)$$

In general, the method indicated in Theorem 1.7 can be summarized in the following way. If letters $\mathbf{S}_{j_1}, \ldots, \mathbf{S}_{j_r}$ are missing from a disjunct $\mathbf{D}_j$, we add as conjuncts to $\mathbf{D}_j$ all of the 2^r possible combinations of $\mathbf{S}_{j_1}, \ldots, \mathbf{S}_{j_r}$ or their denials. For example, to obtain a statement form in full dnf (with respect to A, B, C) logically equivalent to $\neg B$, we construct

$$(\neg B \,\&\, A \,\&\, C) \vee (\neg B \,\&\, A \,\&\, \neg C) \vee (\neg B \,\&\, \neg A \,\&\, C) \vee (\neg B \,\&\, \neg A \,\&\, \neg C)$$

1.9 ADEQUATE SYSTEMS OF CONNECTIVES

As we have remarked earlier, every statement form determines a truth function, and this truth function may be exhibited by means of a truth table. The converse problem suggests itself: For any given truth function, is there a statement form determining it?

There are $2^{(2^n)}$ truth functions of n variables. For, there are 2^n truth assignments to the n variables, and, to each of these assignments, the truth function can associate the value T or the value F.

Example 1.43.

The four truth functions of one variable are

A	$\neg A$	$A \vee \neg A$	$A \,\&\, \neg A$
T	F	T	F
F	T	T	F

Fig. 1-22

Example 1.44.

The sixteen truth functions of two variables are

A	B	$\neg A$	$\neg B$	$A \vee \neg A$	$A \,\&\, \neg A$	$A \vee B$	$A \,\&\, B$
T	T	F	F	T	F	T	T
F	T	T	F	T	F	T	F
T	F	F	T	T	F	T	F
F	F	T	T	T	F	F	F

$A \to B$	$A \leftrightarrow B$	$\neg(A \leftrightarrow B)$	$B \to A$	$\neg A \,\&\, \neg B$	$\neg A \vee \neg B$	$\neg(B \to A)$	$\neg(A \to B)$
T	T	F	T	F	F	F	F
T	F	T	F	F	T	T	F
F	F	T	T	F	T	F	T
T	T	F	T	T	T	F	F

Fig. 1-23

Theorem 1.8. Every truth function is determined by a statement form in the connectives $\neg$, &, $\vee$.

Proof. The given truth function $f(x_1, \ldots, x_n)$ can be exhibited as a "truth table":

x_1	x_2	...	x_n	$f(x_1, x_2, \ldots, x_n)$
T	T	...	T	.
F	T	...	T	.
T	F	...	T	.
F	F	...	T	.
.	.	...	.	.

Fig. 1-24

There are 2^n rows in the table. In each row, the last column indicates the corresponding value $f(x_1, \ldots, x_n)$. In constructing an appropriate statement form, we shall associate the letters $A_1, \ldots, A_n$ with the variables $x_1, \ldots, x_n$.

Case 1: The last column contains only F's. Then the statement form $(A_1 \& \neg A_1) \vee \cdots \vee (A_n \& \neg A_n)$ determines f. (Of course, any contradiction also determines f.)

Case 2: There are some T's in the last column. For $1 \leq i \leq n$ and $1 \leq k \leq 2^n$, let

$$A_{ik} = \begin{cases} A_i & \text{if } A_i \text{ takes the value T in the } k\text{th row} \\ \neg A_i & \text{if } A_i \text{ takes the value F in the } k\text{th row} \end{cases}$$

Let D_k stand for the fundamental conjunction $A_{1k} \& A_{2k} \& \ldots \& A_{nk}$. In an obvious way, D_k is associated with the kth row of the truth table. For, D_k is T under the truth assignment given in the kth row (where A_i is assigned the value given to x_i), and D_k is F under the truth assignment given in any other row. (Notice that, in any other row, say the jth, some A_i will be assigned a value different from its value in the kth row. Hence under the truth assignment corresponding to the jth row, A_{ik} will receive the value F and hence D_k will also receive the value F.) Now let $k_1, \ldots, k_s$ be the rows in which the truth function f has the value T. Let **A** be the statement form $D_{k_1} \vee \cdots \vee D_{k_s}$. Then **A** determines the truth function f. (For the k_lth row, f takes the value T; but D_{k_l} also is T, and therefore so is **A**. For the jth row, where j is different from any of $k_1, \ldots, k_s$, the function f takes the value F; but each D_{k_l} also is F on the jth row, and hence so is **A**.) Notice that **A** is a statement form in the connectives $\neg$, &, $\vee$. ▶

Remark on Theorem 1.8. If the given truth function is not always F (Case 2), the statement form **A** constructed in the proof is in full disjunctive normal form. This gives us a way of constructing a full dnf logically equivalent to a given non-contradictory statement form **C**. Just write down the truth table for **C** and then construct the corresponding statement form **A** as in the proof of Theorem 1.8.

Example 1.45.

Given the truth function

x_1	x_2	$f(x_1, x_2)$
T	T	F
F	T	T
T	F	T
F	F	T

Fig. 1-25

D_2 is $\neg A_1 \& A_2$, D_3 is $A_1 \& \neg A_2$, D_4 is $\neg A_1 \& \neg A_2$. Hence

$$(\neg A_1 \& A_2) \vee (A_1 \& \neg A_2) \vee (\neg A_1 \& \neg A_2)$$

determines the given truth function.

Example 1.46.

Given the truth function

x_1	x_2	x_3	$g(x_1, x_2, x_3)$
T	T	T	T
F	T	T	T
T	F	T	T
F	F	T	F
T	T	F	F
F	T	F	F
T	F	F	F
F	F	F	T

Fig. 1-26

The statement form having g as its truth function is

$$(A_1 \& A_2 \& A_3) \vee (\neg A_1 \& A_2 \& A_3) \vee (A_1 \& \neg A_2 \& A_3) \vee (\neg A_1 \& \neg A_2 \& \neg A_3)$$

Example 1.47.

To find a statement form in full dnf logically equivalent to $(A \vee B) \& (A \vee C \vee \neg B)$, construct the latter's truth table:

A	B	C	$\neg B$	$A \vee B$	$A \vee C \vee \neg B$	$(A \vee B) \& (A \vee C \vee \neg B)$
T	T	T	F	T	T	T
F	T	T	F	T	T	T
T	F	T	T	T	T	T
F	F	T	T	F	T	F
T	T	F	F	T	T	T
F	T	F	F	T	F	F
T	F	F	T	T	T	T
F	F	F	T	F	T	F

Fig. 1-27

Then the method of Theorem 1.8 yields

$$(A \& B \& C) \vee (\neg A \& B \& C) \vee (A \& \neg B \& C) \vee (A \& B \& \neg C) \vee (A \& \neg B \& \neg C)$$

By an *adequate system of connectives* we mean a collection $\mathcal{B}$ of connectives such that every truth function is determined by a statement form in the connectives of $\mathcal{B}$. Thus Theorem 1.8 asserts that $\{\neg, \&, \vee\}$ is an adequate system of connectives.

Corollary 1.9. Each of the following is an adequate system of connectives:

$$(a)\ \{\neg, \&\}, \qquad (b)\ \{\neg, \vee\}, \qquad (c)\ \{\neg, \rightarrow\}$$

Proof.

(*a*) By Theorem 1.8, $\{\neg, \&, \vee\}$ is adequate. But, replacing any statement form $\mathbf{A} \vee \mathbf{B}$ by the logically equivalent statement form $\neg(\neg\mathbf{A} \& \neg\mathbf{B})$, we obtain for any statement form in $\{\neg, \&, \vee\}$ a logically equivalent statement form in $\{\neg, \&\}$.

(*b*) We proceed as in (*a*), but here we replace every $\mathbf{A} \& \mathbf{B}$ by $\neg(\neg\mathbf{A} \vee \neg\mathbf{B})$.

(*c*) We can replace every $\mathbf{A} \vee \mathbf{B}$ by the logically equivalent $(\neg\mathbf{A}) \rightarrow \mathbf{B}$. ▶

There are two binary connectives such that each of them alone forms an adequate system.

Let | be the connective corresponding to the truth-functional operation of *alternative denial*, given by the truth table

A	B	$A \mid B$
T	T	F
F	T	T
T	F	T
F	F	T

Fig. 1-28

$A \mid B$ means "not both A and B". The connective $\mid$ is called the *Sheffer stroke.* $\{\mid\}$ is adequate, since $\neg \mathbf{A}$ is logically equivalent to $\mathbf{A} \mid \mathbf{A}$, and $\mathbf{A} \vee \mathbf{B}$ is logically equivalent to $(\mathbf{A} \mid \mathbf{A}) \mid (\mathbf{B} \mid \mathbf{B})$.

Let $\downarrow$ be the connective corresponding to the truth-functional operation of *joint denial,* given by the truth table

A	B	$A \downarrow B$
T	T	F
F	T	F
T	F	F
F	F	T

Fig. 1-29

$A \downarrow B$ is read "neither A nor B". $\{\downarrow\}$ is adequate, since $\neg \mathbf{A}$ is logically equivalent to $\mathbf{A} \downarrow \mathbf{A}$, and $\mathbf{A} \,\&\, \mathbf{B}$ is logically equivalent to $(\mathbf{A} \downarrow \mathbf{A}) \downarrow (\mathbf{B} \downarrow \mathbf{B})$.

Theorem 1.10. The only one-element adequate systems of binary connectives are $\{\mid\}$ and $\{\downarrow\}$.

Proof. Let $g(x, y)$ be the truth function of a binary connective # forming an adequate system. Clearly, $g(\mathrm{T}, \mathrm{T}) = \mathrm{F}$. For, if $g(\mathrm{T}, \mathrm{T})$ were T, then any statement form in # alone would always take the value T when its statement letters all took the value T, and no such statement form could determine the negation operation. For the same reason (reversing the roles of T and F), $g(\mathrm{F}, \mathrm{F}) = \mathrm{T}$. The situation at this stage is given by Fig. 1-30.

A	B	$A \# B$
T	T	F
F	T	?
T	F	?
F	F	T

Fig. 1-30

Case 1. The second row is F and the third row is T. Then $\mathbf{A} \# \mathbf{B}$ is logically equivalent to $\neg \mathbf{B}$, and all the statement forms in # alone using the letters A and B would be logically equivalent to one of A, B, $\neg A$, $\neg B$. Then $\{\#\}$ would not be adequate.

Case 2. The second row is T and the third row is F. This is handled in exactly the same way as Case 1, since $\mathbf{A} \# \mathbf{B}$ would be logically equivalent to $\neg \mathbf{A}$.

Case 3. The second and third rows are F. Then # is $\downarrow$.

Case 4. The second and third rows are T. Then # is $\mid$. ▶

Solved Problems

1.1. Reduce the following sentences to statement forms.

(a) A necessary condition for x to be prime is that x is odd or $x = 2$.

(b) A sufficient condition for f to be continuous is that f is differentiable.

(c) A necessary and sufficient condition for Jones to be elected is that Jones wins 75 votes.

(d) Grass will grow only if enough moisture is available.

(e) It is raining but the sun is still shining.

(f) He will die today unless medical aid is obtained.

(g) If taxes are increased or government spending decreases, then inflation will not occur this year.

Solution:

(a) $P \to (O \vee D)$, where P is "x is prime", O is "x is odd", and D is "$x = 2$".

(b) $D \to C$, where D is "f is differentiable" and C is "f is continuous".

(c) $E \leftrightarrow V$, where E is "Jones will be elected" and V is "Jones will win 75 votes".

(d) $G \to M$, where G is "grass will grow", and M is "enough moisture is available".

(e) $R \,\&\, S$, where R is "it is raining", and S is "the sun is still shining".

(Note that "but" indicates conjunction, usually with an element of surprise.)

(f) $\neg D \to M$ (or, equivalently, $\neg M \to D$), where D is "he will die today", and M is "medical aid is obtained".

(g) $T \vee G \to \neg I$, where T is "taxes are increased", G is "government spending decreases", and I is "inflation will occur this year".

1.2. Eliminate as many parentheses as possible from:

(a) $\{[(A \vee B) \to (\neg C)] \vee [(((\neg B) \,\&\, C) \,\&\, B)]\}$

(b) $\{[A \,\&\, (\neg(\neg B))] \leftrightarrow [B \leftrightarrow (C \vee B)]\}$

(c) $[(B \leftrightarrow (C \vee B)) \leftrightarrow (A \,\&\, (\neg(\neg B)))]$

Solution:

(a) $[(A \vee B) \to \neg C] \vee [\neg B \,\&\, C \,\&\, B]$

(b) $[A \,\&\, \neg\neg B] \leftrightarrow [B \leftrightarrow (C \vee B)]$

(c) $B \leftrightarrow (C \vee B) \leftrightarrow (A \,\&\, \neg\neg B)$

1.3. Write the truth tables for (a) $(A \vee \neg B) \to (C \,\&\, A)$, (b) $(A \leftrightarrow \neg B) \vee (B \to A)$.

Solution:

(a)

A	B	C	$\neg B$	$A \vee \neg B$	$C \,\&\, A$	$(A \vee \neg B) \to (C \,\&\, A)$
T	T	T	F	T	T	T
F	T	T	F	F	F	T
T	F	T	T	T	T	T
F	F	T	T	T	F	F
T	T	F	F	T	F	F
F	T	F	F	F	F	T
T	F	F	T	T	F	F
F	F	F	T	T	F	F

(*b*)

A	B	$\neg B$	$A \leftrightarrow \neg B$	$B \rightarrow A$	$(A \leftrightarrow \neg B) \vee (B \rightarrow A)$
T	T	F	F	T	T
F	T	F	T	F	T
T	F	T	T	T	T
F	F	T	F	T	T

1.4. Write abbreviated truth tables for
(*a*) $((A \rightarrow B) \rightarrow B) \vee \neg A$, (*b*) $(\neg A \& \neg B) \rightarrow (B \leftrightarrow C)$.

Solution:

(*a*)

$((A \rightarrow B) \rightarrow B) \vee \neg A$
T T T T T T F T
F T T T T T T F
T F F T F T F T
F T F F F T T F

(*b*)

$(\neg A \& \neg B) \rightarrow (B \leftrightarrow C)$
F T F F T T T T T
T F F F T T T T T
F T F T F T F F T
T F T T F F F F T
F T F F T T T F F
T F F F T T T F F
F T F T F T F T F
T F T T F T F T F

1.5. Show that the following are tautologies.
(*a*) $(A \leftrightarrow (A \& \neg A)) \leftrightarrow \neg A$,
(*b*) $((A \rightarrow B) \rightarrow C) \rightarrow ((C \rightarrow A) \rightarrow (D \rightarrow A))$,
(*c*) $(A \rightarrow B) \rightarrow ((B \rightarrow C) \rightarrow (A \rightarrow C))$.

Solution:

(*a*)

A	$\neg A$	$A \& \neg A$	$A \leftrightarrow (A \& \neg A)$	$(A \leftrightarrow (A \& \neg A)) \leftrightarrow \neg A$
T	F	F	F	T
F	T	F	T	T

(*b*) Instead of using a truth table, we show that the statement form cannot be F. Assume that some assignment makes it F. Then $((A \rightarrow B) \rightarrow C)$ is T while $((C \rightarrow A) \rightarrow (D \rightarrow A))$ is F. Since the latter is F, $C \rightarrow A$ is T but $D \rightarrow A$ is F. Since the latter is F, D is T and A is F. Since $C \rightarrow A$ is T and A is F, C must also be F. Since $((A \rightarrow B) \rightarrow C)$ is T and C is F, it follows that $A \rightarrow B$ is F. But this is impossible, since A is F.

(*c*) As in (*b*), we shall show that the statement form is a tautology by proving that the assumption that it is ever F leads to a contradiction. Assume that some assignment makes it F. Then $A \rightarrow B$ is T, while $(B \rightarrow C) \rightarrow (A \rightarrow C)$ is F. Since the latter is F, $B \rightarrow C$ is T and $A \rightarrow C$ is F. Since the latter is F, A is T and C is F. Since $B \rightarrow C$ is T and C is F, it follows that B is F. Since $A \rightarrow B$ is T and B is F, we know that A is F, contradicting the fact that A is T.

1.6. Show that the following are contradictions.
(*a*) $(A \vee B) \& (A \vee \neg B) \& (\neg A \vee B) \& (\neg A \vee \neg B)$,
(*b*) $[(A \& C) \vee (B \& \neg C)] \leftrightarrow [(\neg A \& C) \vee (\neg B \& \neg C)]$.

Solution:

(*a*) Any truth assignment to A and B makes one of the conjuncts false.

(*b*) Let $\mathbf{A}$ stand for the statement form.

A	B	C	$\neg A$	$\neg B$	$\neg C$	$A \& C$	$B \& \neg C$	$\neg A \& C$
T	T	T	F	F	F	T	F	F
F	T	T	T	F	F	F	F	T
T	F	T	F	T	F	T	F	F
F	F	T	T	T	F	F	F	T
T	T	F	F	F	T	F	T	F
F	T	F	T	F	T	F	T	F
T	F	F	F	T	T	F	F	F
F	F	F	T	T	T	F	F	F

$\neg B \& \neg C$	$(A \& C) \vee (B \& \neg C)$	$(\neg A \& C) \vee (\neg B \& \neg C)$	**A**
F	T	F	F
F	F	T	F
F	T	F	F
F	F	T	F
F	T	F	F
F	T	F	F
T	F	T	F
T	F	T	F

1.7. Prove that if **A** and **A** $\rightarrow$ **B** are tautologies, then so is **B**.

Solution:

More generally, any truth assignment making both **A** and **A** $\rightarrow$ **B** true must also make **B** true. For, if **B** were false, then, since **A** is true, **A** $\rightarrow$ **B** would be false by virtue of the truth table for $\rightarrow$.

1.8. Prove:

(*a*) If **A** logically implies **B** and **B** logically implies **C**, then **A** logically implies **C**.

(*b*) If **A** is logically equivalent to **B** and **B** is logically equivalent to **C**, then **A** is logically equivalent to **C**.

(*c*) **A** is logically equivalent to **A**.

(*d*) **A** logically implies **A**.

(*e*) If **A** is logically equivalent to **B**, then **B** is logically equivalent to **A**.

Solution:

(*a*) To show that **A** logically implies **C**, we shall show that, whenever **A** is T, **C** must also be T. Assume **A** T. Since **A** logically implies **B**, **B** must be T. Since **B** logically implies **C**, **C** must also be T.

(*b*) If **A** and **B** always take the same truth value, and **B** and **C** always take the same truth value, then **A** and **C** always take the same truth value.

(*c*) **A** always takes the same truth value as **A**.

(*d*) follows from (*c*).

(*e*) This is clear from the definition of logical equivalence.

1.9. For each of the following, find a logically equivalent statement form in disjunctive normal form:

(*a*) $(A \vee B) \& (\neg B \vee C)$, (*b*) $\neg A \vee (B \rightarrow \neg C)$.

Solution:

(*a*) $(A \vee B) \& (\neg B \vee C)$

$(A \& \neg B) \vee (A \& C) \vee (B \& \neg B) \vee (B \& C)$

$(A \& \neg B) \vee (A \& C) \vee (B \& C)$

(*b*) $\neg A \vee (B \rightarrow \neg C)$

$\neg A \vee (\neg B \vee \neg C)$

$\neg A \vee \neg B \vee \neg C$

1.10. For each of the following, find a logically equivalent statement form in full disjunctive normal form (with respect to all the variables occurring in the statement form):

(*a*) $(A \,\&\, \neg B) \vee (A \,\&\, C)$ (*c*) $B \rightarrow (A \vee \neg C)$

(*b*) $(A \vee B) \leftrightarrow \neg C$ (*d*) $(A \rightarrow B) \rightarrow ((B \rightarrow C) \rightarrow (A \rightarrow C))$

Solution:

(*a*) $(A \,\&\, \neg B) \vee (A \,\&\, C)$

$(A \,\&\, \neg B \,\&\, C) \vee (A \,\&\, \neg B \,\&\, \neg C) \vee (A \,\&\, B \,\&\, C)$

(*b*) $(A \vee B) \leftrightarrow \neg C$

$((A \vee B) \,\&\, \neg C) \vee (\neg(A \vee B) \,\&\, C)$

$((A \,\&\, \neg C) \vee (B \,\&\, \neg C)) \vee (\neg A \,\&\, \neg B \,\&\, C)$

$(A \,\&\, B \,\&\, \neg C) \vee (A \,\&\, \neg B \,\&\, \neg C) \vee (A \,\&\, B \,\&\, \neg C) \vee (\neg A \,\&\, B \,\&\, \neg C) \vee (\neg A \,\&\, \neg B \,\&\, C)$

$(A \,\&\, B \,\&\, \neg C) \vee (A \,\&\, \neg B \,\&\, \neg C) \vee (\neg A \,\&\, B \,\&\, \neg C) \vee (\neg A \,\&\, \neg B \,\&\, C)$

(*c*) $B \rightarrow (A \vee \neg C)$

$\neg B \vee (A \vee \neg C)$

$\neg B \vee A \vee \neg C$

(*d*) $\neg(A \rightarrow B) \vee ((B \rightarrow C) \rightarrow (A \rightarrow C))$

$\neg\neg(A \,\&\, \neg B) \vee (\neg(B \rightarrow C) \vee (A \rightarrow C))$

$(A \,\&\, \neg B) \vee (\neg\neg(B \,\&\, \neg C) \vee \neg A \vee C)$

$(A \,\&\, \neg B) \vee (B \,\&\, \neg C) \vee \neg A \vee C$

$(A \,\&\, \neg B \,\&\, C) \vee (A \,\&\, \neg B \,\&\, \neg C) \vee (A \,\&\, B \,\&\, \neg C) \vee (\neg A \,\&\, B \,\&\, \neg C)$
$\quad \vee (\neg A \,\&\, B \,\&\, C) \vee (\neg A \,\&\, \neg B \,\&\, C) \vee (\neg A \,\&\, \neg B \,\&\, \neg C) \vee (A \,\&\, B \,\&\, C)$

1.11. Two statement forms **A** and **B** in full dnf (with respect to the same statement letters) are logically equivalent if and only if they are essentially the same (i.e. they contain the same fundamental conjunctions except possibly for a change in the order of the conjuncts in each conjunction).

Solution:

Assume that **A** has as one of its disjuncts a fundamental conjunction $\mathbf{B}_1 \,\&\, \ldots \,\&\, \mathbf{B}_k$ (where each $\mathbf{B}_i$ is a literal), no permutation of the conjuncts of which is a disjunct of **B**. Under the truth assignment which assigns T to a statement letter if it is one of the literals $\mathbf{B}_i$ and assigns F to a statement letter if its denial is one of the literals $\mathbf{B}_i$, $\mathbf{B}_1 \,\&\, \ldots \,\&\, \mathbf{B}_k$ is T, and hence **A** is also. But every other essentially different fundamental conjunction is F, and therefore **B** must be F. Thus **A** and **B** could not be logically equivalent.

1.12. By a *fundamental disjunction* we mean either (i) a literal or (ii) a disjunction of two or more literals no two of which involve the same statement letter. One fundamental disjunction **A** is said to be *included* in another **B** if all the literals of **A** are also literals of **B**. A statement form **A** is in *conjunctive normal form* (cnf) if either (i) **A** is a fundamental disjunction or (ii) **A** is a conjunction of two or more fundamental disjunctions of which none is included in another. A statement form **A** in cnf is said to be in *full cnf* (with respect to the statement letters $\mathbf{S}_1, \ldots, \mathbf{S}_k$) if and only if every conjunct of **A** contains all the letters $\mathbf{S}_1, \ldots, \mathbf{S}_k$.

(*a*) Which of the following are in cnf? Which are in full cnf?

(i) $(A \vee B \vee \neg C) \,\&\, (A \vee \neg B)$ (iii) $(A \vee B) \,\&\, (B \vee \neg B)$

(ii) $(A \vee B \vee \neg C) \,\&\, (A \vee B)$ (iv) $\neg A$

(*b*) The denial of a statement form **A** in (full) dnf is logically equivalent to a statement form **B** in (full) cnf obtained by exchanging & and $\vee$ and by changing each literal to its opposite (i.e. omitting the negation sign if it is present or adding it if it is absent). (Example: $\neg((A \& \neg B \& C) \vee (\neg A \& \neg B \& C))$ is logically equivalent to $(\neg A \vee B \vee \neg C) \& (A \vee B \vee \neg C)$.)

(*c*) Any non-tautologous statement form **A** is logically equivalent to a statement form in full cnf (with respect to all statement letters in **A**).

(*d*) For each of the following, find a logically equivalent statement form in cnf (and one in full cnf).

(i) $(A \rightarrow \neg B) \& (A \vee (B \& C))$,

(ii) $(A \& B) \vee (\neg A \& \neg B)$,

(iii) $A \leftrightarrow (B \vee \neg C)$.

(*e*) Given a truth table for a truth function (not always taking the value T), construct a statement form in full cnf determining the given truth function.

Solution:

(*a*) (i) In cnf, but not in full cnf. (ii) Not in cnf, since one conjunct is included in the other. (iii) Not in cnf, since $B \vee \neg B$ is not a fundamental disjunction. (iv) In full cnf.

(*b*) This follows by several applications of De Morgan's Laws (Example 1.28(*a*)).

(*c*) Assume **A** is non-tautologous. Then $\neg$**A** is not a contradiction, and, by Theorem 1.7, $\neg$**A** is logically equivalent to a statement form in full dnf (with respect to all the statement letters in **A**). Hence by part (*b*), $\neg\neg$**A** is logically equivalent to a statement form in full cnf. But **A** is logically equivalent to $\neg\neg$**A**.

(*d*) (i) $(A \rightarrow \neg B) \& (A \vee (B \& C))$

$(\neg A \vee \neg B) \& ((A \vee B) \& (A \vee C))$

$(\neg A \vee \neg B) \& (A \vee B) \& (A \vee C)$ (cnf)

$(\neg A \vee \neg B \vee C) \& (\neg A \vee \neg B \vee \neg C) \& (A \vee B \vee C) \& (A \vee B \vee \neg C) \& (A \vee \neg B \vee C)$ (full cnf)

(ii) $(A \& B) \vee (\neg A \& \neg B)$

$(A \vee \neg A) \& (A \vee \neg B) \& (B \vee \neg A) \& (B \vee \neg B)$

$(A \vee \neg B) \& (B \vee \neg A)$ (full cnf)

(iii) $A \leftrightarrow (B \vee \neg C)$

$(A \rightarrow (B \vee \neg C)) \& ((B \vee \neg C) \rightarrow A)$

$(\neg A \vee B \vee \neg C) \& (\neg(B \vee \neg C) \vee A)$

$(\neg A \vee B \vee \neg C) \& ((\neg B \& C) \vee A)$

$(\neg A \vee B \vee \neg C) \& (\neg B \vee A) \& (C \vee A)$

$(\neg A \vee B \vee \neg C) \& (A \vee \neg B) \& (A \vee C)$ (cnf)

$(\neg A \vee B \vee \neg C) \& (A \vee \neg B \vee C) \& (A \vee \neg B \vee \neg C) \& (A \vee B \vee C)$ (full cnf)

(*e*) Use the same procedure as in the proof of Theorem 1.8, except that we use only the rows ending in F (rather than T), we exchange & and $\vee$ throughout, and we replace each literal by its opposite.

Example.

A	B	C	
T	T	T	F
F	T	T	T
T	F	T	T
F	F	T	T
T	T	F	F
F	T	F	F
T	F	F	T
F	F	F	T

Answer: $(\neg A \vee \neg B \vee \neg C) \& (\neg A \vee \neg B \vee C) \& (A \vee \neg B \vee C)$

1.13. Find a statement form in $\neg$, &, $\vee$ determining the truth function $f(A, B, C)$:

A	B	C	$f(A, B, C)$
T	T	T	T
F	T	T	F
T	F	T	F
F	F	T	F
T	T	F	F
F	T	F	F
T	F	F	T
F	F	F	F

Solution: $(A \& B \& C) \vee (A \& \neg B \& \neg C)$

1.14. Find a statement form in the Sheffer stroke | alone and one in $\downarrow$ alone logically equivalent to the statement form $A \& \neg B$.

Solution:

For the Sheffer stroke,

$$A \& \neg B$$
$$\neg(\neg A \vee B)$$
$$\neg((A \mid A) \vee B)$$
$$\neg\{[(A \mid A) \mid (A \mid A)] \mid (B \mid B)\}$$
$$([(A \mid A) \mid (A \mid A)] \mid (B \mid B)) \mid ([(A \mid A) \mid (A \mid A)] \mid (B \mid B))$$

For $\downarrow$, $\quad A \& \neg B, \quad A \& (B \downarrow B), \quad (A \downarrow A) \downarrow ((B \downarrow B) \downarrow (B \downarrow B))$

1.15. Show that $\{\rightarrow, \vee\}$ is not an adequate system of connectives.

Solution:

If **A** is a statement form in $\rightarrow, \vee$, then **A** takes the value T when the statement letters are T (since $T \rightarrow T = T$ and $T \vee T = T$). Hence negation is not determined by any statement form in $\rightarrow, \vee$.

1.16. Prove that $\{\neg, \leftrightarrow\}$ is not an adequate system of connectives.

Solution:

The eight truth functions in the following diagram are the only ones determined by statement forms in $\neg, \leftrightarrow$. For, if we apply $\neg$ to any of them or if we apply $\leftrightarrow$ to any two of them we obtain another of them.

A	B	$\neg A$	$\neg B$	$A \leftrightarrow A$	$A \leftrightarrow \neg A$	$A \leftrightarrow B$	$A \leftrightarrow \neg B$
T	T	F	F	T	F	T	F
F	T	T	F	T	F	F	T
T	F	F	T	T	F	F	T
F	F	T	T	T	F	T	F

Alternative solution. We shall show that the truth function determined by a statement form in $\neg, \leftrightarrow$ takes T an even number of times. This is clearly true for statement letters, and, when it holds for **A**, it must hold for $\neg$**A**. It remains to show that, if it holds for **A** and **B**, it also holds for **A** $\leftrightarrow$ **B**. Let n be the number of rows in the truth table. n is even (since n is of the form 2^k, where $k \geq 1$). Let j and l be the number of T's of **A** and **B** respectively. Let m be the number of T's of **A** & **B**, and let s be the number of T's of $\neg$**A** & $\neg$**B**. Then $j + l - m = n - s$; hence $j + l - n = m - s$. Since j, l, n are even, it follows that $m - s$ is even, i.e. m and s have the same parity (both odd or both even). Hence $m + s$ is even. But $m + s$ is the number of T's of **A** $\leftrightarrow$ **B**.

1.17. Determine whether the following arguments are correct by representing the sentences as statement forms and checking to see whether the conjunction of the assumptions logically implies the conclusion.

(*a*) Either Arlen is lying or Brewster was in Mexico in April or Crawford was not a blackmailer. If Brewster was not in Mexico in April, then either Arlen is telling the truth or Crawford was a blackmailer. Hence Brewster must have been in Mexico in April.

(*b*) If the budget is not cut, then a necessary and sufficient condition for prices to remain stable is that taxes will be raised. Taxes will be raised only if the budget is not cut. If prices remain stable, then taxes will not be raised. Hence taxes will not be raised.

Solution:

(*a*) Assumptions: $A \vee B \vee \neg C$, $\neg B \rightarrow (\neg A \vee C)$.

Conclusion: B.

Does $(A \vee B \vee \neg C)$ & $(\neg B \rightarrow (\neg A \vee C))$ logically imply B? In other words, is $[(A \vee B \vee \neg C)$ & $(\neg B \rightarrow (\neg A \vee C))] \rightarrow B$ a tautology? Let us try to find a truth assignment making this statement form false. Then the antecedent $(A \vee B \vee \neg C)$ & $(\neg B \rightarrow (\neg A \vee C))$ must be T and the consequent B must be F. Hence $A \vee B \vee \neg C$ is T and $\neg B \rightarrow (\neg A \vee C)$ is T. Since B is F, $\neg B$ is T; and therefore since $\neg B \rightarrow (\neg A \vee C)$ is T, it follows that $\neg A \vee C$ is T. Since $A \vee B \vee \neg C$ is T but B is F, it follows that $A \vee \neg C$ is T. It is clear now that, if we take A to be T, C to be T, and B to be F, then the statement form is F. Therefore the conclusion is not implied by the premises.

(*b*) Assumptions: $\neg B \rightarrow (P \leftrightarrow R)$, $R \rightarrow \neg B$, $P \rightarrow \neg R$.

Conclusion: $\neg R$.

Does $(\neg B \rightarrow (P \leftrightarrow R))$ & $(R \rightarrow \neg B)$ & $(P \rightarrow \neg R)$ logically imply $\neg R$? Let us try to find a truth assignment making the former true and the latter false. Then $\neg B \rightarrow (P \leftrightarrow R)$ is T, $R \rightarrow \neg B$ is T, and $P \rightarrow \neg R$ is T. Since $\neg R$ is F, R is T. But $R \rightarrow \neg B$ is T, and therefore $\neg B$ is T. Hence by the truth of $\neg B \rightarrow (P \leftrightarrow R)$, $(P \leftrightarrow R)$ is T. Since R is T, P must be T. Then since $P \rightarrow \neg R$ is T, $\neg R$ is T, which is impossible. Therefore the argument is correct.

(*a*) and (*b*) can be solved by writing down the truth tables, but the method used above is usually faster.

1.18. Are the following assumptions consistent? f will be continuous (D) if g is bounded (C) or h is linear (E). g is bounded and h is integrable (B) if and only if h is bounded (A) or f is not continuous. If g is bounded, then h is unbounded. If g is unbounded or h is not integrable, then h is linear and f is not continuous.

Solution:

Assumptions: $(C \vee E) \rightarrow D$, $(C \& B) \leftrightarrow (A \vee \neg D)$, $C \rightarrow \neg A$, $(\neg C \vee \neg B) \rightarrow (E \& \neg D)$. Assume that these are all T.

Case 1. C is T. Then A is F. Since $C \vee E$ is T, D is T. Therefore $A \vee \neg D$ is F. Hence C & B is F. Hence B is F, and $\neg C \vee \neg B$ is T. Thus E & $\neg D$ is T, and D is F, which is impossible.

Case 2. C is F. Thus $\neg C$ is T, and therefore so is $\neg C \vee \neg B$. It follows that E & $\neg D$ is T, and therefore E is T and D is F. Since $C \vee E$ is T, then D is T, which is a contradiction. Hence the assumptions are inconsistent.

This and similar problems can also be solved by writing out the complete truth table (which, in this case, has sixteen rows).

1.19. If $\mathbf{A}$ is a statement form in $\neg$, &, $\vee$, and $\mathbf{A}^*$ results from $\mathbf{A}$ by interchanging & and $\vee$ and replacing all statement letters by their denials, show that $\mathbf{A}^*$ is logically equivalent to $\neg\mathbf{A}$.

Solution:

Apply De Morgan's Laws (Example 1.28) to $\neg\mathbf{A}$ until no denials of conjunctions or disjunctions remain. The result is $\mathbf{A}^*$.

1.20. Let $\mathbf{A}$ and $\mathbf{B}$ be statement forms in $\neg$, &, $\vee$. By the *dual* $\mathbf{A}^d$ of $\mathbf{A}$ we mean the statement form obtained from $\mathbf{A}$ by interchanging & and $\vee$. Notice that $(\mathbf{A}^d)^d = \mathbf{A}$.

(*a*) Show that $\mathbf{A}$ is a tautology if and only if $\neg(\mathbf{A}^d)$ is a tautology.

(*b*) Prove that, if $\mathbf{A} \to \mathbf{B}$ is a tautology (i.e. $\mathbf{A}$ logically implies $\mathbf{B}$), then $\mathbf{B}^d \to \mathbf{A}^d$ is a tautology (i.e. $\mathbf{B}^d$ logically implies $\mathbf{A}^d$). (Example: Since $A \& B \to A$ is a tautology, so is $A \to A \vee B$.)

(*c*) Prove: $\mathbf{A} \leftrightarrow \mathbf{B}$ is a tautology (i.e. $\mathbf{A}$ is logically equivalent to $\mathbf{B}$) if and only if $\mathbf{A}^d \leftrightarrow \mathbf{B}^d$ is also a tautology (i.e. $\mathbf{A}^d$ is logically equivalent to $\mathbf{B}^d$). (Example: Since $\neg(A \vee B)$ is logically equivalent to $\neg A \& \neg B$, it follows that $\neg(A \& B)$ is logically equivalent to $\neg A \vee \neg B$.)

Solution:

(*a*) By Problem 1.19, $\neg(\mathbf{A}^d)$ is logically equivalent to $(\mathbf{A}^d)^*$, where $(\mathbf{A}^d)^*$ is obtained from $\mathbf{A}^d$ by exchanging & and $\vee$, and replacing all their statement letters by their denials. But then $(\mathbf{A}^d)^*$ is obtained from $\mathbf{A}$ by replacing all statement letters by their denials; and therefore if $\mathbf{A}$ is a tautology, so is $(\mathbf{A}^d)^*$ (by Theorem 1.1); and conversely if $(\mathbf{A}^d)^*$ is a tautology, so is $\mathbf{A}$. (In this case we substitute for each statement letter its denial and then again use Theorem 1.1 plus the Law of Double Negation (Example 1.24).)

(*b*) Assume $\mathbf{A} \to \mathbf{B}$ is a tautology. Then $\neg\mathbf{A} \vee \mathbf{B}$ is a tautology, and, by part (*a*), $\neg((\neg\mathbf{A} \vee \mathbf{B})^d)$ is a tautology. But $\neg((\neg\mathbf{A} \vee \mathbf{B})^d)$ is $\neg(\neg\mathbf{A}^d \& \mathbf{B}^d)$, which is logically equivalent to $\mathbf{B}^d \to \mathbf{A}^d$.

(*c*) Assume $\mathbf{A} \leftrightarrow \mathbf{B}$ is a tautology. Then $\mathbf{A} \to \mathbf{B}$ and $\mathbf{B} \to \mathbf{A}$ are tautologies. By part (*b*), $\mathbf{B}^d \to \mathbf{A}^d$ and $\mathbf{A}^d \to \mathbf{B}^d$ are also tautologies. Hence $\mathbf{A}^d \leftrightarrow \mathbf{B}^d$ is a tautology. Conversely, if $\mathbf{A}^d \leftrightarrow \mathbf{B}^d$ is a tautology, then, by what we have just proved, $(\mathbf{A}^d)^d \leftrightarrow (\mathbf{B}^d)^d$ is a tautology. But $(\mathbf{A}^d)^d$ is $\mathbf{A}$, and $(\mathbf{B}^d)^d$ is $\mathbf{B}$.

Supplementary Problems

1.21. Write the following sentences as statement forms.

(*a*) A depression will occur if government spending does not increase, and inflation will result only if government spending increases.

(*b*) Jones will lose his job unless he makes good on the deficit, although Jones is the cousin of the boss's wife.

(*c*) Either f is discontinuous or if f is nonlinear, then f is differentiable.

1.22. Assume that the truth values of $\mathbf{A}$, $\mathbf{B}$, $\mathbf{C}$ are T, F, F. Compute the truth values of (*a*) $(\mathbf{A} \to \neg\mathbf{B}) \leftrightarrow ((\mathbf{A} \vee \mathbf{C}) \& \mathbf{B})$, (*b*) $(\mathbf{A} \leftrightarrow (\mathbf{A} \to \mathbf{B})) \vee (\mathbf{A} \to \mathbf{C})$.

1.23. If $A \to B$ is T, what can be deduced about the truth values of

(a) $(A \vee C) \to (B \vee C)$, (b) $(A \& C) \to (B \& C)$, (c) $(\neg A \& B) \leftrightarrow (A \vee B)$?

1.24. In each of the following cases, what further truth values can be deduced from those already given?

(a) $\underset{\mathrm{F}}{\neg A \vee (A \to B)}$, (b) $\underset{\mathrm{T}}{\neg(A \& B) \leftrightarrow (\neg A \vee \neg B)}$, (c) $\underset{\mathrm{F}}{(\neg A \vee B) \to (A \to \neg C)}$.

1.25. Which of the following are statement forms? For the statement forms, determine the principal connectives.

(a) $(((A \vee (\neg B)) \to A) \& (\neg A))$, (b) $((((A \to B) \to A) \to A) \vee B)$, (c) $(\neg(((A \vee B) \vee C) \leftrightarrow (\neg B)))$

1.26. Eliminate as many parentheses as possible from

(a) $(\{[(A \vee (\neg B)) \vee C] \& [A \vee (\neg(\neg B))]\} \& (\neg A))$

(b) $((\neg(A \vee B)) \vee (C \vee (\neg B)))$

(c) $((A \to (\neg(B \vee C))) \to ((\neg A) \to (\neg B)))$

1.27. Write truth tables and abbreviated truth tables for the statement forms of Exercise 1.26.

1.28. Determine which of the following are tautologies, which are contradictions, and which are neither.

(a) $[(A \to B) \to \neg(B \to A)] \leftrightarrow (A \leftrightarrow B)$

(b) $((A \to B) \to B) \to B$

(c) $[(A \to B) \to (C \to D)] \to [E \to \{(D \to A) \to (C \to A)\}]$

(d) $A \leftrightarrow (B \leftrightarrow (A \leftrightarrow (B \leftrightarrow A)))$

(e) $B \& \neg(A \vee B)$

(f) $(A \vee B \vee C) \leftrightarrow [(((A \to B) \to B) \to C) \to C]$

(g) $((A \to B) \to A) \leftrightarrow (B \to (B \to A))$

(h) $(A \to (B \& \neg B)) \to \neg A$

1.29. Show that **A** is logically equivalent to **B** if and only if **A** logically implies **B** and **B** logically implies **A**.

1.30. Show that a statement form logically equivalent to a tautology is a tautology, and a statement form logically equivalent to a contradiction is a contradiction.

1.31. Give an example to show that, if **A** logically implies **B**, then **B** does not necessarily logically imply **A**.

1.32. Of the following pairs **D** and **E**, find those pairs for which **D** is logically equivalent to **E**, those for which **D** logically implies **E**, and those for which **E** logically implies **D**.

	D	**E**
(a)	$\mathbf{C} \leftrightarrow ((\mathbf{A} \& \mathbf{C}) \vee (\mathbf{B} \& \neg \mathbf{C}))$	$\mathbf{B} \to (\mathbf{C} \to \mathbf{A})$
(b)	$\mathbf{A} \vee (\mathbf{B} \leftrightarrow \mathbf{C})$	$(\mathbf{A} \vee \mathbf{B}) \leftrightarrow (\mathbf{A} \vee \mathbf{C})$
(c)	$\neg(\mathbf{B} \vee \mathbf{C})$	$\neg \mathbf{B}$
(d)	$\mathbf{A} \vee (\mathbf{B} \& \mathbf{C})$	$\mathbf{A} \vee \mathbf{B}$
(e)	$\mathbf{A} \leftrightarrow \mathbf{B}$	$\mathbf{B} \leftrightarrow \mathbf{A}$
(f)	$\mathbf{A} \leftrightarrow (\mathbf{B} \leftrightarrow \mathbf{C})$	$(\mathbf{A} \leftrightarrow \mathbf{B}) \leftrightarrow \mathbf{C}$

1.33. Prove: $\mathbf{A}_1 \vee \mathbf{A}_2 \vee \cdots \vee \mathbf{A}_k$ is T if and only if at least one of the $\mathbf{A}_i$'s is T; and $\mathbf{A}_1 \& \mathbf{A}_2 \& \ldots \& \mathbf{A}_k$ is T if and only if all of the $\mathbf{A}_i$'s are T.

1.34. Prove generalizations of De Morgan's Laws and the Distributive Laws, in the sense that the following pairs are logically equivalent:

(a) $\neg(\mathbf{A}_1 \vee \cdots \vee \mathbf{A}_n)$ $\quad \neg\mathbf{A}_1 \& \ldots \& \neg\mathbf{A}_n$

(b) $\neg(\mathbf{A}_1 \& \ldots \& \mathbf{A}_n)$ $\quad \neg\mathbf{A}_1 \vee \cdots \vee \neg\mathbf{A}_n$

(c) $\mathbf{A} \& (\mathbf{B}_1 \vee \cdots \vee \mathbf{B}_n)$ $\quad (\mathbf{A} \& \mathbf{B}_1) \vee \cdots \vee (\mathbf{A} \& \mathbf{B}_n)$

(d) $\mathbf{A} \vee (\mathbf{B}_1 \& \ldots \& \mathbf{B}_n)$ $\quad (\mathbf{A} \vee \mathbf{B}_1) \& \ldots \& (\mathbf{A} \vee \mathbf{B}_n)$

1.35. Which of the following are fundamental conjunctions?

(a) $A \& B \& \neg A$, (b) $B \& \neg A$, (c) $B \& C \& A \& C$.

1.36. Which of the following are in disjunctive normal form?

(a) $(A \& B) \vee (\neg A \& B) \vee (A \& \neg C)$

(b) $(B \& \neg A) \vee (A \& B) \vee (\neg A \& B)$

(c) $\neg A \& \neg B$

(d) $(A \& B) \vee (\neg A \& B) \vee (A \& B \& C)$

(e) $(A \& B) \vee (A \& \neg B) \vee (\neg A \& B) \vee (\neg A \& \neg B)$

1.37. Find a statement form in disjunctive normal form logically equivalent to:

(a) $(A \rightarrow \neg B) \& (B \vee C)$, (b) $(A \vee \neg B) \leftrightarrow C$, (c) $(A \vee B \vee \neg C) \& (B \vee C)$.

1.38. Which of the following are in full dnf (with respect to all the variables occurring in the statement form)? For those not in full dnf, find a logically equivalent statement form in full dnf.

(a) $(A \& B) \vee (\neg A \& \neg B)$

(b) $(A \& B) \vee (\neg A \& C)$

(c) $(A \& B \& C) \vee (A \& B \& \neg C) \vee (A \& \neg B \& C)$

(d) $(A \& B) \vee C$

1.39. Which of the following are in conjunctive normal form? In full cnf? For any not in full cnf (with respect to all its statement letters), find a logically equivalent statement form in full cnf.

(a) $A \vee \neg B$

(b) $(A \vee B) \& (A \vee \neg B) \& (A \vee B \vee C)$

(c) $A \& \neg B$

(d) $(A \vee B) \& (A \vee \neg B)$

(e) $(A \& B) \vee (\neg A \& \neg B)$

1.40. Which statement forms are in both dnf and cnf?

1.41. For each of the following truth tables (a,) (b), (c), construct a corresponding statement form in full dnf and one in full cnf.

A	B	C	(a)	(b)	(c)
T	T	T	F	F	T
F	T	T	F	T	F
T	F	T	F	T	F
F	F	T	T	T	F
T	T	F	T	F	F
F	T	F	F	F	T
T	F	F	F	F	T
F	F	F	T	F	F

1.42. A statement form $\mathbf{A}$ is a tautology if and only if it is logically equivalent to a statement form in full dnf having 2^n disjuncts, where n is the number of statement letters in $\mathbf{A}$.

1.43. Find a statement form in $\neg, \&, \vee$ logically equivalent to the statement forms (*a*) $(A \leftrightarrow (B \vee C)) \rightarrow \neg A$, (*b*) $((A \rightarrow B) \rightarrow B) \rightarrow A$.

1.44. Find a statement form in the Sheffer stroke | and a statement form in $\downarrow$ logically equivalent to the statement forms (*a*) $(A \vee \neg B) \& (A \rightarrow C)$, (*b*) $\neg(A \& \neg B)$.

1.45. Show that $\mathbf{A} + \mathbf{B}$ is logically equivalent to $\neg(\mathbf{A} \leftrightarrow \mathbf{B})$. (Cf. Example 1.2 on page 4.)

1.46. Find a statement form logically equivalent to the denial of $(A \vee (B \& C)) \vee (\neg A \& (B \vee C))$.

1.47. Show that $\{\leftrightarrow, \rightarrow, \&, \vee\}$ is not an adequate system of connectives.

1.48. Show that $\{\neg, +\}$ is not an adequate system of connectives.

1.49. For which binary connectives $\square$ does $\{\neg, \square\}$ form an adequate system of connectives?

1.50. Determine whether the following arguments are correct.

(*a*) If April is rainy, then flowers will bloom in May and mosquitoes will thrive in June. If mosquitoes thrive in June, then malaria will increase in July. If flowers bloom in May, there will be a lot of honey in September. If April is not rainy, then the lawns will be brown this summer. Hence either there will be a lot of honey in September and malaria will increase in July, or the lawns will be brown this summer.

(*b*) If f is continuous, then g or h is differentiable. If g is not differentiable, then f is not continuous and f is bounded. A sufficient condition for g or h to be differentiable is that f be bounded. Hence g is differentiable.

(*c*) If an orange precipitate forms, then either sodium or potassium is present. If sodium is not present, then iron is present. If iron is present and an orange precipitate forms, then potassium is not present. Hence sodium is present.

1.51. Check the consistency of the following sets of assumptions.

(*a*) If the roof needs repair or the house has to be painted, then either the house will be sold or no vacation will be taken this summer. The house will be sold if and only if the roof needs repair and a vacation will be taken this summer. If the house has to be painted, then the house will not be sold or the roof does not need repair. Either a vacation will be taken this summer, or the house has to be painted and the house will be sold.

(*b*) Either devaluation will occur, or, if exports do not decrease, then price controls will be imposed. If devaluation does not occur, then exports will decrease. If price controls are imposed, then exports will not decrease.

1.52.^{D} A computer (called Farfel) has been built to answer any yes-or-no question, but it has been programmed either to answer all questions truthfully or to give incorrect answers to all questions. If we wish to find out whether Fermat's Last Theorem is true, what question should we put to the computer? (*Hint*: Let A stand for "Fermat's Last Theorem is true" and let B stand for "Farfel answers all questions truthfully". Construct a statement form $\mathbf{A}$ such that, if "$\mathbf{A}$?" is put as a question to Farfel, then the answer will be "Yes" if and only if A is true.)

1.53.^{D} Find the duals of statement forms in $\neg, \&, \vee$ which are logically equivalent to $A \rightarrow B$ and $A \leftrightarrow B$, and extend Problems 1.19-1.20 to statement forms in $\neg, \&, \vee, \rightarrow, \leftrightarrow$.

1.54.^{D} Prove that a statement form $\mathbf{A}$ whose only connective is $\leftrightarrow$ is a tautology if and only if every statement letter occurs in $\mathbf{A}$ an even number of times. (*Hint*: Problem 1.32(*e*, *f*).)

Chapter 2

The Algebra of Sets

2.1 SETS

By a *set* we mean any collection of objects.† For example, we may speak of the set of all living Americans, the set of all letters of the English alphabet, or the set of all real numbers less than 4. In most cases, sets will be defined by means of a characteristic property of the objects belonging to the set. In the examples above, we used the properties of being a living American, a letter of the English alphabet, or a real number less than 4.

Notation: For a given property $P(x)$, let $\{x : P(x)\}$ denote the set of all objects x such that $P(x)$ is true.

Example 2.1.

The set of all real roots of the equation $x^4 - 2x^2 - 3 = 0$ is denoted by

$$\{x : x \text{ is a real number } \& \ x^4 - 2x^2 - 3 = 0\}$$

Sometimes we shall define a set merely by listing its elements within braces: $\{a, b, c, \ldots, h\}$. In particular, $\{b\}$ is the set having b as its only member. Such a set $\{b\}$ is called a *singleton*. The set $\{b, c\}$ contains b and c as its only members, and, if $b \neq c$, then $\{b, c\}$ is called an *unordered pair*. Notice that $\{b, c\} = \{c, b\}$.

Example 2.2.

The set of integers strictly between 1 and 5 is equal to $\{2, 3, 4\}$.

Example 2.3.

The set of all real roots of the equation $x^4 - 2x^2 - 3 = 0$ is equal to the set $\{\sqrt{3}, -\sqrt{3}\}$.

We shall extend this method of denoting sets by listing a few elements of the set, followed by dots, in such a way as to indicate the characteristic property of the elements of the set.

Example 2.4.

$\{1, 2, 3, 4, \ldots\}$ is intended to represent the set of positive integers. $\{1, 4, 9, 16, 25, \ldots, n^2, \ldots\}$ is the set of squares of positive integers. {Washington, Adams, Jefferson, Madison, ...} is the set of Presidents of the United States.

Definition: An object x belonging to a set A is said to be a *member* or *element* of A. We shall write $x \in A$ to indicate that x is a member of A. The denial of $x \in A$ will be written $x \notin A$.

Example 2.5. $6 \in \{x : x \text{ is an even integer}\}$, $1 \notin \{x : x \text{ is an even integer}\}$

---

†Synonyms for *set* are *totality, family,* and *class.*

2.2 EQUALITY AND INCLUSION OF SETS. SUBSETS

The sets A and B are equal when and only when A and B have the same members. Equality of A and B is designated in the usual way by $A = B$, and denial of this equality by $A \neq B$.

Example 2.6. $\{x : x^2 = 1 \text{ and } x \text{ is a real number}\} = \{x : x = 1 \text{ or } x = -1\}$

We say that A is a *subset* of B if and only if every member of A is also a member of B. We write $A \subseteq B$ as an abbreviation for: *A is a subset of B.* Sometimes, instead of saying that A is a subset of B, one says that A is *included* in B. The denial of $A \subseteq B$ is written $A \nsubseteq B$.

Example 2.7. $\{1, 3\} \subseteq \{1, 2, 3, 6\}$; $\{b, a\} \subseteq \{c, a, b\}$; $\{1, 2, 4\} \nsubseteq \{1, 2, 5\}$

Obvious properties of the inclusion relation are

Incl (i) $A \subseteq A$ (Reflexivity).

Incl (ii) If $A \subseteq B$ and $B \subseteq C$, then $A \subseteq C$ (Transitivity).

Incl (iii) $A = B$ if and only if $(A \subseteq B \ \& \ B \subseteq A)$.

It is convenient to introduce a special sign for the relation of *proper inclusion.* We shall use $A \subset B$ as an abbreviation for $A \subseteq B \ \& \ A \neq B$. Thus $A \subset B$ if and only if every member of A is a member of B but there is a member of B which is not a member of A. If $A \subset B$, we say that A is a *proper subset* of B. Hence the only subset of B which is not a proper subset of B is B itself. The denial of $A \subset B$ is written $A \not\subset B$.

Some basic properties of proper inclusion are:

PI(i) $A \not\subset A$.

PI(ii) If $A \subset B \ \& \ B \subseteq C$, then $A \subset C$.

PI(iii) If $A \subseteq B \ \& \ B \subset C$, then $A \subset C$.

PI(iv) If $A \subset B$, then $B \nsubseteq A$.

Example 2.8. $\{1, 3\} \subset \{1, 2, 3\}$; $\{1, 3\} \not\subset \{1, 3\}$; $\{1, 4\} \not\subset \{1, 3\}$

2.3 NULL SET. NUMBER OF SUBSETS

Whenever $P(x)$ is a property satisfied by no objects at all, then $\{x : P(x)\}$ is a set having no members. For example, $\{x : x \neq x\}$ is a set with no members. We shall use $\emptyset$ to denote a set with no members. The set $\emptyset$ is called the *null set* or *empty set.* There is precisely one null set, since any two null sets would contain the same members (namely, none at all) and therefore must be equal. The null set is included in every set: $\emptyset \subseteq A$ for all A.

Example 2.9.

The only subset of $\emptyset$ is $\emptyset$ itself.

Example 2.10.

The subsets of $\{x\}$ are $\emptyset$ and $\{x\}$. Thus a singleton has two subsets.

Example 2.11.

If $x \neq y$, the subsets of the unordered pair $\{x, y\}$ are $\emptyset$, $\{x\}$, $\{y\}$ and $\{x, y\}$. Thus a two-element set has four subsets.

Example 2.12.

If x, y and z are three distinct objects, then the subsets of $\{x, y, z\}$ are $\emptyset$, $\{x\}$, $\{y\}$, $\{z\}$, $\{x, y\}$, $\{x, z\}$, $\{y, z\}$ and $\{x, y, z\}$. Thus there are eight subsets of a three-element set.

Let $\mathcal{P}(A)$ denote the set of all subsets of A. Then $\mathcal{P}(A) = \{B : B \subseteq A\}$. Examples 2.9-2.12 suggest the following result.

Theorem 2.1. For any non-negative integer n, if a set A has n elements, then the set $\mathcal{P}(A)$ of all subsets of A has 2^n elements.

First proof: The result is clear when $n = 0$ (Example 2.9). Assume a set A has n elements, where $n > 0$. In choosing an arbitrary subset C of A, there are two possibilities for each element x of A: $x \in C$ or $x \notin C$. Whether one element x is in the subset C is independent of whether any other element y is in C. Hence there are 2^n ways of choosing a subset of A.

Second proof: By induction on n. The case for $n = 0$ is clear (Example 2.9). Assume that the result is true for $n = k$, and assume that A is a set with $k+1$ elements, i.e. $A = \{a_1, \ldots, a_k, a_{k+1}\}$. We must prove that A has 2^{k+1} subsets. Let $B = \{a_1, \ldots, a_k\}$. Since B has k elements, then by inductive hypothesis B has 2^k subsets. Every subset C of B can be thought of as determining two distinct subsets of A, i.e. C itself and C together with the element a_{k+1}. In addition, every subset D of A is determined in this way by precisely one subset C of B, i.e. C is obtained by removing a_{k+1} from D (where, if $a_{k+1} \notin D$, then C is identical with D). Thus the number of subsets of A is twice the number of subsets of B. But since B has 2^k subsets, A has 2^{k+1} subsets. ▶

2.4 UNION

Given sets A and B, their *union* $A \cup B$ consists of all elements of A or B or both. Thus $A \cup B = \{x : x \in A \vee x \in B\}$. Remember that $\vee$ stands for the inclusive "or", i.e. for any sentences **A**, **B**, **A** $\vee$ **B** means **A** or **B** or both.

Example 2.13.

$$\{1, 2, 3\} \cup \{1, 3, 4, 6\} = \{1, 2, 3, 4, 6\}$$

$$\{a\} \cup \{b\} = \{a, b\}$$

$$\{0, 2, 4, 6, 8, \ldots\} \cup \{1, 3, 5, 7, 9, \ldots\} = \{0, 1, 2, 3, 4, 5, \ldots\}$$

If we represent the elements of A and B by points within two circles, then their union consists of all points lying within either of the two circles (see Fig. 2-1). The union operation on sets has the obvious properties:

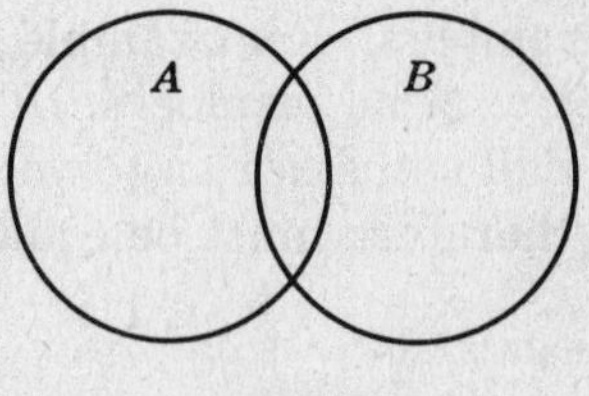

Fig. 2-1

U(i) $A \cup A = A$ (Idempotence)

U(ii) $A \cup B = B \cup A$ (Commutativity)

U(iii) $A \cup \emptyset = A$

U(iv) $(A \cup B) \cup C = A \cup (B \cup C)$ (Associativity)

U(v) $A \cup B = B$ if and only if $A \subseteq B$

U(vi) $A \subseteq A \cup B$ & $B \subseteq A \cup B$

2.5 INTERSECTION

Given sets A and B, their *intersection* $A \cap B$ consists of all objects which are in both A and B. Thus,

$$A \cap B = \{x : x \in A \ \& \ x \in B\}$$

Example 2.14.

$$\{1, 2, 4\} \cap \{1, 3, 4\} = \{1, 4\}$$
$$\{1, 3, 5\} \cap \{2, 4, 6\} = \emptyset$$
$$\{1, 3, 5\} \cap \{0, 2\} = \emptyset$$
$$\{0, 1, 2\} \cap \{0, 3\} = \{0\}$$

Pictorially, we can visualize the intersection as consisting of the shaded area of Fig. 2-2.

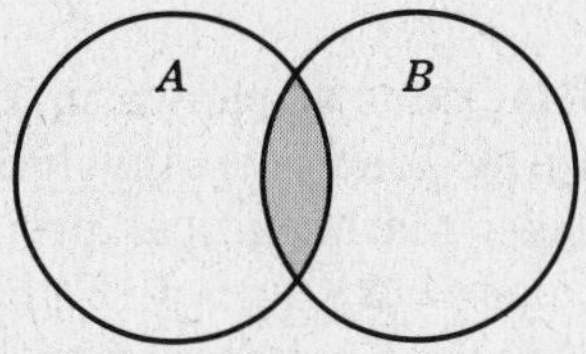

Fig. 2-2

Two sets A and B such that $A \cap B = \emptyset$ are said to be *disjoint*.

The following properties of the intersection operation are evident.

Int (i) $A \cap A = A$ (Idempotence).

Int (ii) $A \cap B = B \cap A$ (Commutativity).

Int (iii) $A \cap \emptyset = \emptyset$.

Int (iv) $(A \cap B) \cap C = A \cap (B \cap C)$ (Associativity).

Int (v) $A \cap B = A$ if and only if $A \subseteq B$.

Int (vi) $A \cap B \subseteq A \ \& \ A \cap B \subseteq B$.

The associative laws for unions and intersections allow us to omit parentheses in writing unions or intersections of three or more sets. Thus we write $A \cap B \cap C$ to stand for either $(A \cap B) \cap C$ or $A \cap (B \cap C)$, since these sets are equal. Similarly $A \cap B \cap C \cap D$ has a unique meaning, since any of the five ways of inserting parentheses yields the same result.

Important relations between unions and intersections are given by the *distributive laws*:

Dist (i) $A \cap (B \cup C) = (A \cap B) \cup (A \cap C)$.

Dist (ii) $A \cup (B \cap C) = (A \cup B) \cap (A \cup C)$.

Property Dist (i) can be verified directly from the definitions by logical manipulations. Thus,

$$\begin{aligned} A \cap (B \cup C) &= \{x : x \in A \ \& \ x \in B \cup C\} \\ &= \{x : x \in A \ \& \ (x \in B \vee x \in C)\} \\ &= \{x : (x \in A \ \& \ x \in B) \vee (x \in A \ \& \ x \in C)\} \\ &= \{x : x \in A \cap B \vee x \in A \cap C\} \\ &= (A \cap B) \cup (A \cap C) \end{aligned}$$

We also can check Dist (i) pictorially. In Fig. 2-3 below, we have vertical lines for $B \cup C$ and horizontal lines for A. Hence $A \cap (B \cup C)$ is represented by the cross-hatched area. In Fig. 2-4 below, the vertical lines indicate $A \cap B$ and the horizontal lines $A \cap C$. The combined area represents $(A \cap B) \cup (A \cap C)$ and is seen to be identical with the cross-hatched area of Fig. 2-3. Dist (ii) may be handled in a similar manner (see Problem 2.3).

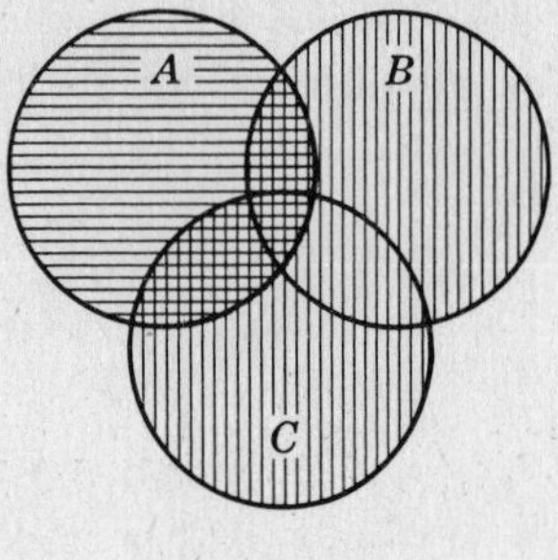

Fig. 2-3

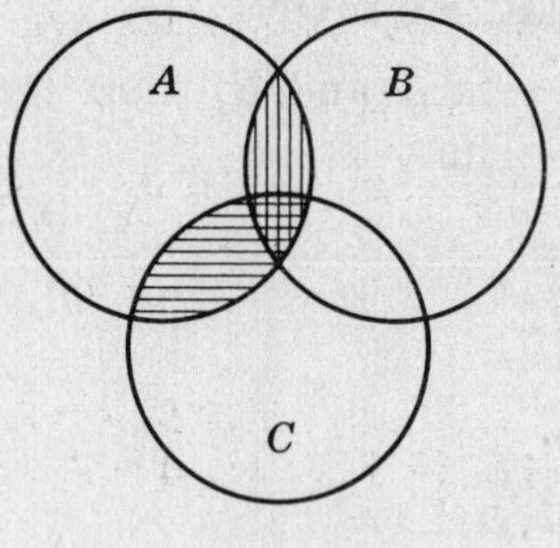

Fig. 2-4

Diagrams as shown in Fig. 2-3 and 2-4 are called *Venn diagrams.* They are useful for verifying identities involving operations on sets, but should not be considered tools of rigorous mathematical proof. Similar pictorial methods can be given for four or more sets (see [112]† and J. F. Randolph, *American Mathematical Monthly,* 1965, pp. 117-127), but this does not seem fruitful enough to warrant consideration here.

Example 2.15. Show that $A \cap (A \cup B) = A$.

By U(vi), $A \subseteq A \cup B$. Hence, by Int (v), $A \cap (A \cup B) = A$.

Example 2.16. Show that $A \cup (A \cap B) = A$.

$$A \cup (A \cap B) = (A \cup A) \cap (A \cup B) = A \cap (A \cup B) = A$$

The first equality is justified by Dist (ii), the second by U(i), and the third by Example 2.15.

The distributive laws have the obvious generalizations:

Dist (i′) $\quad A \cap (B_1 \cup B_2 \cup \cdots \cup B_n) = (A \cap B_1) \cup (A \cap B_2) \cup \cdots \cup (A \cap B_n)$

Dist (ii′) $\quad A \cup (B_1 \cap B_2 \cap \cdots \cap B_n) = (A \cup B_1) \cap (A \cup B_2) \cap \cdots \cap (A \cup B_n)$

These can be proved directly, using mathematical induction.

2.6 DIFFERENCE AND SYMMETRIC DIFFERENCE

By the *difference* $B \sim A$ of B and A we mean the set of all those objects in B which are not in A (the shaded area of Fig. 2-5). Thus,

$$B \sim A = \{x : x \in B \ \& \ x \notin A\}$$

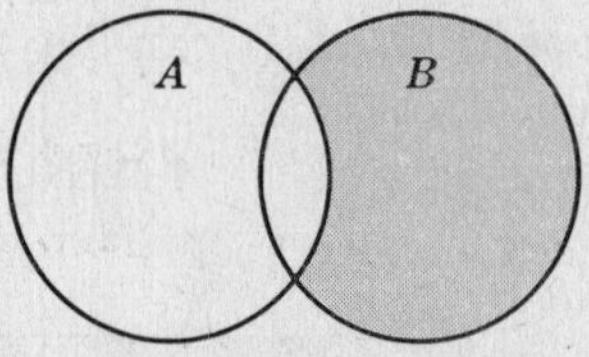

Fig. 2-5

Clearly,

D(i) $\quad B \sim B = \emptyset$

D(ii) $\quad B \sim \emptyset = B$

D(iii) $\quad \emptyset \sim A = \emptyset$

D(iv) $\quad (A \sim B) \sim C = A \sim (B \cup C)$
$\qquad\qquad\qquad = (A \sim C) \sim B$

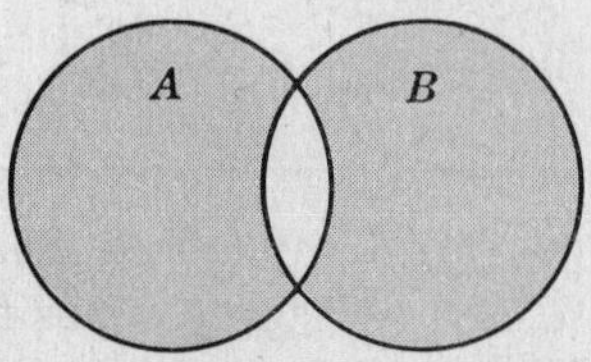

Fig. 2-6

The *symmetric difference* $A \Delta B$ of sets A and B is $(A \sim B) \cup (B \sim A)$ (the shaded area of Fig. 2-6). Fig. 2-6 makes it clear that $A \Delta B = (A \cup B) \sim (A \cap B)$.

†Throughout this book numbers in brackets refer to Bibliography, pages 201-208.

The following properties are easily verified.

SD(i) $A \triangle A = \emptyset$

SD(ii) $A \triangle B = B \triangle A$

SD(iii) $A \triangle \emptyset = A$

Example 2.17.

Let $A = \{0,1,2,3,5\}$, $B = \{0,1,2,3\}$, $C = \{0,1,4,5\}$. Then $B \subset A$, $C \not\subset A$, $A \cap C = \{0,1,5\}$, $B \cap C = \{0,1\}$, $A \sim B = \{5\}$, $A \sim C = \{2,3\}$, $B \sim C = \{2,3\}$, $C \sim B = \{4,5\}$, $A \triangle B = \{5\}$, $A \triangle C = \{2,3,4\}$, $B \triangle C = \{2,3,4,5\}$.

2.7 UNIVERSAL SET. COMPLEMENT

We shall often find it useful to confine our attention to the subsets of some given set X. In such a case, X is called the *universal set* or the *universe* (of discourse).

The union, intersection, difference, and symmetric difference of subsets of X are again subsets of X. When we restrict ourselves to subsets of X, there is still another operation which can be introduced. If $A \subseteq X$, then the *complement* $\bar{A}$ of A is defined to be $X \sim A$. Thus, $\bar{A} = \{x : x \in X \ \&\ x \notin A\}$. *Whenever we use complements, it is assumed that we are dealing only with subsets of some fixed universe* X.

The following assertions are easily verified.

C(i) $\bar{\bar{A}} = A$

C(ii) $\overline{A \cup B} = \bar{A} \cap \bar{B}$ } De Morgan's Laws

C(iii) $\overline{A \cap B} = \bar{A} \cup \bar{B}$ }

C(iv) $A \cap \bar{A} = \emptyset$

C(v) $A \cup \bar{A} = X$

C(vi) $\bar{\emptyset} = X$

C(vii) $\bar{X} = \emptyset$

C(viii) $A \subseteq B$ if and only if $\bar{B} \subseteq \bar{A}$

C(ix) $A = B$ if and only if $\bar{A} = \bar{B}$

C(x) $A \sim B = A \cap \bar{B}$

C(xi) $A \triangle B = (A \cap \bar{B}) \cup (\bar{A} \cap B)$

From C(x) and C(xi) we see that difference and symmetric difference are dispensable in the presence of union, intersection and complement.

Example 2.18.

Let us check C(ii) using definitions and logical transformations.

$$\begin{aligned} \overline{A \cup B} &= \{x : x \in X \ \&\ x \notin A \cup B\} = \{x : x \in X \ \&\ \neg(x \in A \vee x \in B)\} \\ &= \{x : x \in X \ \&\ (x \notin A \ \&\ x \notin B)\} = \{x : (x \in X \ \&\ x \notin A) \ \&\ (x \in X \ \&\ x \notin B)\} \\ &= \{x : x \in X \ \&\ x \notin A\} \cap \{x : x \in X \ \&\ x \notin B\} = \bar{A} \cap \bar{B} \end{aligned}$$

We also may use Venn diagrams to verify the validity of C(ii). Compare Fig. 2-7 and 2-8.

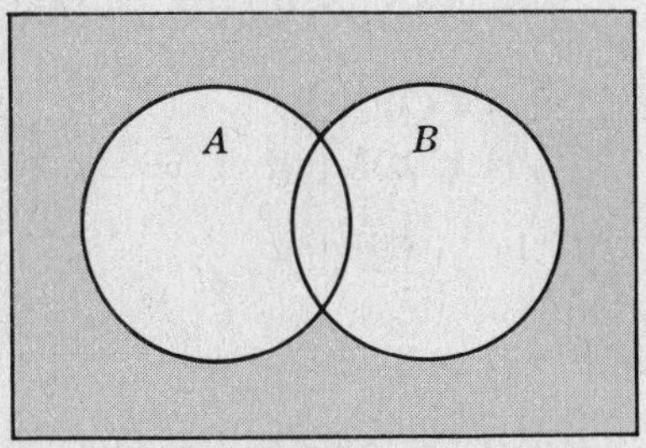

$\overline{A \cup B}$ is the shaded area.

Fig. 2-7

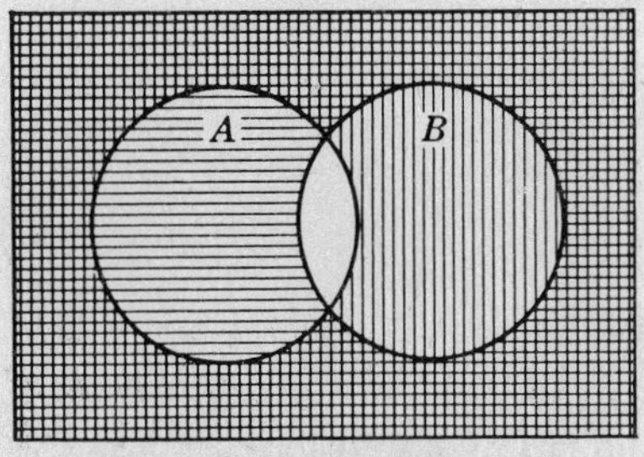

$\bar{A} \cap \bar{B}$ is the cross-hatched area.

Fig. 2-8

De Morgan's Laws C(ii)-C(iii) have the obvious generalizations:

C(ii′) $\overline{A_1 \cup A_2 \cup \cdots \cup A_n} = \bar{A}_1 \cap \bar{A}_2 \cap \cdots \cap \bar{A}_n$

C(iii′) $\overline{A_1 \cap A_2 \cap \cdots \cap A_n} = \bar{A}_1 \cup \bar{A}_2 \cup \cdots \cup \bar{A}_n$

Example 2.19.

$A \subseteq B$ if and only if $A \cap \bar{B} = \emptyset$. In Fig. 2-9, the cross-hatched area is $A \cap \bar{B}$. To say that this is $\emptyset$ is equivalent to saying that A is entirely within B.

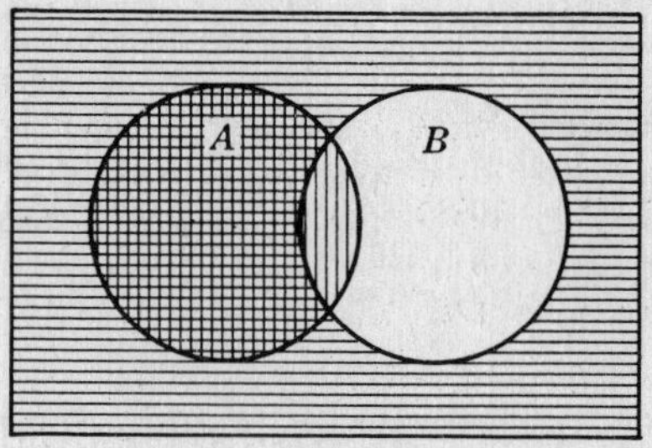

Fig. 2-9

Logical proof:

$$A = A \cap X = A \cap (B \cup \bar{B}) = (A \cap B) \cup (A \cap \bar{B})$$

Hence if $A \cap \bar{B} = \emptyset$, then $A = A \cap B$; therefore, by Int (v), $A \subseteq B$. On the other hand, if $A \subseteq B$, then by Int (v), $A = A \cap B$ and therefore

$$A \cap \bar{B} = (A \cap B) \cap \bar{B} = A \cap (B \cap \bar{B}) = A \cap \emptyset = \emptyset$$

2.8 DERIVATIONS OF RELATIONS AMONG SETS

We have seen two ways of verifying propositions about sets: by means of analogous logical laws, or by pictorial methods (usually Venn diagrams). The first method is the only rigorous one, but the use of diagrams is sometimes quicker and more intuitive.

Example 2.20.

Prove $A \sim (B \cup C) = (A \sim B) \cap (A \sim C)$.

This is clear from Figs. 2-10 and 2-11.

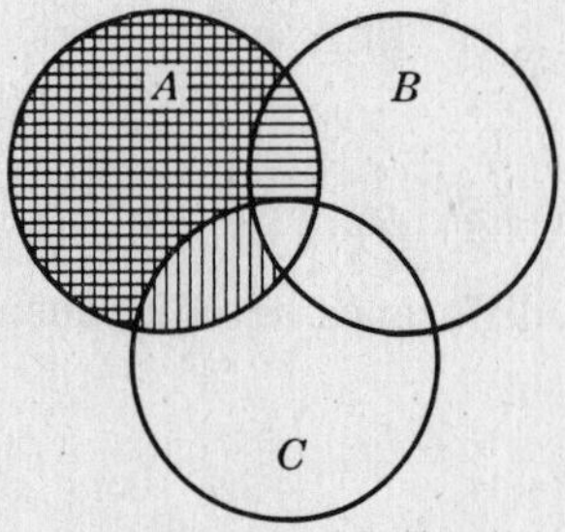

Cross-hatched: $(A \sim B) \cap (A \sim C)$

Fig. 2-10

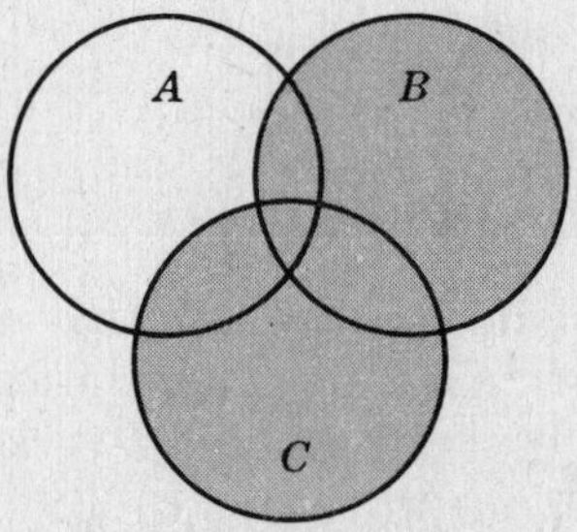

Unshaded: $A \sim (B \cup C)$

Fig. 2-11

More rigorously,

$$\begin{aligned} A \sim (B \cup C) &= \{x : x \in A \ \& \ x \notin (B \cup C)\} \\ &= \{x : x \in A \ \& \ (x \notin B \ \& \ x \notin C)\} \\ &= \{x : (x \in A \ \& \ x \notin B) \ \& \ (x \in A \ \& \ x \notin C)\} \\ &= \{x : x \in A \ \& \ x \notin B\} \cap \{x : x \in A \ \& \ x \notin C\} \\ &= (A \sim B) \cap (A \sim C) \end{aligned}$$

Example 2.21.

Prove: $(A \cup B) \cap \bar{B} = A$ if and only if $A \cap B = \emptyset$.

In Fig. 2-12 below, the cross-hatched part represents $(A \cup B) \cap \bar{B}$ and lies entirely within A. The rest of A is the lens-shaped intersection $A \cap B$. Hence to say that $(A \cup B) \cap \bar{B}$ is identical with A is equivalent to saying that $A \cap B = \emptyset$.

Logical proof:

$$(A \cup B) \cap \bar{B} = (A \cap \bar{B}) \cup (B \cap \bar{B}) \quad \text{(by Dist (ii))}$$
$$= (A \cap \bar{B}) \cup \emptyset \quad \text{(by C(iv))}$$
$$= A \cap \bar{B} \quad \text{(by U(iii))}$$

Hence $(A \cup B) \cap \bar{B} = A$ if and only if $A \cap \bar{B} = A$. But $A \cap \bar{B} = A$ if and only if $A \subseteq \bar{B}$ (Int (v)), which holds if and only if $A \cap B = \emptyset$ (by Example 2.19 and C(i)).

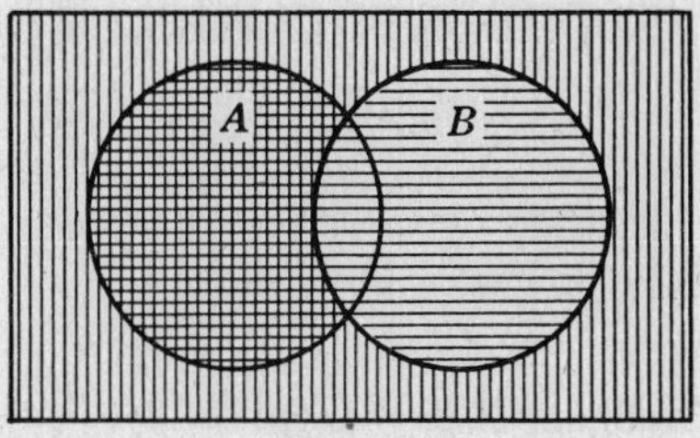

Fig. 2-12

Example 2.22.

Simplify $\overline{A \cap \bar{B}} \cup (B \cap C)$.

$$\overline{A \cap \bar{B}} \cup (B \cap C) = (\bar{A} \cup \bar{\bar{B}}) \cup (B \cap C) \quad \text{(by C(iii))}$$
$$= (\bar{A} \cup B) \cup (B \cap C) \quad \text{(by C(i))}$$
$$= \bar{A} \cup (B \cup (B \cap C)) \quad \text{(by U(iv))}$$
$$= \bar{A} \cup B \quad \text{(by Example 2.16)}$$

In simplifying expressions, we make frequent use of De Morgan's Laws C(ii) and C(iii) for distributing complement bars over smaller expressions, C(i) for eliminating double complements, Examples 2.15 and 2.16, and the distributive laws Dist (i) and Dist (ii).

Example 2.23.

Simplify $(A \cup B \cup C) \cap (\overline{A \cap \bar{B} \cap \bar{C}}) \cap \bar{C}$.

$$(A \cup B \cup C) \cap (\overline{A \cap \bar{B} \cap \bar{C}}) \cap \bar{C} = (A \cup B \cup C) \cap (\bar{A} \cup B \cup C) \cap \bar{C} \quad \text{(De Morgan)}$$
$$= [(A \cup B \cup C) \cap (\bar{A} \cup B \cup C)] \cap \bar{C} \quad \text{(Associativity of } \cap)$$
$$= [(A \cap \bar{A}) \cup (B \cup C)] \cap \bar{C} \quad \text{(Dist (ii))}$$
$$= [B \cup C] \cap \bar{C} = (B \cap \bar{C}) \cup (C \cap \bar{C}) = (B \cap \bar{C}) \cup \emptyset \quad \text{(C(iv), U(iii))}$$
$$= B \cap \bar{C}$$

2.9 PROPOSITIONAL LOGIC AND THE ALGEBRA OF SETS

Every truth-functional operation determines a corresponding operation on sets. For example, denial determines complementation: $\bar{A} = \{x : \neg(x \in A)\}$; conjunction determines the intersection operation: $A \cap B = \{x : x \in A \;\&\; x \in B\}$; and disjunction determines the union operation: $A \cup B = \{x : x \in A \vee x \in B\}$. In general, if # is a connective corresponding to a truth function $f(x_1, \ldots, x_n)$, then we define a corresponding operation @ on sets by $@(A_1, \ldots, A_n) = \{x : \#(x \in A_1, \ldots, x \in A_n)\}$. Thus the set-theoretic operation of symmetric difference corresponds to the exclusive usage of "or".

Example 2.24.

The operation of alternative denial determines the set-theoretic operation $\overline{A \cap B}$, while joint denial determines the operation $\bar{A} \cap \bar{B}$.

In general, a uniform way of determining the set-theoretic operation corresponding to a given truth function is to express the latter in terms of $\neg$, &, $\vee$, and then replace $\neg$, &, $\vee$ by $^{-}$, $\cap$, $\cup$ respectively. The statement letters need not be replaced since they can serve as set variables in the new expression.

2.10 ORDERED PAIRS. FUNCTIONS

If $x \neq y$, then $\{x, y\}$ was called the *unordered pair* of x and y. We say "unordered" because $\{x, y\} = \{y, x\}$. Let us define an *ordered pair* $\langle x, y\rangle$, which is determined by x and y, in that order. By this we mean that the following proposition holds: if $\langle x, y\rangle = \langle u, v\rangle$, then $x = u$ and $y = v$.

Theorem 2.2. $\langle x, y\rangle = \{\{x\}, \{x, y\}\}$ is an adequate definition of an ordered pair.

Proof. Assume $\langle x, y\rangle = \langle u, v\rangle$. We must prove that $x = u$ and $y = v$. We have

$$\{\{x\}, \{x, y\}\} = \{\{u\}, \{u, v\}\} \tag{2.1}$$

Since $\{x\}$ is a member of the left side of equation (*2.1*), it must also be a member of the right side. Hence

$$\{x\} = \{u\} \quad \text{or} \quad \{x\} = \{u, v\}$$

Therefore $x = u$ or $x = u = v$. In either case, $x = u$. Now by (*2.1*),

$$\{x, y\} = \{u\} \quad \text{or} \quad \{x, y\} = \{u, v\}$$

If $\{x, y\} = \{u, v\}$, then $\{x, y\} = \{x, v\}$ since $x = u$. Hence $y = x$ or $y = v$. If $y = x$, then $\{y\} = \{y, v\}$ and $y = v$. In all cases, $y = v$. If $\{x, y\} \neq \{u, v\}$, then $\{x, y\} = \{u\}$ and so $x = y = u$. By (*2.1*),

$$\{u, v\} = \{x\} \quad \text{or} \quad \{u, v\} = \{x, y\}$$

Since $\{u, v\} \neq \{x, y\}$, $\{u, v\} = \{x\}$ and so $u = v = x$. Therefore $y = v$. ▶

Let us recall the definition of a function. A *function* f from A into B is a way of associating an element of B to each element of A. The phrase "way of associating" may be replaced by a more precise notion:

(1) f is a set of ordered pairs such that, if $\langle x, y\rangle \in f$ and $\langle x, z\rangle \in f$, then $y = z$;

(2) for every x in A there exists some y in B such that $\langle x, y\rangle \in f$. (Such an object y must be unique, by virtue of (1); it is denoted in the standard way by $f(x)$.)

We say that f is a function from A *onto* B if f is a function from A into B and every element of B is a value $f(x)$ for some x in A.

Example 2.25.

The function f such that $f(x) = x^2$ for every x in the set A of all integers is a function from A into (but not onto) A. On the other hand, f is a function from A onto the set B of all squares.

A function f is said to be *one-one* if it assigns different values to different arguments, i.e. $f(x) = f(y)$ implies $x = y$.

Example 2.26.

The function f in Example 2.25 is not one-one, since $f(-n) = n^2 = f(n)$ for all integers n. On the other hand, the function g such that $g(x) = x^2$ for all non-negative integers x is a one-one function, since $u^2 = v^2$ implies $u = v$ for all non-negative integers u and v.

A one-one function from A onto B is called a *one-one correspondence* between A and B. For example, the function h such that $h(x) = x + 1$ for all odd integers x is a one-one correspondence between the set of all odd integers and the set of all even integers.

2.11 FINITE, INFINITE, DENUMERABLE, AND COUNTABLE SETS

A *finite* set is a set which is either empty or can be enumerated by the positive integers from 1 up to some integer n. More precisely, A is finite if there is a positive integer n such that there is a one-one correspondence between A and the set of all positive integers less than n. (When $n = 1$, A must be the empty set.)

For example, to justify the assertion that the set of fingers on a hand is finite we set up the correspondence

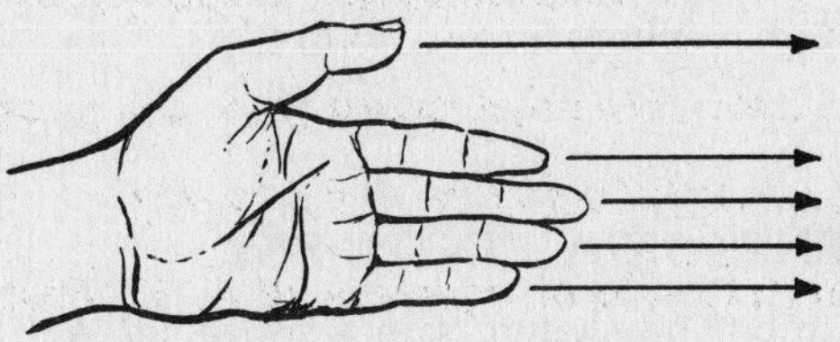

It is clear that a subset of a finite set is finite (and hence that the intersection of any set with a finite set is finite). Also obvious is the fact that the union of two finite sets is finite.

A set is said to be *infinite* if it is not finite. Examples are the set of positive integers, the set of rational numbers, and the set of real numbers. Clearly any set containing an infinite set must also be infinite, and therefore the union of an infinite set with any other set is infinite. However, the intersection of two infinite sets need not be infinite. For example, the set of even integers and the set of odd integers have an empty intersection.

A set A is said to be *denumerable* (or *countably infinite*) if and only if A can be enumerated by the set P of all positive integers, i.e. if there is a one-one correspondence between P and A.

Example 2.27.

(1) The set of positive even integers is denumerable. Here the one-one correspondence is given by $f(n) = 2n$. (2) The set of all integers is denumerable. Here the enumeration is given by $0, 1, -1, 2, -2, 3, -3, \ldots$. The one-one correspondence is $g(n) = \begin{cases} n/2 & \text{if } n \text{ is even} \\ -(n-1)/2 & \text{if } n \text{ is odd} \end{cases}$.

Clearly the union of a finite set and a denumerable set is also denumerable. (Just enumerate the finite set first and continue with the enumeration of the denumerable set, omitting repetitions.) The union of two denumerable sets is again denumerable. (For, if $A = \{a_1, a_2, \ldots\}$ and $B = \{b_1, b_2, \ldots\}$, then $A \cup B = \{a_1, b_1, a_2, b_2, \ldots\}$, where in the latter enumeration we omit any repeated objects.) If we remove a finite number of elements from a denumerable set, the remaining set is still denumerable.

A set is said to be *countable* if and only if it is either finite or denumerable. Obviously, any subset B of a countable set A is also countable. (For, in an enumeration of A, we omit all objects which are not in B. The resulting enumeration of B does or does not terminate. If it does, B is finite. If it does not, B is denumerable.) The union of two countable sets is a countable set. This follows from what has been said above about finite and denumerable sets.

2.12 FIELDS OF SETS

By a *field of sets on X* we mean a non-empty collection $\mathcal{F}$ of subsets of X such that, for any members A and B in $\mathcal{F}$, the sets $A \cup B$, $A \cap B$, and $\bar{A}$ are also in $\mathcal{F}$. Another way of expressing this is to say that $\mathcal{F}$ is closed under the operations of union, intersection and complementation. Since $A \cup B = \overline{\bar{A} \cap \bar{B}}$ and $A \cap B = \overline{\bar{A} \cup \bar{B}}$, it suffices to verify closure under complementation and either union or intersection.

Examples of fields of sets are:

(1) the set $\mathcal{P}(X)$ of all subsets of X;

(2) the set of all finite subsets of X and their complements;

(3) $\{\emptyset, X\}$.

Notice that any field $\mathcal{F}$ of subsets of X must contain both $\emptyset$ and X. For, if $B \in \mathcal{F}$, then $\bar{B} \in \mathcal{F}$ and hence $\emptyset = B \cap \bar{B} \in \mathcal{F}$. Therefore $X = \bar{\emptyset} \in \mathcal{F}$.

2.13 NUMBER OF ELEMENTS IN A FINITE SET

Let $\#(A)$ stand for the number of elements in a finite set A. Clearly

$$\#(A_1 \cup A_2) = \#(A_1) + \#(A_2) - \#(A_1 \cap A_2)$$

For three sets, we have

$$\begin{aligned} \#(A_1 \cup A_2 \cup A_3) = {} & \#(A_1) + \#(A_2) + \#(A_3) \\ & - \#(A_1 \cap A_2) - \#(A_1 \cap A_3) - \#(A_2 \cap A_3) \\ & + \#(A_1 \cap A_2 \cap A_3) \end{aligned}$$

and, for four sets,

$$\begin{aligned} \#(A_1 \cup A_2 \cup A_3 \cup A_4) = {} & \#(A_1) + \#(A_2) + \#(A_3) + \#(A_4) \\ & - \#(A_1 \cap A_2) - \#(A_1 \cap A_3) - \#(A_1 \cap A_4) - \#(A_2 \cap A_3) - \#(A_2 \cap A_4) - \#(A_3 \cap A_4) \\ & + \#(A_1 \cap A_2 \cap A_3) + \#(A_1 \cap A_2 \cap A_4) + \#(A_1 \cap A_3 \cap A_4) + \#(A_2 \cap A_3 \cap A_4) \\ & - \#(A_1 \cap A_2 \cap A_3 \cap A_4) \end{aligned}$$

The general formula for n sets should be clear from the examples for $n = 2, 3, 4$.

Example 2.28.

In a two-party election district consisting of 135 voters, 67 people voted for at least one Democrat and 84 people voted for at least one Republican. How many people voted for candidates of both parties?

$$\#(R \cap D) = \#(R) + \#(D) - \#(R \cup D) = 84 + 67 - 135 = 16$$

Here R is the set of people who voted for at least one Republican and D the set of people who voted for at least one Democratic candidate. Hence $R \cup D$ is the set of all voters and $R \cap D$ is the set of all people who split their ballots.

Example 2.29.

A government committee reported that, among the students using marijuana, LSD or heroin at a certain university, 90% used marijuana, 6% used LSD and 7% heroin, while 4% took marijuana and LSD, 5% marijuana and heroin, 2% heroin and LSD, and 1% took all three. Are the committee's figures consistent?

Note that, if there are n students taking at least one of the drugs, and if H is a set of students, then the percentage in H is $\#(H)/n$. Hence if we let A, B, C be the sets of students taking marijuana, LSD and heroin respectively, and we divide the equation for $\#(A \cup B \cup C)$ by n to obtain the percentages,

$$\%(A \cup B \cup C) = \%(A) + \%(B) + \%(C) - \%(A \cap B) - \%(A \cap C) - \%(B \cap C) + \%(A \cap B \cap C)$$

$$100 = 90 + 6 + 7 - 4 - 5 - 2 + 1 = 93$$

which is impossible. Hence the figures are not consistent.

Solved Problems

2.1. Show that the cancellation law

$$\text{if} \quad A \cup B = A \cup C \quad \text{then} \quad B = C$$

is false by giving a counterexample.

Solution:

$A = C = \{a\}$. $B = \emptyset$.

2.2. Show that parentheses are necessary for writing expressions involving more than one of the operations $\cap$ and $\cup$.

Solution:

Consider $A \cap B \cup C$. This is either $A \cap (B \cup C)$ or $(A \cap B) \cup C$. But these two sets are not necessarily equal. Take $A = \emptyset$ and $B = C \neq \emptyset$. Then $A \cap (B \cup C) = \emptyset$, but $(A \cap B) \cup C = \emptyset \cup C = C$.

2.3. Prove the distributive law Dist (ii), page 33: $A \cup (B \cap C) = (A \cup B) \cap (A \cup C)$.

Solution:

Logical Proof.

$$\begin{aligned} A \cup (B \cap C) &= \{x : x \in A \lor x \in (B \cap C)\} \\ &= \{x : x \in A \lor (x \in B \ \& \ x \in C)\} = \{x : (x \in A \lor x \in B) \ \& \ (x \in A \lor x \in C)\} \\ &= \{x : x \in A \lor x \in B\} \cap \{x : x \in A \lor x \in C\} = (A \cup B) \cap (A \cup C) \end{aligned}$$

Pictorial Proof. In Fig. 2-13, the vertical lines indicate $B \cap C$ and the shaded area is A. In Fig. 2-14, the vertical lines indicate $A \cup B$, the horizontal lines $A \cup C$, and the cross-hatched area $(A \cup B) \cap (A \cup C)$ is identical with the marked area of Fig. 2-13.

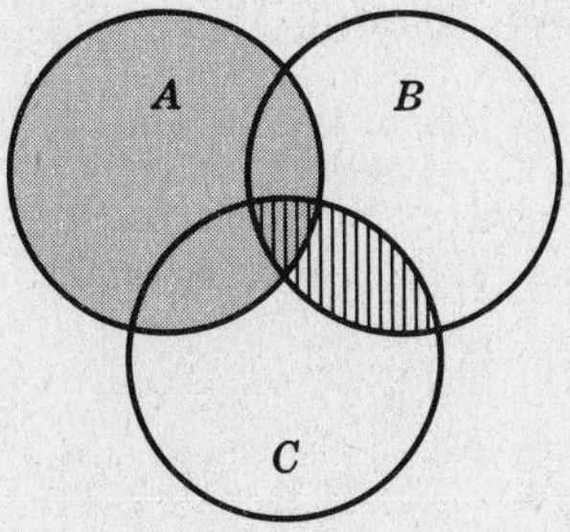

Fig. 2-13

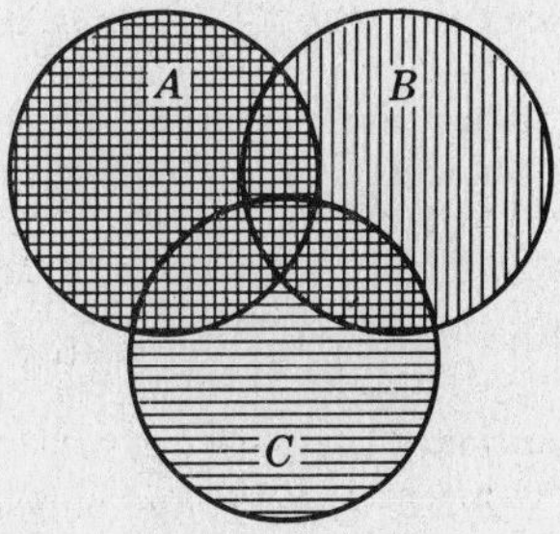

Fig. 2-14

2.4. Prove the generalized distributive law Dist (ii′), page 34:

$$A \cup (B_1 \cap \cdots \cap B_n) = (A \cup B_1) \cap \cdots \cap (A \cup B_n)$$

Solution:

For $n = 1$, the assertion is obvious and the case $n = 2$ is the distributive law Dist (ii). Now using mathematical induction, we assume the result true for $n = k$. Then for $n = k + 1$,

$$\begin{aligned} A \cup (B_1 \cap \cdots \cap B_k \cap B_{k+1}) &= A \cup ((B_1 \cap \cdots \cap B_k) \cap B_{k+1}) \\ &= [A \cup (B_1 \cap \cdots \cap B_k)] \cap [A \cup B_{k+1}] && \text{(by Dist(ii))} \\ &= [(A \cup B_1) \cap \cdots \cap (A \cup B_k)] \cap (A \cup B_{k+1}) && \text{(by the inductive hypothesis)} \\ &= (A \cup B_1) \cap \cdots \cap (A \cup B_k) \cap (A \cup B_{k+1}) \end{aligned}$$

DIFFERENCE AND SYMMETRIC DIFFERENCE

2.5. Using a Venn diagram, determine whether the following conditions are compatible.

(i) $A \cap B = \emptyset$ (iii) $(C \cap A) \sim B = \emptyset$

(ii) $(C \cap B) \sim A = \emptyset$ (iv) $(C \cap A) \cup (C \cap B) \cup (A \cap B) \neq \emptyset$

Solution:

In Fig. 2-15, (iv) says that $E \cup F \cup G \cup H$ is non-empty. (i) says that $E \cup F$ is empty. Hence $G \cup H$ is non-empty. (iii) says that G is empty and (ii) says that H is empty. Hence $G \cup H$ is empty. Therefore conditions (i)-(iv) are inconsistent.

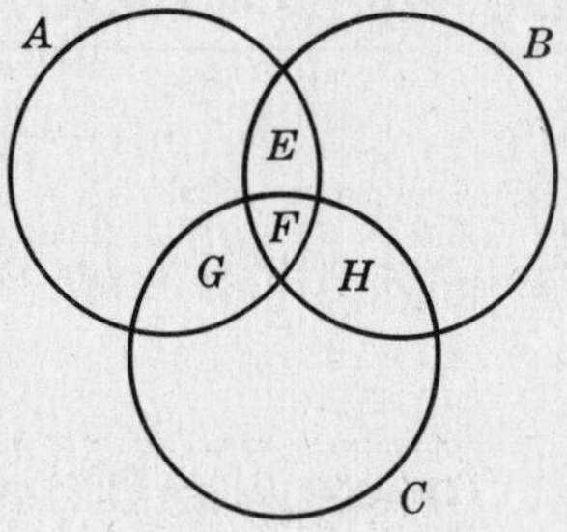

Fig. 2-15

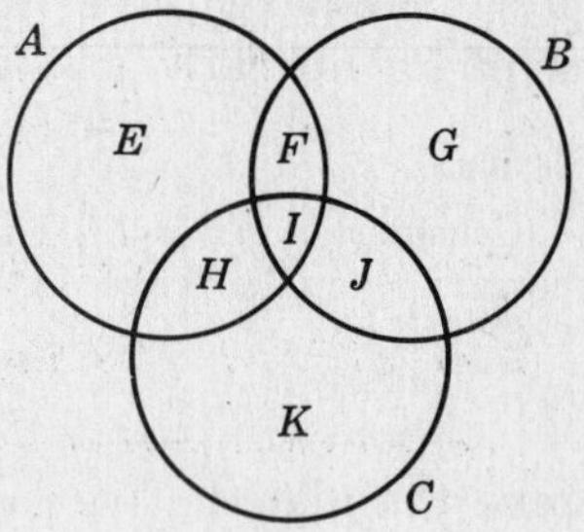

Fig. 2-16

2.6. Show that $A \,\Delta\, (B \,\Delta\, C) = (A \,\Delta\, B) \,\Delta\, C$.

Solution:

In Fig. 2-16, $A \,\Delta\, B = E \cup H \cup G \cup J$ and $C = H \cup I \cup J \cup K$, and so $(A \,\Delta\, B) \,\Delta\, C = E \cup G \cup I \cup K$. $B \,\Delta\, C = F \cup G \cup H \cup K$ and $A = E \cup F \cup H \cup I$, and so $A \,\Delta\, (B \,\Delta\, C) = G \cup K \cup I \cup E$. Thus $(A \,\Delta\, B) \,\Delta\, C = A \,\Delta\, (B \,\Delta\, C)$.

Observe that

$$(A \,\Delta\, B) \,\Delta\, C = \overbrace{(A \cap \bar{B} \cap \bar{C})}^{E} \cup \overbrace{(\bar{A} \cap B \cap \bar{C})}^{G} \cup \overbrace{(\bar{A} \cap \bar{B} \cap C)}^{K} \cup \overbrace{(A \cap B \cap C)}^{I}$$

A logical derivation of this result is rather tedious and is left to the reader. It is easiest to prove by showing $(A \,\Delta\, B) \,\Delta\, C \subseteq A \,\Delta\, (B \,\Delta\, C)$ and $A \,\Delta\, (B \,\Delta\, C) \subseteq (A \,\Delta\, B) \,\Delta\, C$.

2.7. Show that $A \,\Delta\, B = \emptyset$ if and only if $A = B$.

Solution:

$A \,\Delta\, B = \emptyset$ if and only if $(A \sim B) \cup (B \sim A) = \emptyset$,

if and only if $A \sim B = \emptyset$ and $B \sim A = \emptyset$,

if and only if $A \subseteq B$ and $B \subseteq A$,

if and only if $A = B$.

Note: By C(xi), page 35, this result can be restated as

$$A = B \text{ if and only if } (A \cap \bar{B}) \cup (\bar{A} \cap B) = \emptyset$$

2.8. Prove the cancellation law: If $A \,\Delta\, B = A \,\Delta\, C$, then $B = C$.

Solution:

Assume $A \,\Delta\, B = A \,\Delta\, C$. Then $A \,\Delta\, A \,\Delta\, B = A \,\Delta\, A \,\Delta\, C$ (parentheses can be omitted by virtue of Problem 2.6). Since $A \,\Delta\, A = \emptyset$, we obtain: $\emptyset \,\Delta\, B = \emptyset \,\Delta\, C$. But $\emptyset \,\Delta\, D = D$ for any D. Hence $B = C$.

2.9. Prove the distributive law: $A \cap (B \Delta C) = (A \cap B) \Delta (A \cap C)$.

Solution:

$$\begin{aligned} A \cap (B \Delta C) &= A \cap ((B \sim C) \cup (C \sim B)) \\ &= (A \cap (B \sim C)) \cup (A \cap (C \sim B)) \;=\; ((A \cap B) \sim C) \cup ((A \cap C) \sim B) \\ &= ((A \cap B) \sim (A \cap C)) \cup ((A \cap C) \sim (A \cap B)) \;=\; (A \cap B) \Delta (A \cap C) \end{aligned}$$

The problem can also be handled by means of a Venn diagram, as in Problem 2.6.

2.10. Prove C(iii): $\overline{A \cap B} = \bar{A} \cup \bar{B}$, logically and pictorially.

Solution:

Logical Proof.

$$\begin{aligned} \overline{A \cap B} &= \{x : x \in X \;\&\; x \notin (A \cap B)\} \\ &= \{x : x \in X \;\&\; (x \notin A \vee x \notin B)\} \\ &= \{x : (x \in X \;\&\; x \notin A) \vee (x \in X \;\&\; x \notin B)\} \\ &= \{x : x \in X \;\&\; x \notin A\} \cup \{x : x \in X \;\&\; x \notin B\} \;=\; \bar{A} \cup \bar{B} \end{aligned}$$

Pictorial Proof. See Fig. 2-17 and 2-18.

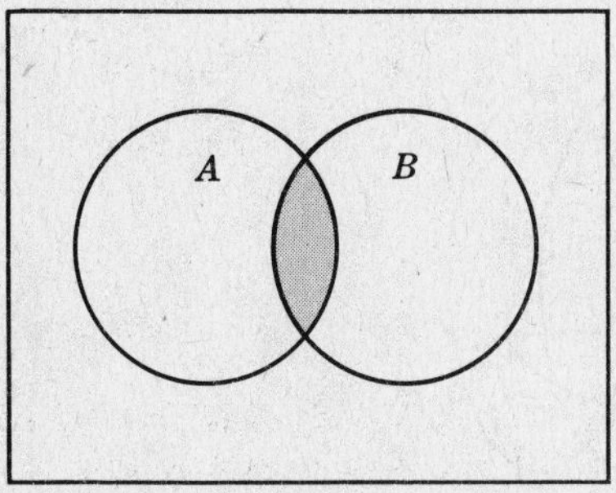

Unshaded region: $\overline{A \cap B}$

Fig. 2-17

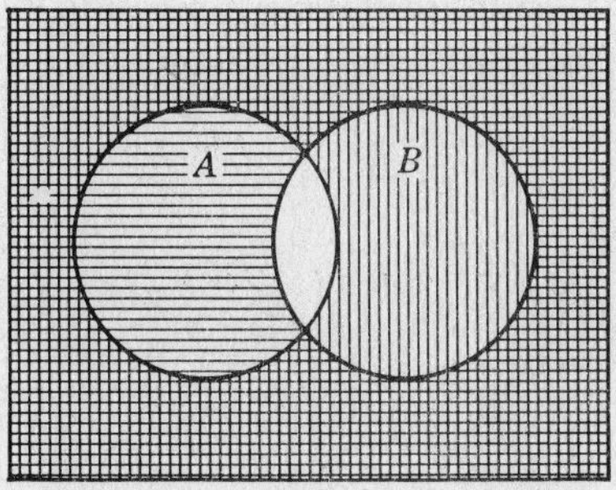

$\bar{A}$: vertical, $\bar{B}$: horizontal
Marked region: $\bar{A} \cup \bar{B}$

Fig. 2-18

2.11. Prove C(viii): $A \subseteq B$ if and only if $\bar{B} \subseteq \bar{A}$.

Solution:

Recall that A and B are subsets of some fixed universe X. Then

$A \subseteq B$ if and only if, for any x in X, if $x \in A$, then $x \in B$,
if and only if, for any x in X, if $x \notin B$, then $x \notin A$,†
if and only if, for any x in X, if $x \in \bar{B}$, then $x \in \bar{A}$,
if and only if $\bar{B} \subseteq \bar{A}$.

2.12. Using mathematical induction prove the generalized De Morgan Law C(iii′):

$$\overline{A_1 \cap \cdots \cap A_n} = \bar{A}_1 \cup \bar{A}_2 \cup \cdots \cup \bar{A}_n$$

Solution:

It is obvious for $n = 1$. The case $n = 2$ is simply C(iii). Assume the result true for $n = k$. Then for $n = k + 1$,

$$\begin{aligned} \overline{A_1 \cap \cdots \cap A_k \cap A_{k+1}} &= \overline{(A_1 \cap \cdots \cap A_k) \cap A_{k+1}} \\ &= \overline{A_1 \cap \cdots \cap A_k} \cup \bar{A}_{k+1} && \text{(by C(iii))} \\ &= (\bar{A}_1 \cup \cdots \cup \bar{A}_k) \cup \bar{A}_{k+1} && \text{(by inductive hypothesis)} \\ &= \bar{A}_1 \cup \cdots \cup \bar{A}_k \cup \bar{A}_{k+1} \end{aligned}$$

†We have used here the logical law of contraposition: $P \rightarrow Q$ is logically equivalent to $\neg Q \rightarrow \neg P$.

2.13. Prove: $A \cup B = \overline{\bar{A} \cap \bar{B}}$ and $A \cap B = \overline{\bar{A} \cup \bar{B}}$.

Solution:

By De Morgan's law C(ii), $\overline{A \cup B} = \bar{A} \cap \bar{B}$. Hence $\overline{\overline{A \cup B}} = \overline{\bar{A} \cap \bar{B}}$. But $\overline{\overline{A \cup B}} = A \cup B$, by C(i). Likewise, by De Morgan's law C(iii), $\overline{A \cap B} = \bar{A} \cup \bar{B}$. Hence $A \cap B = \overline{\overline{A \cap B}} = \overline{\bar{A} \cup \bar{B}}$.

2.14. Prove: (1) $(A \cup B) \cap \bar{A} = B \cap \bar{A}$ (3) $(A \cap B) \cup B = B$

(2) $(A \cap B) \cup \bar{A} = B \cup \bar{A}$ (4) $(A \cup B) \cap B = B$

Solution:

(1) $(A \cup B) \cap \bar{A} = (A \cap \bar{A}) \cup (B \cap \bar{A}) = \emptyset \cup (B \cap \bar{A}) = B \cap \bar{A}$.

(2) $(A \cap B) \cup \bar{A} = (A \cup \bar{A}) \cap (B \cup \bar{A}) = X \cap (B \cup \bar{A}) = B \cup \bar{A}$.

(3) $(A \cap B) \cup B \subseteq B \cup B = B$. Also, $B \subseteq (A \cap B) \cup B$. Hence $(A \cap B) \cup B = B$.

(4) $(A \cup B) \cap B = (A \cap B) \cup (B \cap B) = (A \cap B) \cup B = B$ (by (3)).

2.15. (*a*) Show that the four ellipses in the diagram below form an appropriate Venn diagram for four sets.

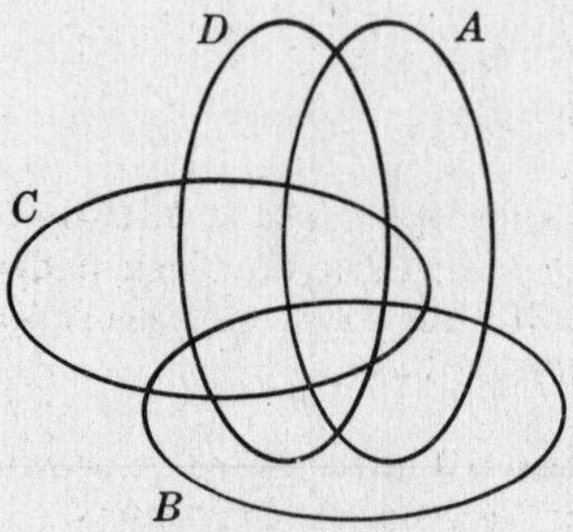

(*b*) Using the diagram of part (*a*), what conclusion can you draw from the following assumptions?

(i) $C \subseteq (B \cap \bar{D}) \cup (D \cap \bar{B})$.

(ii) Everything in both A and C is either in both B and D or in neither B nor D.

(iii) Everything in both B and C is either A or D.

(iv) Everything in both C and D is either in A or B.

Solution:

(*a*) Show that the fifteen regions of the diagram cover all possible cases:

$$A \cap B \cap C \cap \bar{D},\ A \cap B \cap \bar{C} \cap D,\ A \cap \bar{B} \cap C \cap D,\ \bar{A} \cap B \cap C \cap D,\ A \cap B \cap C \cap D,$$
$$A \cap B \cap \bar{C} \cap \bar{D},\ A \cap \bar{B} \cap C \cap \bar{D},\ \bar{A} \cap B \cap C \cap \bar{D},\ A \cap \bar{B} \cap \bar{C} \cap D,\ \bar{A} \cap B \cap \bar{C} \cap D,$$
$$\bar{A} \cap \bar{B} \cap C \cap D,\ A \cap \bar{B} \cap \bar{C} \cap \bar{D},\ \bar{A} \cap B \cap \bar{C} \cap \bar{D},\ \bar{A} \cap \bar{B} \cap C \cap \bar{D},\ \bar{A} \cap \bar{B} \cap \bar{C} \cap D.$$

(*b*) $C = \emptyset$.

2.16. Algebra of Sets and Algebra of Logic. Given a statement form **C** in ¬, &, ∨, let $S(\mathbf{C})$ be the expression obtained from **C** by substituting ¯, ∩, ∪ for ¬, &, ∨ respectively. Example:

$$S((A \vee B) \ \&\ \neg C) = (A \cup B) \cap \bar{C}$$

(*a*) Prove: **A** is logically equivalent to **B** if and only if $S(\mathbf{A}) = S(\mathbf{B})$ holds for all sets (where the statement letters of a statement form **C** are interpreted in $S(\mathbf{C})$ as set variables ranging over all subsets of a fixed universe).

(*b*) Prove that **A** logically implies **B** if and only if $S(\mathbf{A}) \subseteq S(\mathbf{B})$ holds for all sets.

Solution:

(*a*) If we replace each statement letter **L** in **A** by $x \in \mathbf{L}$, then the resulting sentence is equivalent to $x \in S(\mathbf{A})$ (since $x \in W_1 \cap W_2$ if and only if $x \in W_1$ & $x \in W_2$; $x \in W_1 \cup W_2$ if and only if $x \in W_1 \vee x \in W_2$; and $x \in \bar{W}$ if and only if $\neg(x \in W)$). Hence if **A** is logically equivalent to **B**, then $x \in S(\mathbf{A})$ if and only if $x \in S(\mathbf{B})$, which implies that $S(\mathbf{A}) = S(\mathbf{B})$. Conversely, assume that **A** is not logically equivalent to **B**. In general if we are given a truth assignment to the statement letters in an arbitrary statement form **C**, and if we replace statement letters which are T by $\{\emptyset\}$ and statement letters which are F by $\emptyset$, then, under this substitution of sets for statement letters, $S(\mathbf{C}) = \{\emptyset\}$ if **C** is T under the given assignment, and $S(\mathbf{C}) = \emptyset$ if **C** is F under the given assignment. This holds because, under the correspondence associating $\{\emptyset\}$ with T and $\emptyset$ with F, the truth-functional operations correspond to the set-theoretic operations (where sets are restricted to subsets of the universe $\{\emptyset\}$).

$$\neg T = F \qquad \overline{\{\emptyset\}} = \emptyset$$

$$\neg F = T \qquad \bar{\emptyset} = \{\emptyset\}$$

$$T \,\&\, T = T \qquad \{\emptyset\} \cap \{\emptyset\} = \{\emptyset\}$$

$$T \,\&\, F = F \,\&\, T = F \qquad \{\emptyset\} \cap \emptyset = \emptyset \cap \{\emptyset\} = \emptyset$$

$$F \,\&\, F = F \qquad \emptyset \cap \emptyset = \emptyset$$

$$T \vee T = T \vee F = F \vee T = T \qquad \{\emptyset\} \cup \{\emptyset\} = \{\emptyset\} \cup \emptyset = \emptyset \cup \{\emptyset\} = \{\emptyset\}$$

$$F \vee F = F \qquad \emptyset \cup \emptyset = \emptyset$$

Since **A** is not logically equivalent to **B**, there is a truth assignment making one of them T and the other F, say **A** is T and **B** is F. Then under the substitution of $\{\emptyset\}$ for the true statement letters and of $\emptyset$ for the false statement letters, $S(\mathbf{A}) = \{\emptyset\}$ and $S(\mathbf{B}) = \emptyset$. Hence $S(\mathbf{A}) = S(\mathbf{B})$ does not always hold.

Remark: Lurking behind this rather long-winded discussion is what in mathematics is called an "isomorphism" between the structures

$$\langle\{T, F\}, \neg, \&, \vee\rangle \quad \text{and} \quad \langle\{\{\emptyset\}, \emptyset\}, {}^{-}, \cap, \cup\rangle$$

Note that we also have shown that an equation $S(\mathbf{A}) = S(\mathbf{B})$ holds for all sets if and only if it holds in the domain $\{\{\emptyset\}, \emptyset\}$ of all subsets of $\{\emptyset\}$.

(*b*) **A** logically implies **B** if and only if **A & B** is logically equivalent to **A**. By part (*a*), the latter holds if and only if $S(\mathbf{A} \,\&\, \mathbf{B}) = S(\mathbf{A})$ always holds. But $S(\mathbf{A} \,\&\, \mathbf{B}) = S(\mathbf{A}) \cap S(\mathbf{B})$, and $S(\mathbf{A}) \cap S(\mathbf{B}) = S(\mathbf{A})$ if and only if $S(\mathbf{A}) \subseteq S(\mathbf{B})$.

2.17. Define ordered n-tuples (for $n \geqq 3$) by induction:

$$\langle x_1, x_2, \ldots, x_n\rangle = \langle\langle x_1, x_2, \ldots, x_{n-1}\rangle, x_n\rangle$$

Thus $\langle x_1, x_2, x_3\rangle = \langle\langle x_1, x_2\rangle, x_3\rangle$ and $\langle x_1, x_2, x_3, x_4\rangle = \langle\langle\langle x_1, x_2\rangle, x_3\rangle, x_4\rangle$. Prove that if $\langle x_1, x_2, \ldots, x_n\rangle = \langle u_1, u_2, \ldots, u_n\rangle$, then $x_1 = u_1$, $x_2 = u_2$, ..., $x_n = u_n$.

Solution:

We already have proved this result for $n = 2$. Now assume it is true for $n = k \geqq 2$, and we shall prove it must then hold for $k + 1$. We have, by assumption,

$$\langle x_1, x_2, \ldots, x_k, x_{k+1}\rangle = \langle u_1, u_2, \ldots, u_k, u_{k+1}\rangle$$

Hence by definition,

$$\langle\langle x_1, x_2, \ldots, x_k\rangle, x_{k+1}\rangle = \langle\langle u_1, u_2, \ldots, u_k\rangle, u_{k+1}\rangle$$

By the result for $n = 2$, we conclude $x_{k+1} = u_{k+1}$ and $\langle x_1, x_2, \ldots, x_k\rangle = \langle u_1, u_2, \ldots, u_k\rangle$. But the latter equation, by virtue of the inductive hypothesis, implies $x_1 = u_1$, $x_2 = u_2$, ..., $x_k = u_k$.

FINITE, INFINITE, DENUMERABLE, AND COUNTABLE SETS

2.18. Prove that the set W of all ordered pairs of non-negative integers is denumerable.

Solution:

Arrange W in the following infinite array:

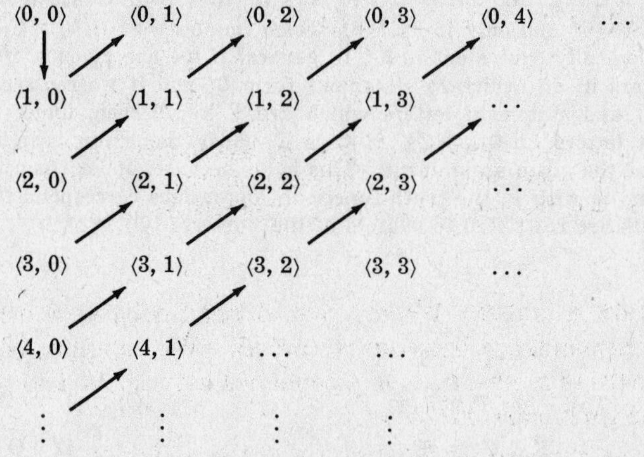

Enumerate the ordered pairs as indicated by the arrows, going up each diagonal from left to right. Notice that the pair $\langle i, j\rangle$ appears in the $[((i+j)(i+j+1)/2)+(j+1)]$th place in the enumeration. This can be seen as follows: all pairs in the same diagonal have the same sum. Adding up all pairs in diagonals preceding the one containing $\langle i, j\rangle$, we obtain

$$1+2+\cdots+(i+j) = (i+j)(i+j+1)/2$$

There are j pairs in the same diagonal as $\langle i, j\rangle$ and preceding $\langle i, j\rangle$.

2.19. Prove that the set of all non-negative rational numbers is denumerable.

Solution:

Every non-negative rational number corresponds to a fraction m/n, where (i) m and n are non-negative integers, (ii) $n \neq 0$, and (iii) m and n have no common integral factors other than ± 1. We can associate the ordered pair $\langle m, n\rangle$ with m/n, and use the enumeration given in Problem 2.18, merely omitting those pairs $\langle m, n\rangle$ which do not satisfy conditions (i)-(iii).

2.20. The set A of all real roots of all nonzero polynomials with integral coefficients (such roots are called *real algebraic numbers*) is denumerable.

Solution:

Any nonzero polynomial has only a finite number of roots. First list the finite set of all real roots of all polynomials of degree at most one whose coefficients are in magnitude ≤ 1 (i.e. whose coefficients are either $0, 1$, or -1). Then list the finite set of all real roots of all polynomials of degree ≤ 2 whose coefficients are in magnitude ≤ 2, etc. In general, at the nth step we list the finite set of all real roots of all polynomials of degree $\leq n$ whose coefficients are in magnitude $\leq n$. Of course, we omit repetitions. In this way, we obtain an enumeration of all real algebraic numbers. That the set A is not finite follows from the fact that all integers belong to A.

2.21. Show that the set of all real numbers is not countable.

Solution:

Let R_1 be the set of all real numbers x such that $0 \leq x < 1$. It suffices to show that R_1 is not countable, since any subset of a countable set is countable. Every x in R_1 is representable as a unique infinite decimal

$$x = .a_1a_2a_3\ldots$$

where the infinite decimal does not end with an infinite string of 9's. (Thus although a decimal such as .1362000... is also representable as .1361999..., we shall use only the first representation.) Assume now that R_1 can be enumerated:

$$x_1 = .a_{11}a_{12}a_{13}\ldots$$
$$x_2 = .a_{21}a_{22}a_{23}\ldots$$
$$\ldots\ldots\ldots\ldots\ldots\ldots$$
$$x_k = .a_{k1}a_{k2}a_{k3}\ldots$$
$$\ldots\ldots\ldots\ldots\ldots\ldots$$

Construct a decimal $y = .b_1b_2b_3\ldots$ as follows:

$$b_i = \begin{cases} 0 & \text{if } a_{ii} \neq 0 \\ 1 & \text{if } a_{ii} = 0 \end{cases}$$

Thus, for all i, $a_{ii} \neq b_i$. But then, y is in R_1 and is different from all of the numbers $x_1, x_2, \ldots$ (since the decimal representation of y differs from that of x_i in the ith place). This contradicts the assumption that the sequence $x_1, x_2, \ldots$ exhausts R_1.

2.22. Given two sets A and B. We say that A *has the same cardinality as* B if there is a one-one correspondence between A and B. We say that A *has smaller cardinality than* B if there is a one-one correspondence between A and a subset of B but A does not have the same cardinality as B.

Prove that, for any set A, A has smaller cardinality than the set $\mathcal{P}(A)$ of all subsets of A (Cantor's Theorem).

Solution:

(1) There is a one-one correspondence between A and a subset of $\mathcal{P}(A)$. Namely, to each element x of A associate the set $\{x\}$ in $\mathcal{P}(A)$. Clearly if x and y are distinct elements of A, $\{x\} \neq \{y\}$.

(2) We must show that there is no one-one correspondence f between A and $\mathcal{P}(A)$. Assume, on the contrary, that there is such a one-one correspondence f. Let $C = \{x : x \in A \;\&\; x \notin f(x)\}$. Thus C consists of all elements x of A such that x is not a member of the corresponding subset $f(x)$ of A. But $C \subseteq A$. Hence $C \in \mathcal{P}(A)$. So there must be an element y in A such that $f(y) = C$. Then by definition of C, $y \in C$ if and only if $y \notin f(y)$. Since $f(y) = C$, it follows that $y \in C$ if and only if $y \notin C$. But either $y \in C$ or $y \notin C$. Hence $y \in C \;\&\; y \notin C$, which is a contradiction.

FIELD OF SETS

2.23. Prove that the collection $\mathcal{F}$ of all subsets B of X such that either B or $\bar{B}$ is countable is a field of sets.

Solution:

Assume $B \in \mathcal{F}$. Then either $\bar{\bar{B}}$ or $\bar{B}$ is countable. Hence $\bar{B} \in \mathcal{F}$. Assume now that A is also in $\mathcal{F}$.

Case 1: B is countable. Then $A \cap B$ is countable. Hence $A \cap B \in \mathcal{F}$.

Case 2: A is countable. Then $A \cap B$ is countable. Hence $A \cap B \in \mathcal{F}$.

Case 3: $\bar{B}$ is countable and $\bar{A}$ is countable. Hence $\bar{A} \cup \bar{B}$ is countable. But $\overline{A \cap B} = \bar{A} \cup \bar{B}$. Therefore $A \cap B \in \mathcal{F}$.

2.24. Let X be the set of all integers, and let k be a fixed integer. Let $\mathcal{G}$ be the collection of all subsets B of X such that, for any u in B, both $u+k$ and $u-k$ are also in B. (This means that a shift of k units does not alter B.) Show that $\mathcal{G}$ is a field of sets.

Solution:

Let $B \in \mathcal{G}$. Assume $u \in \bar{B}$. Hence $u \notin B$. So $u - k \notin B$. (For, if $u - k \in B$, then $u = (u-k) + k \in B$.) Also, $u + k \in \bar{B}$. (For, if $u + k \in B$, then $u = (u+k) - k \in B$.) Thus

$\bar{B} \in \mathcal{G}$. Assume now that A and B are in $\mathcal{G}$. Let us consider $A \cap B$. Assume $u \in A \cap B$. Then $u \in A$ & $u \in B$. Hence $u \pm k \in A$ & $u \pm k \in B$. Therefore $u \pm k \in A \cap B$. Thus, $A \cap B \in \mathcal{G}$.

Additional question: How many elements does $\mathcal{G}$ have?

NUMBER OF ELEMENTS IN A FINITE SET

2.25. Derive the equality

$$\#(A \cup B \cup C) = \#(A) + \#(B) + \#(C) - \#(A \cap B) - \#(A \cap C) - \#(B \cap C) + \#(A \cap B \cap C) \quad (1)$$

for arbitrary finite sets A, B and C.

Solution:

Take any element x in $A \cup B \cup C$. If x is in precisely one of the sets A, B, C, then x is counted once on the right side of (*1*). (For example, if $x \in B \cap \tilde{A} \cap \tilde{C}$, then x is counted only in $\#(B)$.) If x belongs to precisely two of the sets A, B, C, then x will be counted twice in the positive sense on the right side and once in the negative sense. (For example, if $x \in A \cap \tilde{B} \cap C$, then x is counted twice in the positive sense in $\#(A)$ and $\#(C)$, and x is subtracted once in $\#(A \cap C)$.) Lastly, if x belongs to $A \cap B \cap C$, then x is counted in every term on the right side, four times in the positive sense and three times in the negative sense. Thus the net effect of the right side of (*1*) is to count the number of elements in $A \cup B \cup C$.

2.26. If a boating club of 75 members admitted only owners of sailboats or powerboats, and if 48 members owned sailboats and 33 members owned powerboats, how many members owned both sailboats and powerboats?

Solution:

Let A = the set of all members owning sailboats, and B = the set of all members owning powerboats.

$$\#(A \cup B) = \#(A) + \#(B) - \#(A \cap B)$$

$$75 = 48 + 33 - \#(A \cap B)$$

Hence $\#(A \cap B) = 6$.

2.27. Among 50 students taking examinations in mathematics, physics and chemistry, 37 passed mathematics, 24 physics and 43 chemistry; at most 19 passed mathematics and physics, at most 29 mathematics and chemistry, and at most 20 physics and chemistry. What is the largest possible number that could have passed all three?

Solution:

Let M, P, C stand for the collections of students passing mathematics, physics and chemistry, respectively.

$$\#(M \cup P \cup C) = \#(M) + \#(P) + \#(C) - \#(M \cap P) - \#(M \cap C) - \#(P \cap C) + \#(M \cap P \cap C)$$

$$50 \geq 37 + 24 + 43 - \#(M \cap P) - \#(M \cap C) - \#(P \cap C) + \#(M \cap P \cap C)$$

Hence

$$\#(M \cap P \cap C) \leq \#(M \cap P) + \#(M \cap C) + \#(P \cap C) - 54$$

$$\leq 19 + 29 + 20 - 54 = 14$$

Supplementary Problems

2.28. List all subsets of $\{\emptyset, \{\emptyset\}\}$.

2.29. $\{x : x \text{ is a real number } \& \ x^2 < 0\} \ = \ ?$

2.30. Prove: $B \cup C = \emptyset$ if and only if $B = \emptyset$ & $C = \emptyset$.

2.31. Prove: If $A \cup B = A$ for all sets A, then $B = \emptyset$.

2.32. Show with an example that $A \cup B \cap C$ requires parentheses to be unambiguous.

2.33. Prove: If $B \subseteq A$ & $C \subseteq A$, then $B \cup C \subseteq A$; and if $A \subseteq B$ & $A \subseteq C$, then $A \subseteq B \cap C$.

2.34. Prove: If $C \subseteq A$, then $B \cap C \subseteq A$ and $C \subseteq A \cup B$.

2.35. Prove: $\mathcal{P}(A) \cap \mathcal{P}(B) = \mathcal{P}(A \cap B)$.

2.36. Is $\mathcal{P}(A) \cup \mathcal{P}(B) = \mathcal{P}(A \cup B)$? If not, give a counterexample.

2.37. $\{x : x \text{ is an integral multiple of } 2\} \cap \{x : x \text{ is an integral multiple of } 3\} \ = \ ?$

2.38. Disprove the cancellation law: If $A \cap B = A \cap C$, then $B = C$.

In Problems 2.39-2.54, determine whether the given equation is always true, using rigorous logical methods and also, if possible, Venn diagrams. If an equation is not always true, specify a counterexample.

2.39. $(A \cup B) \cap (B \cup C) \cap (C \cup A) = (A \cap B) \cup (B \cap C) \cup (C \cap A)$.

2.40. $A \sim (B \sim C) = (A \sim B) \cup (A \cap B \cap C)$.

2.41. $A \cap (B \sim C) = (A \cap B) \sim (A \cap C)$.

2.42. $A \cup (B \sim C) = (A \cup B) \sim (A \cup C)$.

2.43. $A \sim (B \cap C) = (A \sim B) \cup (A \sim C)$.

2.44. $A \sim (A \cap B) = A \sim B$.

2.45. $A \cap B = A \sim (A \sim B)$.

2.46. $A \cup B = A \cup (B \sim A)$.

2.47. $(A \sim C) \cup (B \sim C) = (A \cup B) \sim C$.

2.48. $A \cup (B \,\Delta\, C) = (A \cup B) \,\Delta\, (A \cup C)$.

2.49. $A \sim (B \,\Delta\, C) = (A \sim B) \,\Delta\, (A \sim C)$.

2.50. $\overline{A \sim B} = \bar{A} \cup B$.

2.51. $\bar{A} \sim \bar{B} = B \sim A$.

2.52. $A \,\Delta\, B = \bar{A} \,\Delta\, \bar{B}$.

2.53. $\overline{A \,\Delta\, B} = (\bar{A} \cap \bar{B}) \cup (A \cap B)$.

2.54. $\overline{A \,\Delta\, B} = A \,\Delta\, \bar{B} = \bar{A} \,\Delta\, B$.

2.55. Let I = the set of integers = $\{\ldots, -2, -1, 0, 1, 2, \ldots\}$; N = the set of non-negative integers = $\{0, 1, 2, \ldots\}$; Np = the set of non-positive integers = $\{0, -1, -2, \ldots\}$; E = the set of even integers; P = the set of prime numbers. Find: $N \cap (Np)$, $I \sim N$, $I \sim (Np)$, $N \sim (Np)$, $N \cup (Np)$, $N \,\Delta\, (Np)$, $I \sim E$, $E \cap P$.

2.56. Prove that $A = B$ is equivalent to each of the following conditions.

(i) $A \sim B = B \sim A$.

(ii) $A \cup B = A \cap B$.

(iii) $A \cup C = B \cup C$ & $A \cap C = B \cap C$.

2.57. Prove that $A \subseteq B$ is equivalent to each of the following conditions.

(i) $A \triangle B = B \sim A$, (ii) $\bar{A} \cup B = X$.

2.58. Prove: $A \triangle B = A \cup B$ if and only if $A \cap B = \emptyset$.

2.59. Prove: $A \subseteq \bar{A}$ if and only if $A = \emptyset$, and $\bar{A} \subseteq A$ if and only if $A = X$.

2.60. Using Venn diagrams, determine the compatibility of the following conditions.

(i) $(A \cap B) \cup C = A \sim B$ and $C \cap A = B \cap A$.

(ii) $A \sim (B \sim C) \subseteq C \cup B$ and $A \cap B \cap C = \emptyset$ and $C \sim B \subseteq A$.

(iii) $(B \sim A) \cap C \neq \emptyset$ and $C \sim A \subseteq C \sim B$.

2.61. Prove the following identities.

(i) $$\begin{aligned} A_1 \cup A_2 \cup \cdots \cup A_n &= (A_1 \sim A_2) \cup (A_2 \sim A_3) \cup \cdots \cup (A_{n-1} \sim A_n) \\ &\qquad \cup (A_n \sim A_1) \cup (A_1 \cap A_2 \cap \cdots \cap A_n) \\ &= A_1 \cup (A_2 \sim A_1) \cup (A_3 \sim (A_1 \cup A_2)) \cup (A_4 \sim (A_1 \cup A_2 \cup A_3)) \cup \cdots \\ &\qquad \cup (A_n \sim (A_1 \cup A_2 \cup \cdots \cup A_{n-1})) \end{aligned}$$

(ii) $A_1 \cap A_2 \cap \cdots \cap A_n = A_1 \sim [(A_1 \sim A_2) \cup (A_1 \sim A_3) \cup \cdots \cup (A_1 \sim A_n)]$.

(iii) $(A_1 \sim B_1) \cap (A_2 \sim B_2) \cap \cdots \cap (A_n \sim B_n) = (A_1 \cap A_2 \cap \cdots \cap A_n) \sim (B_1 \cup B_2 \cup \cdots \cup B_n)$.

2.62. Simplify the following expressions.

(i) $\overline{\overline{(A \cap B) \cup C} \cap \bar{B}}$, (ii) $((A \cup B) \cap \bar{A}) \cup \overline{B \cap A}$.

2.63. Find the set-theoretic operations corresponding to the truth functions for $\rightarrow$ and $\leftrightarrow$.

2.64. Determine whether each of the following sets of ordered pairs is a function.

(i) $\{\langle x, y\rangle : x$ and y are human beings and x is the father of $y\}$.

(ii) $\{\langle x, y\rangle : x$ and y are human beings and y is the father of $x\}$.

(iii) $\{\langle x, y\rangle : x$ and y are real numbers and $x^2 + y^2 = 1\}$.

(iv) $\{\langle x, y\rangle : (x = 1 \ \& \ y = 2) \vee (x = -1 \ \& \ y = 0)\}$.

2.65. For each of the following functions f from the set of integers I into I, determine whether f is a function from I onto I and also whether f is one-one.

(i) $f(x) = 2x + 1$ (iii) $f(x) = \begin{cases} x+1 & \text{if } x \text{ is even} \\ x-1 & \text{if } x \text{ is odd} \end{cases}$

(ii) $f(x) = -x$ (iv) $f(x) = x^2 - 3x + 5$

2.66. Prove: The set of all rational numbers is denumerable.

2.67. Prove: The set of all irrational real numbers is not countable.

2.68. By a *left-open interval* of the set R of real numbers, we mean either an interval

$$(a, b] = \{x : a < x \leq b\}$$

or an infinite interval of the form

$$(a, \infty) = \{x : a < x\} \quad \text{or} \quad (-\infty, b] = \{x : x \leq b\}$$

Let $\mathcal{F}$ be the collection of sets of real numbers consisting of $\emptyset$, R, and all unions of a finite number of left-open intervals. Show that $\mathcal{F}$ is a field of sets.

2.69. (1) For finite sets A, B, C, D, derive the formula

$$\begin{aligned}\#(A\cup B\cup C\cup D) &= \#(A)+\#(B)+\#(C)+\#(D)\\ &\quad -\#(A\cap B)-\#(A\cap C)-\#(A\cap D)-\#(B\cap C)-\#(B\cap D)-\#(C\cap D)\\ &\quad +\#(A\cap B\cap C)+\#(A\cap B\cap D)+\#(A\cap C\cap D)+\#(B\cap C\cap D)\\ &\quad -\#(A\cap B\cap C\cap D)\end{aligned}$$

(2) Generalize the result of (1) to any finite sets $A_1, \ldots, A_n$.

2.70. In an advertising survey of a hundred coffee and tea drinkers, 70 people were found to drink coffee at times, and 30 people drank both tea and coffee. How many people sometimes drank tea?

2.71. Among Americans taking vacations last year, 90% took vacations in the summer, 65% in the winter, 10% in the spring, 7% in the autumn, 55% in the winter and summer, 8% in the spring and summer, 6% in the autumn and summer, 4% in the winter and spring, 4% in the winter and autumn, 3% in the spring and autumn, 3% in the summer, winter and spring, 3% in the summer, winter and autumn, 2% in the summer, autumn and spring, and 2% in the winter, spring and autumn. What percentage took vacations during every season?

2.72. If A and B both contain n elements, prove that $A-B$ and $B-A$ have an equal number of elements. Show that this is no longer the case when A and B are infinite.

2.73. (*a*) Show that the maximum number of sets obtainable from A and B by applying the union and difference operations is eight. (*b*) Show that the maximum number of sets obtainable from A, B, C by applying the union and difference operations is $128 = 2^7$. (*Hint*: How many regions appear in the Venn diagram for A, B, C?) (*c*) Generalize (*b*) to the case of n sets $A_1, \ldots, A_n$.

2.74. (*a*) Prove: $(A\cap B\cap C)\cup(A\cap B\cap D)\cup(A\cap C\cap D)\cup(B\cap C\cap D)$

$= (A\cup B)\cap(A\cup C)\cap(A\cup D)\cap(B\cup C)\cap(B\cup D)\cap(C\cup D)$.

(*b*) Prove: $(A\cap B)\cup(A\cap C)\cup(A\cap D)\cup(B\cap C)\cup(B\cap D)\cup(C\cap D)$

$= (A\cup B\cup C)\cap(A\cup B\cup D)\cap(A\cup C\cup D)\cap(B\cup C\cup D)$.

(*c*)^D Prove the following generalization of (*a*), (*b*) and Problem 2.41: Given n sets $A_1, \ldots, A_n$. Let $k \leq n$. Show that the union U of all intersections of k of the sets $A_1, \ldots, A_n$ is equal to the intersection I of all unions of $n-k+1$ of the sets $A_1, \ldots, A_n$. (*Note*: In Problem 2.41, $n=3$, $k=2$; in (*a*), $n=4$, $k=3$; in (*b*), $n=4$, $k=2$.) *Hint*: Prove $U\subseteq I$ and $I\subseteq U$.

Chapter 3

Boolean Algebras

3.1 OPERATIONS

An *n-ary operation on a set Y* is defined to be any function f which, to each n-tuple $\langle y_1, \ldots, y_n \rangle$ of elements $y_1, \ldots, y_n$ in Y, assigns an element $f(y_1, \ldots, y_n)$ in Y. A more traditional way of asserting that f is an n-ary operation on Y is to say that Y is *closed* under the function f.

Example 3.1.

Addition, multiplication and subtraction are binary operations on the set of integers. (We use "binary" instead of "2-ary".) The function f such that $f(x) = x - 1$ for every integer x is a singulary operation on the set of all integers. (We use "singulary" instead of "1-ary".)

Example 3.2.

The subtraction function $x - y$ is not a binary operation on the set of non-negative integers, because the value $x - y$ is not always a non-negative integer. The division function x/y is not a binary operation on the set of positive integers. (Why?)

3.2 AXIOMS FOR A BOOLEAN ALGEBRA

By a *Boolean algebra* we mean a set B together with two binary operations $\wedge$ and $\vee$ on B, a singulary operation $'$ on B, and two specific elements 0 and 1 of B such that the following axioms hold.

(1) For any x and y in B, $x \vee y = y \vee x$ } Commutative Laws

(2) For any x and y in B, $x \wedge y = y \wedge x$ } Commutative Laws

(3) For any x, y, z in B, $x \wedge (y \vee z) = (x \wedge y) \vee (x \wedge z)$ } Distributive Laws

(4) For any x, y, z in B, $x \vee (y \wedge z) = (x \vee y) \wedge (x \vee z)$ } Distributive Laws

(5) For any x in B, $x \vee 0 = x$.

(6) For any x in B, $x \wedge 1 = x$.

(7) For any x in B, $x \vee x' = 1$.

(8) For any x in B, $x \wedge x' = 0$.

(9) $0 \neq 1$.

A Boolean algebra will be designated by a sextuple $\langle B, \wedge, \vee, ', 0, 1 \rangle$. Sometimes one refers to the set B as a Boolean algebra, but this is just a loose misuse of language.

Example 3.3.

(*a*) The two-element Boolean algebra

$$\mathcal{B}_0 = \langle \{\emptyset, \{\emptyset\}\}, \cap, \cup, ^-, \emptyset, \{\emptyset\} \rangle$$

where $B = \{\emptyset, \{\emptyset\}\}$; $\wedge$ = the ordinary set-theoretic intersection operation $\cap$; $\vee$ = the ordinary set-theoretic union operation $\cup$; $'$ = the ordinary set-theoretic operation of complementation; $0 = \emptyset$; and $1 = \{\emptyset\}$. In Chapter 2, we have verified properties (1)-(9). Of course, we first must note that $\cap$, $\cup$ and $^-$ are operations on $\{\emptyset, \{\emptyset\}\}$.

(*b*) The Boolean algebra of all subsets of a non-empty set A, under the usual operations of intersection, union, and complementation, and with $\emptyset$ and A as the distinguished elements 0 and 1: $\langle \mathcal{P}(A), \cap, \cup, ^-, \emptyset, A\rangle$. When there is no danger of confusion, we shall refer to this Boolean algebra simply as $\mathcal{P}(A)$. Part (*a*) is a special case of (*b*) when $A = \{\emptyset\}$. (Notice that we have omitted the case where $A = \emptyset$; in this case, $0 = \emptyset = A = 1$, violating Axiom (9).)

Example 3.4. (This example should be omitted by those not familiar with elementary number theory.)

Let B be the set of all positive integers which are integral divisors of 70. Thus,

$$B = \{1, 2, 5, 7, 10, 14, 35, 70\}$$

For any x and y in B, let $x \wedge y$ be the greatest common divisor of x and y, let $x \vee y$ be the least common multiple of x and y, and let $x' = 70/x$. (For example, $5 \wedge 14 = 1$, $5 \vee 14 = 70$, $10 \wedge 35 = 5$, $10 \vee 35 = 70$, $5' = 14$, $10' = 7$.) Then $\langle B, \wedge, \vee, ', 1, 70\rangle$ is a Boolean algebra. Verification of Axioms (1)-(9) uses elementary properties of *greatest common divisor* and *least common multiple.*

Example 3.5.

It seems evident that a set of sentences closed under the operations of conjunction, disjunction and negation should form a Boolean algebra. However, this is not quite so. For example, $A \& B$ and $B \& A$ are not equal, but only logically equivalent. Thus we should have to replace the equality sign = in Axioms (1)-(9) by the relation of logical equivalence. In addition, 0 could be any sentence of the form $\mathbf{A} \& \neg \mathbf{A}$, and 1 could be any sentence of the form $\mathbf{A} \vee \neg \mathbf{A}$. If we wish to retain the equality sign = with its usual meaning (i.e. identity), then we may proceed as follows. By the *statement bundle* $[\mathbf{A}]$ determined by a statement form $\mathbf{A}$, we mean the set of all statement forms which are logically equivalent to $\mathbf{A}$. Then, it is clear that: (i) $[\mathbf{A}] = [\mathbf{B}]$ if and only if $\mathbf{A}$ is logically equivalent to $\mathbf{B}$; (ii) if $[\mathbf{A}] \neq [\mathbf{B}]$, then $[\mathbf{A}] \cap [\mathbf{B}] = \emptyset$. If K_1 and K_2 are statement bundles, it is obvious that if $\mathbf{A}_1$ and $\mathbf{B}_1$ are statement forms in K_1 and $\mathbf{A}_2$ and $\mathbf{B}_2$ are statement forms in K_2, then $\mathbf{A}_1 \& \mathbf{A}_2$ is logically equivalent to $\mathbf{B}_1 \& \mathbf{B}_2$, $\mathbf{A}_1 \vee \mathbf{A}_2$ is logically equivalent to $\mathbf{B}_1 \vee \mathbf{B}_2$, and $\neg \mathbf{A}_1$ is logically equivalent to $\neg \mathbf{B}_1$. Therefore if we take an arbitrary statement form $\mathbf{C}_1$ from K_1 and an arbitrary statement form $\mathbf{C}_2$ from K_2, then we may define $K_1 \& K_2$ to be $[\mathbf{C}_1 \& \mathbf{C}_2]$, $K_1 \vee K_2$ to be $[\mathbf{C}_1 \vee \mathbf{C}_2]$, and K_1' to be $[\neg \mathbf{C}_1]$. If B is taken to be the set of all statement bundles, 0 is taken to be $[A \& \neg A]$, and 1 is taken to be $[A \vee \neg A]$, then $\langle B, \&, \vee, ', 0, 1\rangle$ is a Boolean algebra. Verification of Axioms (1)-(9) reduces to well-known properties of the algebra of logic. (For example, to check Axiom (1), we consider any statement bundles K_1 and K_2, and we take any statement forms $\mathbf{C}_1$ and $\mathbf{C}_2$ in K_1 and K_2 respectively. Then $K_1 \vee K_2 = [\mathbf{C}_1 \vee \mathbf{C}_2]$ and $K_2 \vee K_1 = [\mathbf{C}_2 \vee \mathbf{C}_1]$. But $[\mathbf{C}_1 \vee \mathbf{C}_2] = [\mathbf{C}_2 \vee \mathbf{C}_1]$, since $\mathbf{C}_1 \vee \mathbf{C}_2$ is logically equivalent to $\mathbf{C}_2 \vee \mathbf{C}_1$.)

Terminology: $x \wedge y$ is called the *meet* of x and y.

$x \vee y$ is called the *join* of x and y.

x' is called the *complement* of x.

0 is called the *zero element.*

1 is called the *unit element.*

If it is necessary to distinguish the meet, join, complement, zero element and unit element of a Boolean algebra $\mathcal{B}$ from those of another Boolean algebra, we shall add the subscript $\mathcal{B}$: $\wedge_{\mathcal{B}}, \vee_{\mathcal{B}}, '_{\mathcal{B}}, 0_{\mathcal{B}}, 1_{\mathcal{B}}$.

Unless something is said to the contrary, we shall assume in what follows that $\mathcal{B} = \langle B, \wedge, \vee, 0, 1\rangle$ is an arbitrary Boolean algebra.

Theorem 3.1. *Uniqueness of the complement*: If $x \vee y = 1$ and $x \wedge y = 0$, then $y = x'$.

Proof. First,

$$\begin{aligned} y &= y \vee 0 && \text{by Axiom (5)} \\ &= y \vee (x \wedge x') && \text{by Axiom (8)} \\ &= (y \vee x) \wedge (y \vee x') && \text{by Axiom (4)} \\ &= (x \vee y) \wedge (y \vee x') && \text{by Axiom (1)} \\ &= 1 \wedge (y \vee x') && \text{by hypothesis} \\ &= (y \vee x') \wedge 1 && \text{by Axiom (2)} \\ &= y \vee x' && \text{by Axiom (6)} \end{aligned}$$

Second, $x' = x' \vee 0$ by Axiom (5)
$= x' \vee (x \wedge y)$ by hypothesis
$= (x' \vee x) \wedge (x' \vee y)$ by Axiom (4)
$= (x \vee x') \wedge (x' \vee y)$ by Axiom (1)
$= 1 \wedge (x' \vee y)$ by Axiom (7)
$= (x' \vee y) \wedge 1$ by Axiom (2)
$= x' \vee y$ by Axiom (6)
$= y \vee x'$ by Axiom (1)
$= y$ by the first part above ▶

Corollary 3.2. For any z in B, $(z')' = z$. (Notation: We shall denote $(z')'$ by z'', $((z')')'$ by z''', etc.)

Proof. First, $z' \vee z = z \vee z'$ by Axiom (1)
$= 1$ by Axiom (7)

Second, $z' \wedge z = z \wedge z'$ by Axiom (2)
$= 0$ by Axiom (8)

Hence by Theorem 3.1, taking x to be z' and y to be z, we obtain $z = z''$. ▶

Theorem 3.3. *Idempotence*: For any x in B,

(i) $x \wedge x = x$, (ii) $x \vee x = x$

Proof.

(i) $x = x \wedge 1$ by Axiom (6)
$= x \wedge (x \vee x')$ by Axiom (7)
$= (x \wedge x) \vee (x \wedge x')$ by Axiom (3)
$= (x \wedge x) \vee 0$ by Axiom (8)
$= x \wedge x$ by Axiom (5)

(ii) $x = x \vee 0$ by Axiom (5)
$= x \vee (x \wedge x')$ by Axiom (8)
$= (x \vee x) \wedge (x \vee x')$ by Axiom (4)
$= (x \vee x) \wedge 1$ by Axiom (7)
$= x \vee x$ by Axiom (6) ▶

Definition: By the *dual* of a proposition concerning a Boolean algebra B, we mean the proposition obtained by substituting $\vee$ for $\wedge$, $\wedge$ for $\vee$, 0 for 1, and 1 for 0, i.e. by exchanging $\wedge$ and $\vee$, and exchanging 0 and 1.

Example 3.6.

The dual of $x \wedge (y \vee z) = (x \wedge y) \vee (x \wedge z)$ is $x \vee (y \wedge z) = (x \vee y) \wedge (x \vee z)$, and vice versa. The dual of $x \vee x' = 1$ is $x \wedge x' = 0$, and vice versa.

It is obvious that if **B** is the dual of **A**, then **A** is the dual of **B**.

Theorem 3.4. *Duality Principle* (Proof-theoretic version): If a proposition **A** is derivable from Axioms (1)-(9), then the dual of **A** is also derivable from Axioms (1)-(9).

Proof. The dual of each of Axioms (1)-(9) is again an axiom: (1) and (2) are duals of each other, and so are the pairs (3)-(4), (5)-(6), and (7)-(8). (9) is its own dual. Thus if in a proof of **A** we replace every proposition by its dual, the result is again a proof (since axioms are replaced by axioms), but this new proof is now a proof of the dual of **A**. ▶

The proof of Theorem 3.3 is twice as long as it need be. Since $x \vee x = x$ is the dual of $x \wedge x = x$, it would have sufficed to prove $x \wedge x = x$ and then cite the Duality Principle to obtain $x \vee x = x$. As a matter of fact, the proof of the Duality Principle is illustrated in the proof of Theorem 3.3: the proof of (ii) is obtained by taking the duals of the propositions in the proof of (i).

Theorem 3.5. For all x, y, z in B,

(i) $x \wedge 0 = 0$

(ii) $x \vee 1 = 1$

(iii) $x \wedge (x \vee y) = x$ } Absorption Laws

(iv) $x \vee (x \wedge y) = x$ } Absorption Laws

(v) $[y \wedge x = z \wedge x \;\&\; y \wedge x' = z \wedge x'] \rightarrow y = z$

(vi) $x \vee (y \vee z) = (x \vee y) \vee z$ } Associative Laws

(vii) $x \wedge (y \wedge z) = (x \wedge y) \wedge z$ } Associative Laws

(viii) $(x \vee y)' = x' \wedge y'$ } De Morgan's Laws

(ix) $(x \wedge y)' = x' \vee y'$ } De Morgan's Laws

(x) $x \vee y = (x' \wedge y')'$

(xi) $x \wedge y = (x' \vee y')'$

(xii) $x \wedge y' = 0 \leftrightarrow x \wedge y = x$

(xiii) $0' = 1$

(xiv) $1' = 0$

(xv) $x \wedge (x' \vee y) = x \wedge y$

(xvi) $x \vee (x' \wedge y) = x \vee y$

Proof. From now on, we usually will not cite the particular axioms or theorems being used in a proof.

(i) $x \wedge 0 = (x \wedge 0) \vee 0 = (x \wedge 0) \vee (x \wedge x') = (x \wedge x') \vee (x \wedge 0) = x \wedge (x' \vee 0) = x \wedge x' = 0$

(ii) is the dual of (i).

(iii) $x \wedge (x \vee y) = (x \vee 0) \wedge (x \vee y) = x \vee (0 \wedge y) = x \vee 0 = x.$

(iv) is the dual of (iii).

(v) Assume $y \wedge x = z \wedge x \;\&\; y \wedge x' = z \wedge x'$. Then

$$\begin{aligned} y &= y \wedge 1 = y \wedge (x \vee x') = (y \wedge x) \vee (y \wedge x') \\ &= (z \wedge x) \vee (z \wedge x') = z \wedge (x \vee x') = z \wedge 1 = z \end{aligned}$$

(vi) We shall use (v), replacing y by $x \vee (y \vee z)$ and z by $(x \vee y) \vee z$. Thus to apply (v) we must show

(*a*) $(x \vee (y \vee z)) \wedge x = ((x \vee y) \vee z) \wedge x$ and (*b*) $(x \vee (y \vee z)) \wedge x' = ((x \vee y) \vee z) \wedge x'$

To prove (*a*): $(x \vee (y \vee z)) \wedge x = x \wedge (x \vee (y \vee z)) = x$ by (iii). Also,

$$\begin{aligned} ((x \vee y) \vee z) \wedge x &= x \wedge ((x \vee y) \vee z) \\ &= [x \wedge (x \vee y)] \vee [x \wedge z] \\ &= x \vee (x \wedge z) \quad \text{by (iii)} \\ &= x \quad \text{by (iv)} \end{aligned}$$

Thus $(x \vee (y \vee z)) \wedge x = x = ((x \vee y) \vee z) \wedge x$

To prove (*b*):
$$\begin{aligned}(x \vee (y \vee z)) \wedge x' &= x' \wedge (x \vee (y \vee z)) \\ &= (x' \wedge x) \vee (x' \wedge (y \vee z)) \\ &= 0 \vee (x' \wedge (y \vee z)) = x' \wedge (y \vee z)\end{aligned}$$

Also,
$$\begin{aligned}((x \vee y) \vee z) \wedge x' &= x' \wedge [(x \vee y) \vee z] = (x' \wedge (x \vee y)) \vee (x' \wedge z) \\ &= [(x' \wedge x) \vee (x' \wedge y)] \vee (x' \wedge z) \\ &= [0 \vee (x' \wedge y)] \vee [x' \wedge z] \\ &= (x' \wedge y) \vee (x' \wedge z) = x' \wedge (y \vee z)\end{aligned}$$

Thus $(x \vee (y \vee z)) \wedge x' = x' \wedge (y \vee z) = ((x \vee y) \vee z) \wedge x'$

(vii) is the dual of (vi).

(viii) To prove $(x \vee y)' = x' \wedge y'$, we use the uniqueness of the complement (Theorem 3.1). We must show

$$(c)\ (x \vee y) \wedge (x' \wedge y') = 0 \quad \text{and} \quad (d)\ (x \vee y) \vee (x' \wedge y') = 1$$

To prove (*c*):
$$\begin{aligned}(x \vee y) \wedge (x' \wedge y') &= (x' \wedge y') \wedge (x \vee y) \\ &= [(x' \wedge y') \wedge x] \vee [(x' \wedge y') \wedge y] \\ &= [x \wedge (x' \wedge y')] \vee [x' \wedge (y' \wedge y)] \\ &= [(x \wedge x') \wedge y'] \vee [x' \wedge (y \wedge y')] \\ &= [0 \wedge y'] \vee [x' \wedge 0] = 0 \vee 0 = 0\end{aligned}$$

To prove (*d*):
$$\begin{aligned}(x \vee y) \vee (x' \wedge y') &= [(x \vee y) \vee x'] \wedge [(x \vee y) \vee y'] \\ &= [x' \vee (x \vee y)] \wedge [x \vee (y \vee y')] \\ &= [(x' \vee x) \vee y] \wedge [x \vee 1] \\ &= [(x \vee x') \vee y] \wedge 1 = (x \vee x') \vee y \\ &= 1 \vee y = y \vee 1 = 1\end{aligned}$$

(ix) is the dual of (viii).

(x) By (viii), $(x \vee y)' = x' \wedge y'$. Hence $(x \vee y)'' = (x' \wedge y')'$. But $(x \vee y)'' = x \vee y$, by Corollary 3.2.

(xi) is the dual of (x).

(xii) $x = x \wedge 1 = x \wedge (y \vee y') = (x \wedge y) \vee (x \wedge y')$. Therefore $x \wedge y' = 0$ implies $x = x \wedge y$. Conversely, assume $x = x \wedge y$. Then

$$\begin{aligned}x \wedge y' &= 0 \vee (x \wedge y') = (x \wedge x') \vee (x \wedge y') \\ &= x \wedge (x' \vee y') = x \wedge (x \wedge y)' = x \wedge x' = 0\end{aligned}$$

(xiii) Since $0 \vee 1 = 1$ and $0 \wedge 1 = 0$, we obtain $0' = 1$ by Theorem 3.1.

(xiv) is the dual of (xiii).

(xv) $x \wedge (x' \vee y) = (x \wedge x') \vee (x \wedge y) = 0 \vee (x \wedge y) = x \wedge y$.

(xvi) is the dual of (xv). ▶

3.3 SUBALGEBRAS

It is clear that a Boolean algebra $\langle B, \wedge, \vee, ', 0, 1\rangle$ has a unique zero element 0 and a unique unit element 1. For, assume that z is also a possible zero element; in particular, $x = x \vee z$ for all x in B. Hence if we let $x = 0$, $0 = 0 \vee z$. But $0 \vee z = z \vee 0 = z$. Thus $0 = z$. Likewise, if u were a possible unit element, then $1 = 1 \wedge u = u \wedge 1 = u$.

The subset $\{0,1\}$ is closed† under the operations $\wedge, \vee, '$. For,

$$\begin{cases} 0 \vee 1 = 1 = 1 \vee 0 \\ 1 \vee 1 = 1 \\ 0 \vee 0 = 0 \end{cases} \qquad \begin{cases} 0 \wedge 1 = 0 = 1 \wedge 0 \\ 1 \wedge 1 = 1 \\ 0 \wedge 0 = 0 \end{cases} \qquad \begin{cases} 0' = 1 \\ 1' = 0 \end{cases}$$

Thus if we let $\wedge_{\{0,1\}}, \vee_{\{0,1\}}, '_{\{0,1\}}$ denote the restrictions of the operations $\wedge, \vee, '$ to the set $\{0,1\}$, then $\mathcal{B}^* = \langle\{0,1\}, \wedge_{\{0,1\}}, \vee_{\{0,1\}}, '_{\{0,1\}}, 0, 1\rangle$ is itself a Boolean algebra. Notice that all we had to observe was that $\{0,1\}$ is closed under the operations $\wedge, \vee, '$; it is easy to check that Axioms (1)-(9) are then automatically satisfied.

More generally, if A is any non-empty subset of B closed under the operations $\wedge, \vee, '$, then $\langle A, \wedge_A, \vee_A, '_A, 0, 1\rangle$ is a Boolean algebra, where $\wedge_A, \vee_A, '_A$ are the restrictions of the operations $\wedge, \vee, '$ to the set A. Observe that 0 and 1 must belong to A. For, if $x \in A$, then $x' \in A$, and thus we obtain $0 = x \wedge x' \in A$ and $1 = x \vee x' \in A$. The Boolean algebra $\langle A, \wedge_A, \vee_A, '_A, 0, 1\rangle$ is called a *subalgebra* of $\mathcal{B}$. In particular, the Boolean algebra $\mathcal{B}^*$ determined by $\{0,1\}$ is a subalgebra of $\mathcal{B}$. In fact, $\mathcal{B}^*$ is the "smallest" subalgebra of $\mathcal{B}$, since $\mathcal{B}^*$ is a subalgebra of any other subalgebra of $\mathcal{B}$.

To show that a subset A of B is closed under $\wedge, \vee, '$, it suffices to show that A is closed either under $\wedge$ and $'$, or under $\vee$ and $'$. For, if A is closed under $\wedge$ and $'$, then, for any x, y in A, $x \vee y = (x' \wedge y')' \in A$. Likewise, if A is closed under $\vee$ and $'$, then, for any x, y in A, $x \wedge y = (x' \vee y')' \in A$.

Example 3.7.

Let $\mathcal{B}$ be the Boolean algebra $\mathcal{P}(K)$ of all subsets of an infinite set K under the usual set-theoretic operations of intersection, union and complementation, and with $\emptyset$ and K as the zero element and unit element, respectively. Let A be the set of all subsets of K which are either finite or cofinite (i.e. the complement of a finite set). Then A is closed under intersection, union and complement, and therefore A determines a subalgebra of $\mathcal{B}$. In general, the subalgebras are the fields of subsets of K.

3.4 PARTIAL ORDERS

In a Boolean algebra $\mathcal{B}$, we define a binary relation $\leq$ on B by stipulating that

$$x \leq y \quad \text{if and only if} \quad x \wedge y = x\text{††}$$

Theorem 3.6. $x \leq y$ if and only if $x \vee y = y$.

Proof. Assume $x \leq y$. Then

$$x \vee y = (x \wedge y) \vee y = y$$

(using Theorem 3.5(iv)). Conversely, if $x \vee y = y$, then

$$x \wedge y = x \wedge (x \vee y) = x$$

(using Theorem 3.4(iii)). ▶

Example 3.8.

In a Boolean algebra $\mathcal{P}(A)$, the relation $x \leq y$ is equivalent to $x \subseteq y$, for any subsets x and y of A.

---

†Recall that A is said to be closed under the operations $\wedge, \vee, '$ if, for any x and y in A, the objects $x \wedge y$, $x \vee y$, and x' are also in A.

††The symbol $\leq$ should not be confused with the symbol for the usual ordering of integers or of real numbers. If necessary, use $\leq_{\mathcal{B}}$ instead of $\leq$.

Example 3.9.

In the Boolean algebra $\mathcal{B}$ associated with the propositional calculus (Example 3.5), if **A** and **B** are statement forms, then $[\mathbf{A}] \leq [\mathbf{B}]$ if and only if $[\mathbf{A}] \wedge [\mathbf{B}] = [\mathbf{A}]$. But $[\mathbf{A}] \wedge [\mathbf{B}] = [\mathbf{A} \,\&\, \mathbf{B}]$. Thus $[\mathbf{A}] \leq [\mathbf{B}]$ if and only if $[\mathbf{A} \,\&\, \mathbf{B}] = [\mathbf{A}]$, i.e. if and only if $\mathbf{A} \,\&\, \mathbf{B}$ and **A** are logically equivalent. Clearly $\mathbf{A} \,\&\, \mathbf{B}$ is logically equivalent to **A** if and only if **A** logically implies **B**. Thus $[\mathbf{A}] \leq [\mathbf{B}]$ if and only if **A** logically implies **B** (or, equivalently, if and only if $\mathbf{A} \rightarrow \mathbf{B}$ is a tautology).

Theorem 3.7. (PO 1) $x \leq x$ (Reflexivity)

(PO 2) $(x \leq y \;\&\; y \leq z) \rightarrow x \leq z$ (Transitivity)

(PO 3) $(x \leq y \;\&\; y \leq x) \rightarrow x = y$ (Anti-symmetry)

Proof.

(PO 1) $x \wedge x = x$.

(PO 2) Assume $x \wedge y = x$ and $y \wedge z = y$. Then

$$x \wedge z = (x \wedge y) \wedge z = x \wedge (y \wedge z) = x \wedge y = x$$

(PO 3) Assume $x \wedge y = x$ and $y \wedge x = y$. Then

$$x = x \wedge y = y \wedge x = y$$ ▶

In general, a binary relation R on a set A is any subset of $A \times A$, i.e. any set of ordered pairs $\langle u, v\rangle$ such that $u \in A \;\&\; v \in A$. For example, the relation of fatherhood on the set of human beings is the set of all ordered pairs $\langle x, y\rangle$ such that x and y are people and x is the father of y. In accordance with tradition, one often writes xRy instead of $\langle x, y\rangle \in R$.

A binary relation R on a set A satisfying the analogues of (PO 2) – (PO 3),

(2) $(xRy \;\&\; yRz) \rightarrow xRz$,

(3) $(xRy \;\&\; yRx) \rightarrow x = y$,

is called a *partial order* on A.

A partial order on a set A is said to be *reflexive* if and only if xRx holds for all x in A, while R is said to be *irreflexive* on A if and only if xRx is false for all x in A. For example, the ordinary relation $\leq$ on the set of integers is reflexive while the relation $<$ on the set of integers is irreflexive. In Theorem 3.7 we have seen that the binary relation $\leq$ on a Boolean algebra $\mathcal{B}$ is a reflexive partial order.

If $\leq$ is a reflexive partial order on a set A, we can define $x < y$ to mean that $x \leq y \;\&\; x \neq y$. Then we have

Theorem 3.8. (i) $\neg(x < x)$

(ii) $(x < y \;\&\; y \leq z) \rightarrow x < z$

(iii) $(x \leq y \;\&\; y < z) \rightarrow x < z$

(iv) $(x < y \;\&\; y < z) \rightarrow x < z$

(v) $\neg(x < y \;\&\; y < x)$

(vi) $<$ is an irreflexive partial order on A.

Of course, given an irreflexive partial order $<$ on A, we can define a reflexive partial order $\leq$ on A as follows: $x \leq y \leftrightarrow (x < y$ or $x = y)$.

Let $\leq$ be a partial order on a set A. An element z of A is said to be an *upper bound* of a subset $Y \subseteq A$ if $y \leq z$ for all y in Y. An element z of A is said to be a *least upper bound* (lub) of a subset $Y \subseteq A$ if and only if

(1) z is an upper bound of Y,

(2) $z \leq w$ for every upper bound w of Y.

Clearly, by (PO 3), a subset Y of A has at most one lub.

Similarly, an element z of A is said to be a *lower bound* of a subset $Y \subseteq A$ if and only if $z \leq y$ for all y in Y; z is called a *greatest lower bound* (glb) of Y if and only if

(3) z is a lower bound of Y,

(4) $w \leq z$ for every lower bound w of Y.

Again, by (PO 3), Y has at most one glb.

Example 3.10.

The usual order relation $\leq$ on the set I of integers is a partial order on I. Any non-empty subset of I having an upper bound (respectively, lower bound) must have a lub (respectively, glb), which is, in fact, the greatest element (respectively, smallest element) of the set. However, there are non-empty subsets which have no lub, e.g. the set I itself or the set of even integers.

Example 3.11.

The usual order relation $\leq$ on the set R of real numbers is a partial order on R. Any non-empty subset of R having an upper bound (respectively, lower bound) must have a lub (respectively, glb).

Example 3.12.

The usual order relation $\leq$ on the set Q of rational numbers is a partial order on Q. However, in this case, there exist non-empty subsets of Q which are bounded above but do not have a lub. An example is the set of all positive rational numbers x such that $x^2 < 2$. (This is just another way of saying that $\sqrt{2}$ is not rational.)

Examples 3.10-3.12 possess the additional property of *connectedness*:

(Conn): For any x and y in A, $x \leq y$ or $y \leq x$.

A partial order satisfying (Conn) is called a *total order* (synonyms: simple order, linear order). Not all partial orders are total orders.

Example 3.13.

The partial order $\subseteq$ determined by the Boolean algebra of all subsets of $\{0, 1\}$ is not connected, for we have neither $\{0\} \subseteq \{1\}$ nor $\{1\} \subseteq \{0\}$.

A partial order on a finite set A can be indicated by a diagram in which the elements of A are pictured as points, and a point x has the relation to some point y if and only if y can be reached from x by following a sequence of zero or more upward arrows. The order relation $\subseteq$ in the Boolean algebra of all subsets of $\{0, 1\}$ is pictured in Fig. 3-1, and the order relation in the Boolean algebra of all subsets of $\{0, 1, 2\}$ is shown in Fig. 3-2.

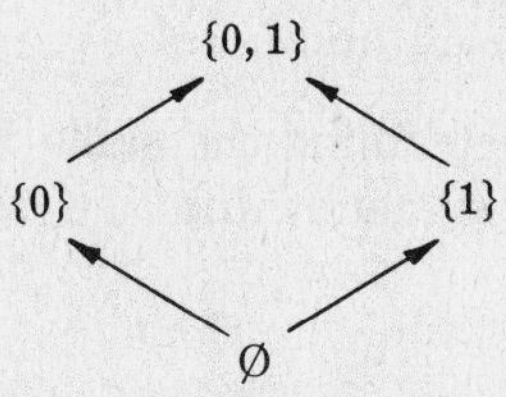

Fig. 3-1

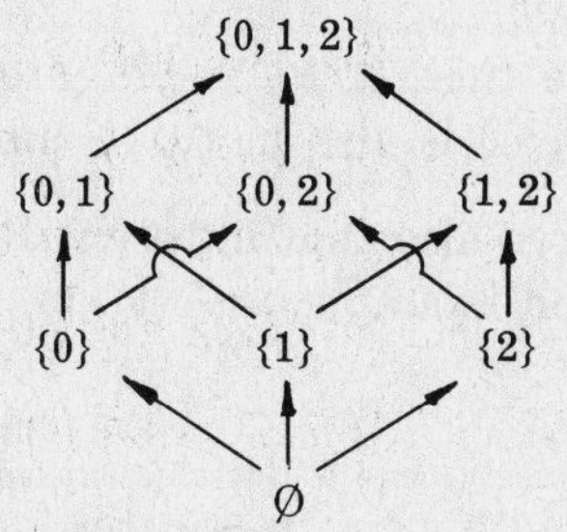

Fig. 3-2

The partial order $\leq$ determined by a Boolean algebra $\mathcal{B}$ has a special property (L) not possessed by all partial orders.

Theorem 3.9. (L) For any x and y, $\{x, y\}$ has both a lub (namely, $x \vee y$) and a glb (namely, $x \wedge y$).

Proof. $x \leq x \vee y$ (since $x \wedge (x \vee y) = x$), and similarly, $y \leq x \vee y$. Thus $x \vee y$ is an upper bound of $\{x, y\}$. Now assume w is any upper bound of $\{x, y\}$. This means that $x \leq w$ and $y \leq w$, i.e. $x \wedge w = x$ and $y \wedge w = y$. Then $(x \vee y) \wedge w = (x \wedge w) \vee (y \wedge w) = x \vee y$, i.e. $x \vee y \leq w$. Thus $x \vee y$ is the lub of $\{x, y\}$. The proof for $x \wedge y$ is left to the reader. ▶

3.5 BOOLEAN EXPRESSIONS AND FUNCTIONS. NORMAL FORMS

By a *Boolean expression* we mean any expression built up from the variables $x, y, z, x_1, y_1, z_1, x_2, y_2, z_2, \ldots$ by applying the operations $\wedge, \vee, '$ a finite number of times. In other words, all variables are Boolean expressions, and if τ and σ are Boolean expressions, so are $(\tau \wedge \sigma)$, $(\tau \vee \sigma)$ and (τ')†.

Example 3.14.

The following are Boolean expressions:

$$((x \vee (y')) \wedge (z_2')),\quad ((y \wedge z) \vee ((x') \wedge y)),\quad ((y \wedge (z \vee x))'),\quad ((((y')') \wedge z) \wedge y),\quad (((y')') \wedge (z \wedge y))$$

We shall use the same conventions for omitting parentheses as were used for statement forms in Chapter 1 (cf. page 5). For this purpose, the symbols $\wedge, \vee, '$ are to correspond to &, $\vee$, ¬.

Example 3.15.

Using the conventions for omitting parentheses, we can write the Boolean expressions of Example 3.14 as follows:

$$(x \vee y') \wedge z_2',\quad (y \wedge z) \vee (x' \wedge y),\quad (y \wedge (z \vee x))',\quad y'' \wedge z \wedge y,\quad y'' \wedge (z \wedge y)$$

Given a Boolean algebra $\mathcal{B} = \langle B, \wedge, \vee, ', 0, 1\rangle$ and a Boolean expression $\tau(u_1, \ldots, u_k)$ having its variables among $u_1, \ldots, u_k$, we can determine a corresponding *Boolean function* $\tau^{\mathcal{B}}(u_1, \ldots, u_k)$: for each k-tuple $\langle b_1, \ldots, b_k\rangle$ of elements of B, $\tau^{\mathcal{B}}(b_1, \ldots, b_k)$ is the element of B obtained by assigning the values $b_1, \ldots, b_k$ to $u_1, \ldots, u_k$ respectively, and interpreting the symbols $\wedge, \vee, '$ to mean the corresponding operations in $\mathcal{B}$. (In order to make the corresponding function unique, we always shall list the variables $u_1, \ldots, u_k$ in the order in which they occur in the list $x, y, z, x_1, y_1, z_1, x_2, y_2, z_2, \ldots$. For example, $y \vee x'$ determines the function $f(x, y) = y \vee x'$; thus $f(1, 0) = 0$ and $f(0, 1) = 1$.)

Example 3.16.

The Boolean expression $x \vee y'$ determines the following function $f(x, y)$ with respect to the two-element Boolean algebra $\mathcal{B}_0$.

$$f(0,0) = 1, \quad f(0,1) = 0, \quad f(1,0) = 1, \quad f(1,1) = 1$$

Notice that, if $b_1, \ldots, b_n$ are in $\{0, 1\}$ and $\tau(u_1, \ldots, u_n)$ is a Boolean expression, then $\tau^{\mathcal{B}}(b_1, \ldots, b_n)$ is also in $\{0, 1\}$, since $\{0, 1\}$ is closed under $\wedge$, $\vee$ and $'$.

Observe also that different Boolean expressions may determine the same Boolean function. For example, $x \wedge (y \vee z)$ and $(x \wedge y) \vee (x \wedge z)$ always determine the same Boolean functions.

---

†More precisely, σ is a Boolean expression if and only if there is a finite sequence $\tau_1, \ldots, \tau_n$ such that τ_n is σ, and, if $1 \leq i \leq n$, then either τ_i is a variable or there exist $j, k < i$ such that τ_i is $(\tau_j \wedge \tau_k)$ or τ_i is $(\tau_j \vee \tau_k)$ or τ_i is (τ_j').

Theorem 3.10. Given a Boolean expression $\tau(u)$, which may contain other variables $u_1, \ldots, u_k$ as well as u. Then the equation

$$\tau(u) = [\tau(0) \wedge u'] \vee [\tau(1) \wedge u]$$

is derivable from the axioms for Boolean algebras.

Proof. See Problem 3.7, page 66. ▶

Now we shall present a normal form theorem for Boolean algebras which is a generalization of the disjunctive normal theorem for propositional logic (Theorem 1.6). The following notation will be convenient:

For any expression τ, $\tau^i = \begin{cases} \tau & \text{if } i = 1 \\ \tau' & \text{if } i = 0 \end{cases}$.

The symbol $\sum$, with appropriate indices, will be used to indicate repeated use of $\vee$. In particular, $\sum_{\alpha=0}^{1} \sigma(\alpha)$ stands for $\sigma(0) \vee \sigma(1)$, while $\sum_{\alpha_1=0}^{1} \sum_{\alpha_2=0}^{1} \sigma(\alpha_1, \alpha_2)$ stands for $\sigma(0,0) \vee \sigma(0,1) \vee \sigma(1,0) \vee \sigma(1,1)$.

Theorem 3.11. (Disjunctive Normal Form) For any Boolean expression $\tau(u_1, \ldots, u_k)$, the equation

$$\tau(u_1, \ldots, u_k) = \sum_{\alpha_1=0}^{1} \sum_{\alpha_2=0}^{1} \cdots \sum_{\alpha_k=0}^{1} \tau(\alpha_1, \alpha_2, \ldots, \alpha_k) \wedge u_1^{\alpha_1} \wedge u_2^{\alpha_2} \wedge \cdots \wedge u_k^{\alpha_k}$$

is derivable from the axioms for Boolean algebra (and therefore the corresponding equation, with τ replaced by $\tau^{\mathcal{B}}$, holds in any Boolean algebra $\mathcal{B}$).

Proof. See Problem 3.8, page 67. ▶

Example 3.17.

When $k = 1$, Theorem 3.11 reads

$$\tau(u) = \sum_{\alpha=0}^{1} \tau(\alpha) \wedge u^{\alpha} = (\tau(0) \wedge u') \vee (\tau(1) \wedge u)$$

When $k = 2$, we obtain

$$\begin{aligned} \tau(u_1, u_2) &= \sum_{\alpha_1=0}^{1} \sum_{\alpha_2=0}^{1} \tau(\alpha_1, \alpha_2) \wedge u_1^{\alpha_1} \wedge u_2^{\alpha_2} \\ &= (\tau(0,0) \wedge u_1' \wedge u_2') \vee (\tau(0,1) \wedge u_1' \wedge u_2) \vee (\tau(1,0) \wedge u_1 \wedge u_2') \vee (\tau(1,1) \wedge u_1 \wedge u_2) \end{aligned}$$

Example 3.18.

When τ is $x \vee y$, Theorem 3.11 states

$$\begin{aligned} x \vee y &= [(0 \vee 0) \wedge x' \wedge y'] \vee [(1 \vee 0) \wedge x \wedge y'] \vee [(0 \vee 1) \wedge x' \wedge y] \vee [(1 \vee 1) \wedge x \wedge y] \\ &= [0 \wedge x' \wedge y'] \vee [1 \wedge x \wedge y'] \vee [1 \wedge x' \wedge y] \vee [1 \wedge x \wedge y] \\ &= [x \wedge y'] \vee [x' \wedge y] \vee [x \wedge y] \end{aligned}$$

Example 3.19.

The representation of $(x \vee y) \wedge (x' \vee y')$ in disjunctive normal form is

$$\begin{aligned} (x \vee y) \wedge (x' \vee y') &= (0 \wedge x' \wedge y') \vee (1 \wedge x' \wedge y) \vee (1 \wedge x \wedge y') \vee (0 \wedge x \wedge y) \\ &= (x' \wedge y) \vee (x \wedge y') \end{aligned}$$

Corollary 3.12. Let $\tau(u_1, \ldots, u_k)$ and $\sigma(u_1, \ldots, u_k)$ be Boolean expressions, and let $\mathcal{B}$ be some Boolean algebra. If the Boolean functions $\tau^{\mathcal{B}}(u_1, \ldots, u_k)$ and $\sigma^{\mathcal{B}}(u_1, \ldots, u_k)$ are equal, then:

(a) the equation $\tau(u_1, \ldots, u_k) = \sigma(u_1, \ldots, u_k)$ is provable from the axioms for Boolean algebras;

(b) $\tau^{\mathcal{C}} = \sigma^{\mathcal{C}}$ for *all* Boolean algebras $\mathcal{C}$.

Proof. (a) Since $\{0,1\} \subseteq B$, $\tau^{\mathcal{B}}(b_1, \ldots, b_k) = \sigma^{\mathcal{B}}(b_1, \ldots, b_k)$ whenever $b_1, \ldots, b_k \in \{0,1\}$. But $\tau^{\mathcal{B}}(b_1, \ldots, b_k) = \sigma^{\mathcal{B}}(b_1, \ldots, b_k)$ holds if and only if the corresponding equation $\tau(b_1, \ldots, b_k) = \sigma(b_1, \ldots, b_k)$ can be proved from the axioms for Boolean algebras. For, the equations $0 \wedge 0 = 0$, $0 \wedge 1 = 1 \wedge 0 = 0$, $1 \wedge 1 = 1$, etc., are all derivable from these axioms, and the values $\tau^{\mathcal{B}}(b_1, \ldots, b_k)$ and $\sigma^{\mathcal{B}}(b_1, \ldots, b_k)$ are computable from these equations. Hence by Theorem 3.11, $\tau(u_1, \ldots, u_k) = \sigma(u_1, \ldots, u_k)$ is derivable from the axioms.

(b) is an immediate consequence of (a). ▶

The remarkable thing about Corollary 3.12 is that, if an equation holds for one Boolean algebra (in particular, if it holds for the two-element Boolean algebra $\mathcal{B}_2$), then it holds for all Boolean algebras. To mathematicians it probably would not have been surprising if we had only asserted that, if an equation holds for all Boolean algebras, then it is provable from the axioms. This latter assertion follows, in fact, from the completeness theorem for first-order logic (see Corollary 2.15(*a*), page 68 of [135]).

3.6 ISOMORPHISMS

A function Φ is called an *isomorphism* from a Boolean algebra $\mathcal{B} = \langle B, \wedge_{\mathcal{B}}, \vee_{\mathcal{B}}, '^{\mathcal{B}}, 0_{\mathcal{B}}, 1_{\mathcal{B}} \rangle$ into a Boolean algebra $\mathcal{C} = \langle C, \wedge_{\mathcal{C}}, \vee_{\mathcal{C}}, '^{\mathcal{C}}, 0_{\mathcal{C}}, 1_{\mathcal{C}} \rangle$ if and only if

(a) Φ is a one-one function from B into C,

(b) for any x, y in B,

$$\begin{aligned} \Phi(x \wedge_{\mathcal{B}} y) &= \Phi(x) \wedge_{\mathcal{C}} \Phi(y) \\ \Phi(x \vee_{\mathcal{B}} y) &= \Phi(x) \vee_{\mathcal{C}} \Phi(y) \\ \Phi(x'^{\mathcal{B}}) &= (\Phi(x))'^{\mathcal{C}} \end{aligned}$$

Such a function Φ is called an isomorphism from $\mathcal{B}$ *onto* $\mathcal{C}$ if, in addition, Φ is a function from B onto C.

Theorem 3.13. Let Φ be an isomorphism from a Boolean algebra $\mathcal{B}$ into (respectively, onto) a Boolean algebra $\mathcal{C}$ (with the notation given above). Then

(a) $\Phi(0_{\mathcal{B}}) = 0_{\mathcal{C}}$ and $\Phi(1_{\mathcal{B}}) = 1_{\mathcal{C}}$.

(b) It is not necessary to assume that

$$\Phi(x \vee_{\mathcal{B}} y) = \Phi(x) \vee_{\mathcal{C}} \Phi(y) \quad \text{for all } x, y \text{ in } B$$

Alternatively, we could omit the assumption that

$$\Phi(x \wedge_{\mathcal{B}} y) = \Phi(x) \wedge_{\mathcal{C}} \Phi(y)$$

(c) If Θ is an isomorphism from $\mathcal{C}$ into (respectively, onto) a Boolean algebra $\mathcal{D} = \langle D, \wedge_{\mathcal{D}}, \vee_{\mathcal{D}}, '^{\mathcal{D}}, 0_{\mathcal{D}}, 1_{\mathcal{D}} \rangle$, then the composite mapping† $\Theta \circ \Phi$ is an isomorphism from $\mathcal{B}$ into (respectively, onto) $\mathcal{D}$.

†The *composite mapping* (or *composition*) $\Theta \circ \Phi$ is the function defined on the domain B of Φ such that $(\Theta \circ \Phi)(x) = \Theta(\Phi(x))$ for each x in B. The *inverse* Φ^{-1} is the function whose domain is the range $\Phi[B]$ of Φ (here, $\Phi[B] = \{\Phi(x) : x \in B\}$) and such that, for any y in $\Phi[B]$, $(\Phi^{-1})(y)$ is the unique x in B such that $\Phi(x) = y$.

(*d*) The inverse mapping Φ^{-1} is an isomorphism from the subalgebra of $\mathcal{C}$ determined by $\Phi[B]$ onto $\mathcal{B}$, and, in particular, if Φ is onto $\mathcal{C}$, then Φ^{-1} is an isomorphism from $\mathcal{C}$ onto $\mathcal{B}$.

Proof.

(*a*)
$$\Phi(0_{\mathcal{B}}) = \Phi(x \wedge_{\mathcal{B}} x'^{\mathcal{B}}) = \Phi(x) \wedge_{\mathcal{C}} \Phi(x'^{\mathcal{B}}) = \Phi(x) \wedge_{\mathcal{C}} (\Phi(x))'^{\mathcal{C}} = 0_{\mathcal{C}}$$
$$\Phi(1_{\mathcal{B}}) = \Phi(0_{\mathcal{B}}'^{\mathcal{B}}) = (\Phi(0_{\mathcal{B}}))'^{\mathcal{C}} = 0_{\mathcal{C}}'^{\mathcal{C}} = 1_{\mathcal{C}}$$

(*b*)
$$\Phi(x \vee_{\mathcal{B}} y) = \Phi((x'^{\mathcal{B}} \wedge_{\mathcal{B}} y'^{\mathcal{B}})'^{\mathcal{B}}) = (\Phi(x'^{\mathcal{B}} \wedge_{\mathcal{B}} y'^{\mathcal{B}}))'^{\mathcal{C}}$$
$$= (\Phi(x'^{\mathcal{B}}) \wedge_{\mathcal{C}} \Phi(y'^{\mathcal{B}}))'^{\mathcal{C}} = ((\Phi(x))'^{\mathcal{C}} \wedge_{\mathcal{C}} (\Phi(y))'^{\mathcal{C}})'^{\mathcal{C}} = \Phi(x) \vee_{\mathcal{C}} \Phi(y)$$

(*c*) First, $\Theta \circ \Phi$ is one-one. (If $x \neq y$, then $\Phi(x) \neq \Phi(y)$ and therefore $\Theta(\Phi(x)) \neq \Theta(\Phi(y))$.)

Second, $(\Theta \circ \Phi)(x'^{\mathcal{B}}) = \Theta(\Phi(x'^{\mathcal{B}})) = \Theta((\Phi(x))'^{\mathcal{C}}) = (\Theta(\Phi(x)))'^{\mathcal{D}} = ((\Theta \circ \Phi)(x))'^{\mathcal{D}}$

Lastly,
$$(\Theta \circ \Phi)(x \wedge_{\mathcal{B}} y) = \Theta(\Phi(x \wedge_{\mathcal{B}} y)) = \Theta(\Phi(x) \wedge_{\mathcal{C}} \Phi(y))$$
$$= \Theta(\Phi(x)) \wedge_{\mathcal{D}} \Theta(\Phi(y)) = (\Theta \circ \Phi)(x) \wedge_{\mathcal{D}} (\Theta \circ \Phi)(y)$$

(*d*) Assume $z, w \in \Phi[B]$. Then $z = \Phi(x)$ and $w = \Phi(y)$ for some x and y in B. Hence $x = \Phi^{-1}(z)$ and $y = \Phi^{-1}(w)$. First, if $z \neq w$, then $x \neq y$ (for, if $x = y$, then $z = \Phi(x) = \Phi(y) = w$). Thus Φ^{-1} is one-one. Second, $\Phi(x \vee_{\mathcal{B}} y) = \Phi(x) \vee_{\mathcal{C}} \Phi(y) = z \vee_{\mathcal{C}} w$. Hence $\Phi^{-1}(z \vee_{\mathcal{C}} w) = x \vee_{\mathcal{B}} y = \Phi^{-1}(z) \vee_{\mathcal{B}} \Phi^{-1}(w)$. Third, $\Phi(x'^{\mathcal{B}}) = (\Phi(x))'^{\mathcal{C}} = z'^{\mathcal{C}}$. Hence $\Phi^{-1}(z'^{\mathcal{C}}) = x'^{\mathcal{B}} = (\Phi^{-1}(z))'^{\mathcal{B}}$. ▶

We say that $\mathcal{B}$ *is isomorphic with* $\mathcal{C}$ if and only if there is an isomorphism from $\mathcal{B}$ onto $\mathcal{C}$. From Theorem 3.13(*d, c*) it follows that, if $\mathcal{B}$ is isomorphic with $\mathcal{C}$, then $\mathcal{C}$ is isomorphic with $\mathcal{B}$, and if, in addition, $\mathcal{C}$ is isomorphic with $\mathcal{D}$, then $\mathcal{B}$ is isomorphic with $\mathcal{D}$. Isomorphic Boolean algebras have, in a certain sense, the same Boolean structure. More precisely, this means that any property (formulated in the language of Boolean algebras) holding for one Boolean algebra also holds for any isomorphic Boolean algebra.†

Example 3.20.

Consider the two-element subalgebra $\mathcal{C} = \{0_{\mathcal{B}}, 1_{\mathcal{B}}\}$ of any Boolean algebra $\mathcal{B}$. Let $\mathcal{D}$ be the Boolean algebra whose elements are the integers 0 and 1 and whose operations are:

($\wedge_{\mathcal{D}}$): ordinary multiplication, i.e. $0 \wedge_{\mathcal{D}} 0 = 0 \wedge_{\mathcal{D}} 1 = 1 \wedge_{\mathcal{D}} 0 = 0$, $1 \wedge_{\mathcal{D}} 1 = 1$.

($\vee_{\mathcal{D}}$): addition modulo 2, i.e. $0 \vee_{\mathcal{D}} 0 = 0$, $0 \vee_{\mathcal{D}} 1 = 1 \vee_{\mathcal{D}} 0 = 1$, $1 \vee_{\mathcal{D}} 1 = 0$.

($'^{\mathcal{D}}$): the function $1 - x$, i.e. $0'^{\mathcal{D}} = 1$ and $1'^{\mathcal{D}} = 0$.

Then the function Φ on $\{0_{\mathcal{B}}, 1_{\mathcal{B}}\}$ such that $\Phi(0_{\mathcal{B}}) = 0$ and $\Phi(1_{\mathcal{B}}) = 1$ is an isomorphism of $\mathcal{C}$ onto $\mathcal{D}$.

3.7 BOOLEAN ALGEBRAS AND PROPOSITIONAL LOGIC

A statement form **A** and a Boolean expression τ are said to correspond if τ arises from **A** by replacing $\neg$, &, $\vee$ by $'$, $\wedge$, $\vee$ (respectively) and by replacing the statement letters $A, B, C, A_1, B_1, C_1, \ldots$ by $x, y, z, x_1, y_1, z_1, \ldots$ (respectively).

†For a rigorous formulation of this assertion, see page 90 of [135].

Example 3.21.

$A \vee (B \,\&\, \neg C)$ corresponds to $x \vee (y \wedge z')$. $\neg(\neg A \,\&\, (B_1 \vee A))$ corresponds to $(x' \wedge (y_1 \vee x))'$.

The statement form corresponding to a Boolean expression τ will be denoted SF(τ).

Theorem 3.14. The equation $\tau = \sigma$ holds for all Boolean algebras if and only if SF(τ) is logically equivalent to SF(σ). Hence we have a decision procedure to determine whether $\tau = \sigma$ holds for all Boolean algebras.

Proof. By Corollary 3.12, $\tau = \sigma$ holds for all Boolean algebras if and only if $\tau = \sigma$ holds for the two-element Boolean algebra $\mathcal{C} = \langle\{F, T\}, \&_{\{F,T\}}, \vee_{\{F,T\}}, '^{\{F,T\}}, F, T\rangle$, where the operations $\&_{\{F,T\}}, \vee_{\{F,T\}}, '^{\{F,T\}}$ have the obvious meanings given by the usual truth tables. These are given in detail in Problem 2.16, page 44. It is clear that $\tau = \sigma$ holds for $\mathcal{C}$ if and only if SF(τ) and SF(σ) always take the same truth values. (For, an assignment of truth values, T or F, to the statement letters in SF(τ) and SF(σ) corresponds to substitution of the same truth values for the corresponding variables in τ and σ.) ▶

Example 3.22.

Consider the equation $x \wedge (y \vee z) = (x \wedge y) \vee (x \wedge z)$. The corresponding statement forms are $A \,\&\, (B \vee C)$ and $(A \,\&\, B) \vee (A \,\&\, C)$. To check that these statement forms are logically equivalent, we substitute T and F for A, B, C in all possible ways and verify that the outcomes are the same. For example, if A is F, B is T, and C is F, then $A \,\&\, (B \vee C)$ and $(A \,\&\, B) \vee (A \,\&\, C)$ both are F. The computation we make to determine this is essentially the same as the one we make to see that $x \wedge (y \vee z) = F = (x \wedge y) \vee (x \wedge z)$ when x is F, y is T, and z is F. (Namely, $F \wedge (T \vee F) = F \wedge T = F$ and $(F \wedge T) \vee (F \wedge F) = F \vee F = F$.)

Solved Problems

3.1. In a Boolean algebra, let $x \sim y$ be defined as $x \wedge y'$. Prove:

(a) $x \vee y = x \vee (y \sim x)$

(b) $x \sim (x \sim y) = x \wedge y$

(c) A non-empty subset A determines a subalgebra if and only if A is closed under $\sim$ and $'$.

(d) $x' = 1 \sim x$

(e) $x \leq y \leftrightarrow x \sim y = 0$ (i.e. $x \leq y \leftrightarrow x \wedge y' = 0$)

(f) $x \leq 0 \leftrightarrow x = 0$

(g) $x \wedge y = 0 \leftrightarrow x \sim y = x$

(h) $x \wedge (y \sim z) = (x \wedge y) \sim (x \wedge z)$

(i) Does $x \vee (y \sim z) = (x \vee y) \sim (x \vee z)$ hold?

Solution:

(a) $x \vee (y \sim x) = x \vee (y \wedge x') = (x \vee y) \wedge (x \vee x') = (x \vee y) \wedge 1 = x \vee y$

(b) $x \sim (x \sim y) = x \wedge (x \sim y)' = x \wedge (x \wedge y')' = x \wedge (x' \vee y'')$

$= x \wedge (x' \vee y) = (x \wedge x') \vee (x \wedge y) = 0 \vee (x \wedge y) = x \wedge y$

(c) If A is a subalgebra and if x and y are in A, then $x \sim y = x \wedge y' \in A$. Conversely, if A is closed under $\sim$ and $'$, and if x and y are in A, then $x \sim y \in A$, and therefore $x \wedge y = x \sim (x \sim y) \in A$. Since A is closed under $\wedge$ and $'$, A is a subalgebra.

(d) $1 \sim x = 1 \wedge x' = x'$

(e) This is Theorem 3.5(xii).

(f) $x \sim x = x \wedge x' = 0$. Now use part (e) with $y = 0$.

(g) $x \sim y = x \;\leftrightarrow\; x \wedge y' = x$

$\leftrightarrow\; x \wedge y'' = 0$ (Theorem 3.5(xii))

$\leftrightarrow\; x \wedge y = 0$

(h) $x \wedge (y \sim z) = x \wedge (y \wedge z') = x \wedge y \wedge z'$. On the other hand,

$$(x \wedge y) \sim (x \wedge z) = (x \wedge y) \wedge (x \wedge z)' = (x \wedge y) \wedge (x' \vee z')$$
$$= y \wedge (x \wedge (x' \vee z')) = y \wedge (x \wedge z') = x \wedge y \wedge z'$$

(i) No. $1 \vee (1 \sim 1) = 1 \vee 0 = 1$. However, $(1 \vee 1) \sim (1 \vee 1) = 1 \sim 1 = 0$.

3.2. In our axiom system for Boolean algebras, prove that Axiom (9), $0 \neq 1$, is equivalent (in the presence of the other axioms) to the assertion that the Boolean algebra contains more than one element.

Solution:

Clearly, if $0 \neq 1$, then there is more than one element. Conversely, assume $0 = 1$. Then for any x, $x = x \wedge 1 = x \wedge 0 = 0$. (Notice that in proving results about Boolean algebras we have not used Axiom (9).) Thus every element is equal to 0, and the Boolean algebra contains just one element.

3.3. Let n be an integer greater than 1. Let B be the set of positive integers which are divisors of n. If x and y are in B, define $x' = n/x$, $x \wedge y$ = the greatest common divisor (gcd) of x and y, $x \vee y$ = least common multiple (lcm) of x and y. (This is a generalization of Example 3.5.)

Show that $\langle B, \wedge, \vee, ', 1, n\rangle$ is a Boolean algebra if and only if n is square-free (i.e. n is not divisible by any square greater than 1).

Solution:

Remember that the zero $0_{\mathcal{B}}$ and unit $1_{\mathcal{B}}$ of the algebra are the integers 1 and n respectively. Axioms (1)-(6) and (9) represent simple properties of integers and of greatest common divisors and least common multiples (cf., for example, [129]). However, Axioms (7) and (8) hold if and only if, for all x in B, x and n/x have no factors in common (other than 1), and this condition is equivalent to n being square-free. (Example: if $n = 60$, which is not square-free, $6' = 10$ and $6 \vee 6' = \text{lcm}(6, 10) = 30 \neq 60 = 1_{\mathcal{B}}$, $6 \wedge 6' = \text{gcd}(6, 10) = 2 \neq 1 = 0_{\mathcal{B}}$.)

SUBALGEBRAS

3.4. In the Boolean algebra of all divisors of 70 (see Example 3.4), find all subalgebras.

Solution:

We must find all subsets A of $\{1, 2, 5, 7, 10, 14, 35, 70\}$ closed under $\wedge$ and $'$. Remember that $x \wedge y = \text{gcd}(x, y)$ and $x' = 70/x$.

$$A_1 = \{1, 70\} = \{0_{\mathcal{B}}, 1_{\mathcal{B}}\}$$
$$A_2 = \{1, 2, 35, 70\}$$
$$A_3 = \{1, 5, 14, 70\}$$
$$A_4 = \{1, 7, 10, 70\}$$
$$A_5 = \{1, 2, 5, 7, 10, 14, 35, 70\}$$

3.5. (*a*) Given a subset D of a Boolean algebra $\mathcal{B}$, show that the intersection† of all subalgebras of $\mathcal{B}$ containing D as a subset is itself a subalgebra of $\mathcal{B}$ (called the *subalgebra generated by* D).

(*b*) What is the subalgebra generated by the empty set $\emptyset$?

(*c*) If $D = \{b\}$, what is the subalgebra generated by D?

Solution:

(*a*) Let C be the intersection of all subalgebras containing D. Clearly, if x and y are in C, then $x \wedge y$ and x' are in all subalgebras containing D and hence also are in C.

(*b*) $\{(0_{\mathcal{B}}, 1_{\mathcal{B}}\}$ is a subalgebra containing $\emptyset$ as a subset and is contained in all other subalgebras. Hence $\{0_{\mathcal{B}}, 1_{\mathcal{B}}\}$ is the subalgebra generated by $\emptyset$.

(*c*) $\{0_{\mathcal{B}}, 1_{\mathcal{B}}, b, b'\}$ is a subalgebra containing $\{b\}$ and is contained in every subalgebra containing $\{b\}$. Therefore $\{0_{\mathcal{B}}, 1_{\mathcal{B}}, b, b'\}$ is the subalgebra generated by $\{b\}$.

BOOLEAN EXPRESSIONS AND FUNCTIONS. NORMAL FORMS

3.6. If D is a subset of a Boolean algebra $\mathcal{B}$, show that the subalgebra $\mathcal{C}$ generated by D consists of the set C of all values obtained by substituting elements of D for the variables in all Boolean functions.

Solution:

Every such value, being obtained from elements of D by $\wedge, \vee, '$, must belong to every subalgebra containing D. On the other hand, the set C of all such values clearly forms a subalgebra containing D. Hence C is the intersection of all subalgebras containing D.

3.7. Prove Theorem 3.10: Given a Boolean expression $\tau(u)$, which may contain other variables $u_1, \ldots, u_k$ as well as u. Then the equation

$$\tau(u) \;=\; [\tau(0) \wedge u'] \vee [\tau(1) \wedge u]$$

is derivable from the axioms for Boolean algebras.

Solution:

We shall use induction on the number m of occurrences of $\wedge, \vee, '$ in τ. If $m = 0$, then τ is either u or u_i (for some i). If τ is u, then $\tau(0) = 0$ and $\tau(1) = 1$. Thus

$$\tau(u) \;=\; u \;=\; [0 \wedge u'] \vee [1 \wedge u] \;=\; [\tau(0) \wedge u'] \vee [\tau(1) \wedge u]$$

If τ is u_i, then $\tau(0) = \tau(1) = u_i$. Hence

$$\tau(u) \;=\; u_i \;=\; u_i \wedge (u' \vee u) \;=\; (u_i \wedge u') \vee (u_i \wedge u) \;=\; [\tau(0) \wedge u'] \vee [\tau(1) \wedge u]$$

Now let $m > 0$ and assume that the result is true for all expressions with fewer than m occurrences of $\wedge, \vee, '$.

Case 1. $\tau(u) = [\sigma(u)]'$. Now, by inductive hypothesis,

$$\sigma(u) \;=\; [\sigma(0) \wedge u'] \vee [\sigma(1) \wedge u]$$

Hence

$$\begin{aligned}
\tau(u) &= (\sigma(u))' \;=\; ([\sigma(0) \wedge u'] \vee [\sigma(1) \wedge u])' \\
&= [\sigma(0) \wedge u']' \wedge [\sigma(1) \wedge u]' \;=\; [\sigma(0)' \vee u''] \wedge [\sigma(1)' \vee u'] \\
&= [\tau(0) \vee u] \wedge [\tau(1) \vee u'] \\
&= [\tau(0) \wedge \tau(1)] \vee [\tau(0) \wedge u'] \vee [\tau(1) \wedge u] \vee [u \wedge u'] \\
&= [\tau(0) \wedge \tau(1)] \vee [\tau(0) \wedge u'] \vee [\tau(1) \wedge u] \\
&= [(\tau(0) \wedge \tau(1)) \wedge (u \vee u')] \vee [\tau(0) \wedge u'] \vee [\tau(1) \wedge u] \\
&= [\tau(0) \wedge \tau(1) \wedge u] \vee [\tau(0) \wedge \tau(1) \wedge u'] \vee [\tau(0) \wedge u'] \vee [\tau(1) \wedge u] \\
&= [(\tau(0) \wedge \tau(1) \wedge u) \vee (\tau(1) \wedge u)] \vee [(\tau(0) \wedge \tau(1) \wedge u') \vee (\tau(0) \wedge u')] \\
&= [\tau(1) \wedge u] \vee [\tau(0) \wedge u']
\end{aligned}$$

†The *intersection* of a collection of sets is the set of all objects belonging to every set in the collection.

Case 2. $\tau(u) = \sigma(u) \vee \rho(u)$. Then the inductive hypothesis holds for σ and ρ. Hence

$$\begin{aligned}
\tau(u) &= \sigma(u) \vee \rho(u) \\
&= [(\sigma(0) \wedge u') \vee (\sigma(1) \wedge u)] \vee [(\rho(0) \wedge u') \vee (\rho(1) \wedge u)] \\
&= [(\sigma(0) \wedge u') \vee (\rho(0) \wedge u')] \vee [(\sigma(1) \wedge u) \vee (\rho(1) \wedge u)] \\
&= [(\sigma(0) \vee \rho(0)) \wedge u'] \vee [(\sigma(1) \vee \rho(1)) \wedge u] \\
&= [\tau(0) \wedge u'] \vee [\tau(1) \wedge u]
\end{aligned}$$

Case 3. $\tau(u) = \sigma(u) \wedge \rho(u)$. This is similar to Case 2 and is left to the reader.

3.8. Prove Theorem 3.11: For any Boolean expression $\tau(u_1, \ldots, u_k)$, the equation

$$\tau(u_1, \ldots, u_k) = \sum_{\alpha_1=0}^{1} \sum_{\alpha_2=0}^{1} \cdots \sum_{\alpha_k=0}^{1} [\tau(\alpha_1, \alpha_2, \ldots, \alpha_k) \wedge u_1^{\alpha_1} \wedge u_2^{\alpha_2} \wedge \cdots \wedge u_k^{\alpha_k}]$$

is derivable from the axioms for Boolean algebras (and therefore the corresponding equation for $\tau^{\mathcal{B}}$ holds for any Boolean algebra $\mathcal{B}$).

Solution:

We shall use induction on k. The case $k = 1$ is an immediate consequence of Theorem 3.10. Now assume that the result holds for k and we shall prove it for an expression $\tau(u_1, u_2, \ldots, u_{k+1})$. By Theorem 3.10,

$$\tau(u_1, \ldots, u_{k+1}) = [\tau(0, u_2, \ldots, u_{k+1}) \wedge u_1'] \vee [\tau(1, u_2, \ldots, u_{k+1}) \wedge u_1]$$

But, by inductive hypothesis,

$$\tau(0, u_2, \ldots, u_{k+1}) = \sum_{\alpha_2=0}^{1} \cdots \sum_{\alpha_{k+1}=0}^{1} [\tau(0, \alpha_2, \ldots, \alpha_{k+1}) \wedge u_2^{\alpha_2} \wedge \cdots \wedge u_{k+1}^{\alpha_{k+1}}]$$

and

$$\tau(1, u_2, \ldots, u_{k+1}) = \sum_{\alpha_2=0}^{1} \cdots \sum_{\alpha_{k+1}=0}^{1} [\tau(1, \alpha_2, \ldots, \alpha_{k+1}) \wedge u_2^{\alpha_2} \wedge \cdots \wedge u_{k+1}^{\alpha_{k+1}}]$$

Hence

$$\begin{aligned}
\tau(u_1, u_2, \ldots, u_{k+1}) &= \left\{\left(\sum_{\alpha_2=0}^{1} \cdots \sum_{\alpha_{k+1}=0}^{1} [\tau(0, \alpha_2, \ldots, \alpha_{k+1}) \wedge u_2^{\alpha_2} \wedge \cdots \wedge u_{k+1}^{\alpha_{k+1}}]\right) \wedge u_1'\right\} \\
&\quad \vee \left\{\left(\sum_{\alpha_2=0}^{1} \cdots \sum_{\alpha_{k+1}=0}^{1} [\tau(1, \alpha_2, \ldots, \alpha_{k+1}) \wedge u_2^{\alpha_2} \wedge \cdots \wedge u_{k+1}^{\alpha_{k+1}}]\right) \wedge u_1\right\} \\
&= \left(\sum_{\alpha_2=0}^{1} \cdots \sum_{\alpha_{k+1}=0}^{1} [\tau(0, \alpha_2, \ldots, \alpha_{k+1}) \wedge u_1' \wedge u_2^{\alpha_2} \wedge \cdots \wedge u_{k+1}^{\alpha_{k+1}}]\right) \\
&\quad \vee \left(\sum_{\alpha_2=0}^{1} \cdots \sum_{\alpha_{k+1}=0}^{1} [\tau(1, \alpha_2, \ldots, \alpha_{k+1}) \wedge u_1 \wedge u_2^{\alpha_2} \wedge \cdots \wedge u_{k+1}^{\alpha_{k+1}}]\right) \\
&= \sum_{\alpha_1=0}^{1} \sum_{\alpha_2=0}^{1} \cdots \sum_{\alpha_{k+1}=0}^{1} [\tau(\alpha_1, \alpha_2, \ldots, \alpha_{k+1}) \wedge u_1^{\alpha_1} \wedge u_2^{\alpha_2} \wedge \cdots \wedge u_{k+1}^{\alpha_{k+1}}]
\end{aligned}$$

3.9. Show that in any Boolean algebra $\mathcal{B}$ there are 2^{2^n} different Boolean functions of n variables.

Solution:

By the Disjunctive Normal Form Theorem (Theorem 3.11), the equation

$$\tau(u_1, \ldots, u_n) = \sum_{\alpha_1=0}^{1} \cdots \sum_{\alpha_n=0}^{1} [\tau(\alpha_1, \ldots, \alpha_n) \wedge u_1^{\alpha_1} \wedge \cdots \wedge u_n^{\alpha_n}]$$

is derivable. Hence the function determined by τ depends only on the 2^n values $\tau(\alpha_1, \ldots, \alpha_n)$, where each α_i is either 0 or 1. Each such value is 0 or 1. Hence there are 2^{2^n} different Boolean functions.

3.10. If D is a finite subset of a Boolean algebra $\mathcal{B}$, show that the subalgebra $\mathcal{C}$ generated by D is also finite.

Solution:

Let D have n elements. By Problem 3.6, the elements of $\mathcal{C}$ are the values obtained by substituting elements of D for the variables in all Boolean functions $\tau^{\mathcal{B}}$. Clearly, we may confine our attention to functions $\tau^{\mathcal{B}}$ of at most n variables, since variables for which the same element of D is substituted may be identified. By Problem 3.9, there are 2^{2^n} such functions. Hence since each of the n variables may be replaced by any of the elements of D, we obtain at most $2^{2^n} \cdot n^n$ possible elements in $\mathcal{C}$.

3.11. If τ and σ are Boolean expressions such that

$$\tau(c_1, \ldots, c_n) = 1 \rightarrow \sigma(c_1, \ldots, c_n) = 1$$

for all elements $c_1, \ldots, c_n$ of some Boolean algebra $\mathcal{C}$, then $\tau \leqq \sigma$ is derivable from the axioms for Boolean algebras.

Solution:

Let $\rho(u_1, \ldots, u_n) = \tau \wedge \sigma'$. If $u_1, \ldots, u_n$ are given values 0 or 1, then:

(i) if τ takes the value 0, so does ρ;

(ii) if τ takes the value 1, then, by assumption, so does σ, and therefore σ' assumes the value 0, and so does ρ. Hence $\rho(u_1, \ldots, u_n) = 0$ for all values of $u_1, \ldots, u_n$ in the subalgebra $\{0, 1\}$. Hence by Corollary 3.12, the equation $\rho(u_1, \ldots, u_n) = 0$ is derivable from the axioms for Boolean algebras. (Although 0 itself is not officially a Boolean expression, one can use the expression $u_1 \wedge u_1'$ instead of 0 so as to fit into the formulation of Corollary 3.12.) Thus $\tau \wedge \sigma' = 0$ is derivable. Hence $\tau \leqq \sigma$ is derivable (by Theorem 3.5(xii)).

3.12. (Conjunctive Normal Form.) We shall use $\prod$ to indicate repeated application of $\wedge$. Thus $\prod_{\alpha=0}^{1} \sigma(\alpha)$ denotes $\sigma(0) \wedge \sigma(1)$. Given a Boolean expression $\tau(u_1, \ldots, u_n)$ having its variables among $u_1, \ldots, u_n$, show that the equation

$$\tau(u_1, \ldots, u_n) \;=\; \prod_{\alpha_1=0}^{1} \cdots \prod_{\alpha_n=0}^{1} (\tau(\alpha_1, \ldots, \alpha_n) \vee u_1^{\alpha_1'} \vee \cdots \vee u_n^{\alpha_n'}) \tag{1}$$

is derivable from the axioms for Boolean algebras and therefore holds in every Boolean algebra. Also, write equation (*1*) for the cases $n = 1$ and $n = 2$.

Solution:

$(\tau(u_1, \ldots, u_n))'$ is a Boolean expression, and, by the Disjunctive Normal Form Theorem, the equation

$$\tau(u_1, \ldots, u_n)' \;=\; \sum_{\alpha_1=0}^{1} \cdots \sum_{\alpha_n=0}^{1} [\tau(\alpha_1, \ldots, \alpha_n)' \wedge u_1^{\alpha_1} \wedge \cdots \wedge u_n^{\alpha_n}] \tag{2}$$

is derivable. Taking the complements of both sides of (*2*) and applying De Morgan's Laws, we obtain (*1*). In the case $n = 1$, we obtain

$$\tau(u) \;=\; (\tau(0) \vee u) \wedge (\tau(1) \vee u')$$

For $n = 2$, we obtain

$$\tau(u_1, u_2) \;=\; (\tau(0,0) \vee u_1 \vee u_2) \wedge (\tau(0,1) \vee u_1 \vee u_2') \wedge (\tau(1,0) \vee u_1' \vee u_2) \wedge (\tau(1,1) \vee u_1' \vee u_2')$$

3.13. Write the Boolean expression $(x \wedge (y' \vee z)) \vee z'$ in both disjunctive and conjunctive normal forms.

Solution:

Disjunctive:

$$(x' \wedge y' \wedge z') \vee (x' \wedge y \wedge z') \vee (x \wedge y' \wedge z') \vee (x \wedge y' \wedge z) \vee (x \wedge y \wedge z') \vee (x \wedge y \wedge z)$$

Conjunctive: $$(x \vee y \vee z') \wedge (x \vee y' \vee z')$$

Sometimes, instead of using the theorems on disjunctive and conjunctive normal forms, it is easier to find the appropriate expression by using known laws for Boolean algebras. Thus

$$(x \wedge (y' \vee z)) \vee z' \;=\; (x \vee z') \wedge ((y' \vee z) \vee z') \;=\; x \vee z'$$

Then, $x \vee z' \;=\; (x \vee z') \vee (y \wedge y') \;=\; (x \vee y \vee z') \wedge (x \vee y' \vee z')$.

3.14. Given a Boolean algebra $\mathcal{B}$. (i) Show that the set of all Boolean functions $\tau^{\mathcal{B}}$ is a Boolean algebra $\mathcal{F}$. (ii) Prove that $\mathcal{F}$ is isomorphic to the Boolean algebra of statement bundles (cf. Example 3.6). (iii) Show that the set of all Boolean functions $u^{\mathcal{B}}$, where u is a variable, is a set of generators D of $\mathcal{F}$ (i.e. the subalgebra generated by D is the whole algebra $\mathcal{F}$).

Solution:

(i) The operations of $\wedge, \vee, '$ on Boolean functions are defined in the obvious way. The zero element is $(x \wedge x')^{\mathcal{B}}$ and the unit element is $(x \vee x')^{\mathcal{B}}$. The straightforward verification of Axioms (1)-(9) is left to the reader.

(ii) For each Boolean function $\tau^{\mathcal{B}}$, let $\Psi(\tau^{\mathcal{B}})$ be the statement bundle containing the corresponding statement form $\mathrm{SF}(\tau)$ defined in Section 3.7. This is a well-defined function, for if $\tau^{\mathcal{B}} = \sigma^{\mathcal{B}}$, then, by Corollary 3.12, $\tau = \sigma$ holds for all Boolean algebras, and therefore by Theorem 3.14, $\mathrm{SF}(\tau)$ and $\mathrm{SF}(\sigma)$ belong to the same statement bundle. That the mapping Ψ is one-one follows from the "if" part of Theorem 3.14. The fact that Ψ preserves the Boolean operation can be checked easily by the reader.

(iii) Every Boolean function $\tau^{\mathcal{B}}$ belongs to every subalgebra containing the Boolean functions $u^{\mathcal{B}}$, since $\tau^{\mathcal{B}}$ is obtained from the functions $u^{\mathcal{B}}$ in the same way that τ is built up from the corresponding variables.

3.15. (Boolean Algebra and the Algebra of Sets.) For any Boolean expression τ, form the corresponding set-theoretic expression $\mathrm{Set}(\tau)$ by replacing $\wedge, \vee, '$ by $\cap, \cup, \bar{\ }$. Show that $\tau = \sigma$ holds for all Boolean algebras if and only if $\mathrm{Set}(\tau) = \mathrm{Set}(\sigma)$ holds in all fields of sets.

Solution:

Use Problem 2.16(*a*) and Theorem 3.14.

Supplementary Problems

3.16. Prove the generalized Distributive and De Morgan's Laws:

(a) $x \wedge (y_1 \vee \cdots \vee y_n) = (x \wedge y_1) \vee \cdots \vee (x \wedge y_n)$

(b) $x \vee (y_1 \wedge \cdots \wedge y_n) = (x \vee y_1) \wedge \cdots \wedge (x \vee y_n)$

(c) $(x_1 \vee \cdots \vee x_n)' = x_1' \wedge \cdots \wedge x_n'$

(d) $(x_1 \wedge \cdots \wedge x_n)' = x_1' \vee \cdots \vee x_n'$

3.17. For any Boolean algebra, prove:

(a) $x = 0 \leftrightarrow y = (x \wedge y') \vee (x' \wedge y)$ (Poretzky's Law)

(b) $x \vee y = x \vee z \;\&\; x' \vee y = x' \vee z \rightarrow y = z$

(c) $x \vee y = 0 \leftrightarrow x = 0 \;\&\; y = 0$

(d) $x \wedge y = 1 \leftrightarrow x = 1 \;\&\; y = 1$

(e) $(x \sim y) \vee (y \sim x) = 0 \leftrightarrow x = y$

SUBALGEBRAS

3.18. Show by an example that a subset of a Boolean algebra containing 0 and 1 and closed under $\wedge$ and $\vee$ need not be a subalgebra.

3.19. If A determines a subalgebra of a Boolean algebra $\mathcal{B}$ (i.e. A is closed under $\wedge, \vee, '$) and if $b \in B \sim A$, show that the subalgebra generated by $A \cup \{b\}$ consists of all elements of the form $(a_1 \wedge b) \vee (a_2 \wedge b')$, where $a_1 \in A$ and $a_2 \in A$.

3.20. Prove that, in any Boolean algebra,

(a) $x \leq y' \leftrightarrow x \wedge y = 0$

(b) $x \leq y \leftrightarrow x' \vee y = 1$

BOOLEAN EXPRESSIONS AND FUNCTIONS. NORMAL FORMS

3.21. Simplify the following Boolean expressions.

(a) $(x \vee y) \wedge (x \vee z) \wedge (x' \wedge y)'$

(b) $[x \vee (y \wedge (z \vee x'))]'$

(c) $(x' \wedge y)' \vee (x \wedge y')$

3.22. Prove that two Boolean expressions either determine the same Boolean function in all Boolean algebras or they never determine the same Boolean function.

3.23. Prove that, for any two disjunctive (conjunctive) normal forms in n variables τ and σ, $\tau = \sigma$ holds in all Boolean algebras if and only if τ and σ are precisely the same (i.e. the identity

$$\sum_{\alpha_1=0}^{1} \cdots \sum_{\alpha_n=0}^{1} [\tau(\alpha_1, \ldots, \alpha_n) \wedge u_1^{\alpha_1} \wedge \cdots \wedge u_n^{\alpha_n}] = \sum_{\alpha_1=0}^{1} \cdots \sum_{\alpha_n=0}^{1} [\sigma(\alpha_1, \ldots, \alpha_n) \wedge u_1^{\alpha_1} \wedge \cdots \wedge u_n^{\alpha_n}]$$

holds in all Boolean algebras if and only if $\tau(\alpha_1, \ldots, \alpha_n) = \sigma(\alpha_1, \ldots, \alpha_n)$ for all $\alpha_1, \ldots, \alpha_n$ chosen from $\{0, 1\}$).

3.24. Let $\tau(u)$ be a Boolean expression. Prove:

(a) $\tau(\tau(0)) = \tau(0) \wedge \tau(1) \leq \tau(u) \leq \tau(0) \vee \tau(1) = \tau(\tau(1))$

(b) $\tau(u_1 \vee u_2) \vee \tau(u_1 \wedge u_2) = \tau(u_1) \vee \tau(u_2)$

3.25. Show that the dual of $x \leq y$ is $x \geq y$.

Chapter 4

Switching Circuits and Logic Circuits

4.1 SWITCHING CIRCUITS

A *switch* is a device which is attached to a point in an electric circuit and which may assume either of two states, *closed* or *open*. In the closed state the switch allows current to flow through the point, whereas in the open state no current can flow through the point. We shall indicate a switch by means of the symbol , where A denotes a sentence such that the switch is closed when A is true and open when A is false. We say that two points are connected by a switching circuit if and only if they are connected by wires (lines) on which a finite number of switches are located.

Example 4.1.

In Fig. 4-1 points x and y are connected by a switching circuit. The four switches are said to be *in parallel*. Clearly, current flows between x and y if and only if $A \vee B \vee C \vee D$ is true. This example may be generalized to the case of any finite number of switches connected in parallel. Current flows through the circuit of Fig. 4-2 if and only if the sentence $A_1 \vee A_2 \vee \cdots \vee A_n$ is true.

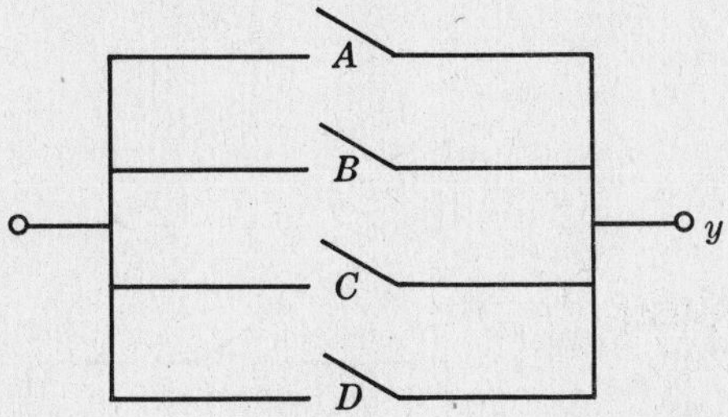

Fig. 4-1

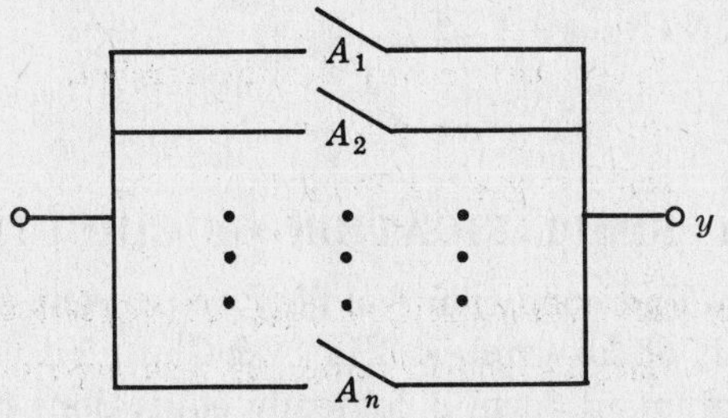

Fig. 4-2

Example 4.2.

In the switching circuit of Fig. 4-3 current can flow between the points x and y if and only if $A \& B$ is true. The two switches are said to be *in series*. This case may be generalized to the case of any finite number of switches connected in series. The condition for current flow through the circuit of Fig. 4-4 is $A_1 \& A_2 \& A_3 \& \ldots \& A_n$.

Fig. 4-3

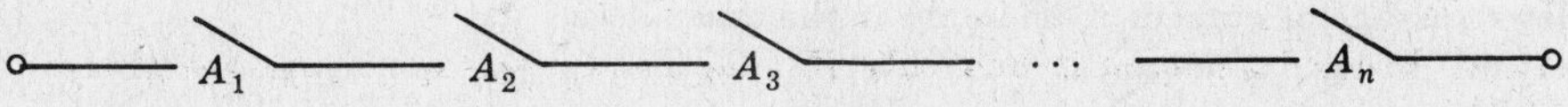

Fig. 4-4

Example 4.3.

In the switching circuit of Fig. 4-5 below, current can flow if and only if $(A \& C) \vee (\neg A \vee B)$ is true.

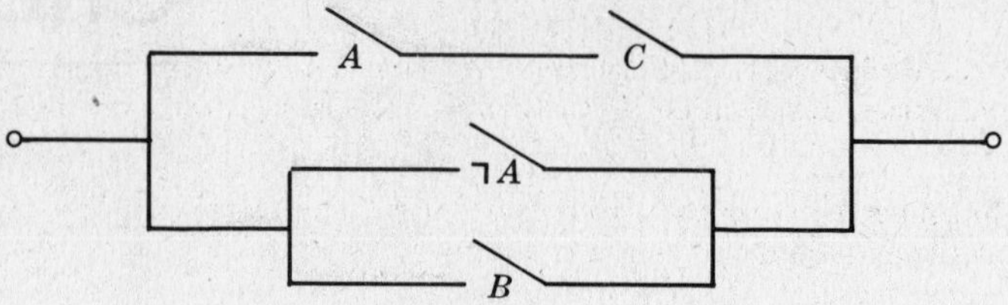

Fig. 4-5

Example 4.3 shows that we may combine switches in parallel and in series in the same circuit. Such a circuit is called a *series-parallel* switching circuit. More precisely, if A is any sentence, then 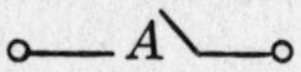

 is a series-parallel switching circuit, and if $S, S_1, \ldots, S_n$ are series-parallel switching circuits, we may form a new series-parallel switching circuit by replacing any switch in S by either

or

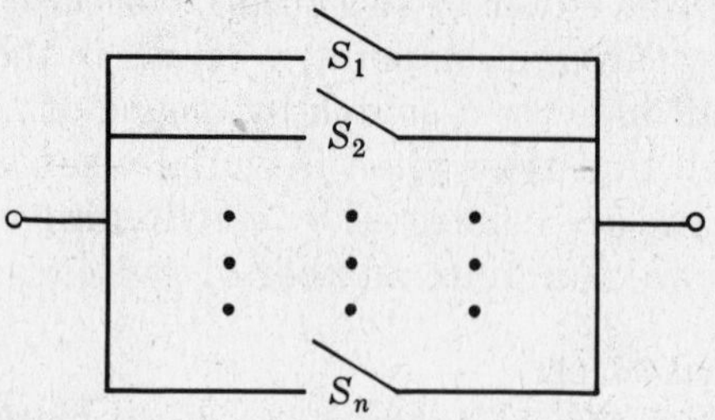

Clearly, a condition for flow of current through a series-parallel switching circuit can be written down by means of conjunctions and disjunctions, starting from the expressions representing the closure of the individual switches. In Example 4.3, this condition was $(A \& C) \vee (\neg A \vee B)$.

4.2 SIMPLIFICATION OF CIRCUITS

The condition for flow of current through the circuit of Example 4.3 is $(A \& C) \vee (\neg A \vee B)$. The latter statement form is logically equivalent to the statement form $((A \& C) \vee \neg A) \vee B$, which in turn is logically equivalent to $C \vee \neg A \vee B$. Hence the circuit of Fig. 4-5 may be replaced by the circuit of Fig. 4-6.

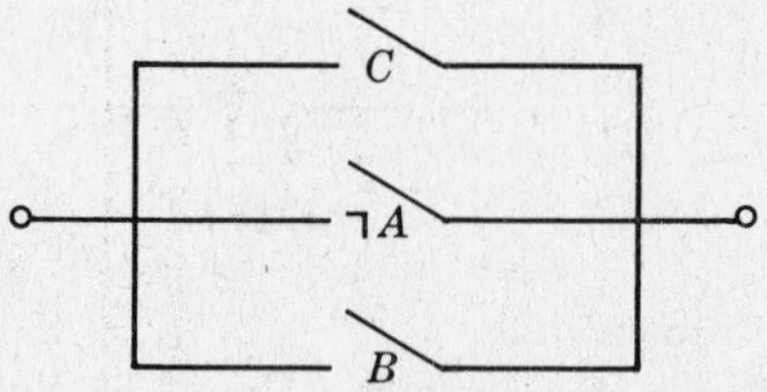

Fig. 4-6

The circuit of Fig. 4-6 is clearly a simplification of that of Fig. 4-5, since it involves fewer switches.

Example 4.4.

A condition for current flow through the circuit of Fig. 4-7 is $(A \& B \& \neg C) \vee (\neg C \& \neg A)$. However, this is logically equivalent to $\neg C \& [(A \& B) \vee \neg A]$, which in turn is logically equivalent to $\neg C \& (B \vee \neg A)$. Hence an equivalent, but simpler, circuit is that of Fig. 4-8. (The two circuits are *equivalent* in the sense that one allows passage of current if and only if the other does.)

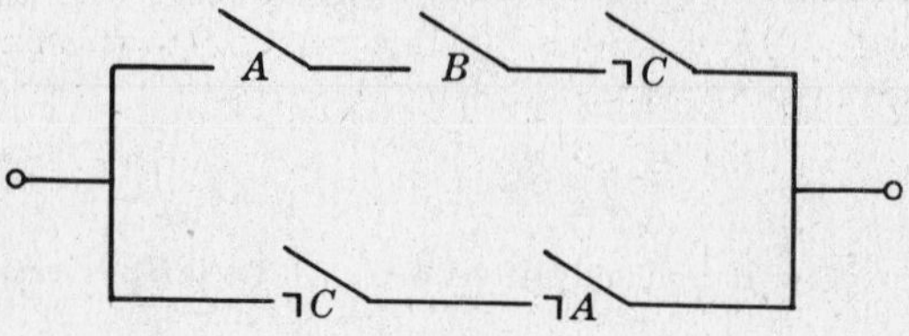

Fig. 4-7

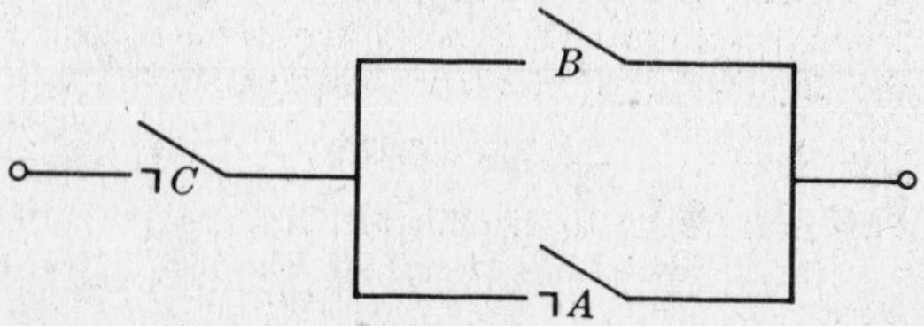

Fig. 4-8

Example 4.5.

A committee of three decides questions by majority vote. Each member can press a button to signify a "Yes" vote. Let us construct a switching circuit which will pass current when and only when a majority votes "Yes".

Let A stand for "member 1 approves", B for "member 2 approves", and C for "member 3 approves". Then a necessary and sufficient condition for a majority vote is

$$(A \& B) \vee (A \& C) \vee (B \& C)$$

A corresponding circuit is shown in Fig. 4-9. However, the given statement form is logically equivalent to $(A \& (B \vee C)) \vee (B \& C)$, having the simpler circuit of Fig. 4-10.

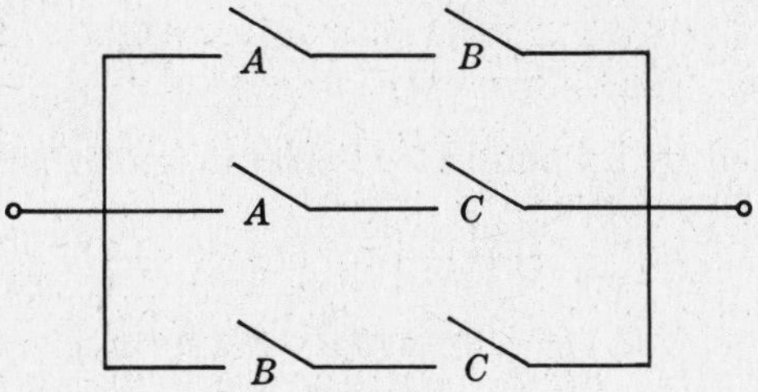

Fig. 4-9

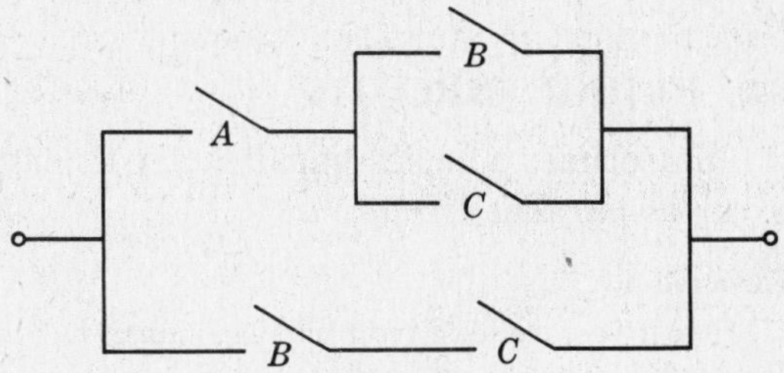

Fig. 4-10

Example 4.6.

A light in a room is to be controlled independently by three wall switches (not to be confused with switches of a circuit), located at the three entrances of the room. This means that flicking any one of the wall switches changes the state of the light (*on* to *off*, and *off* to *on*). Let us design a circuit which allows current to flow to the light under the required conditions.

Let A stand for "wall switch 1 is up", B for "wall switch 2 is up", and C for "wall switch 3 is up". In the truth table of Fig. 4-11, we wish to construct a statement form $f(A,B,C)$ for the required switching circuit.

A	B	C	$f(A,B,C)$
T	T	T	
F	T	T	
T	F	T	
F	F	T	
T	T	F	
F	T	F	
T	F	F	
F	F	F	

Fig. 4-11

	A	B	C	$f(A,B,C)$
(1)	T	T	T	T
(2)	F	T	T	F
(4)	T	F	T	F
(3)	F	F	T	T
(6)	T	T	F	F
(7)	F	T	F	T
(5)	T	F	F	T
(8)	F	F	F	F

Fig. 4-12

The requirement on $f(A,B,C)$ is that its truth value should change whenever the truth value of one of A, B, C changes. We arbitrarily assign the value T to $f(A,B,C)$ when A, B, C are all T (the first row); thus the light will be on when all wall switches are up. Then we proceed down the truth table, changing the truth value of $f(A,B,C)$ whenever the truth value of precisely one of A, B, C changes. We have indicated such a procedure in Fig. 4-12 by writing to the left of each row a number showing at what step the truth value for that row has been determined. Another way of describing the assignment of truth values is to note that T is assigned when an odd number of statement letters have the value T. We find the resulting statement form by the method developed in the proof of Theorem 1.8; this amounts to forming the disjunction of the truth assignments in the rows to which a T is attached:

$$(A \& B \& C) \vee (\neg A \& \neg B \& C) \vee (\neg A \& B \& \neg C) \vee (A \& \neg B \& \neg C)$$

This is logically equivalent to

$$[A \& ((B \& C) \vee (\neg B \& \neg C))] \vee [\neg A \& ((\neg B \& C) \vee (B \& \neg C))]$$

having the circuit shown in Fig. 4-13 below.

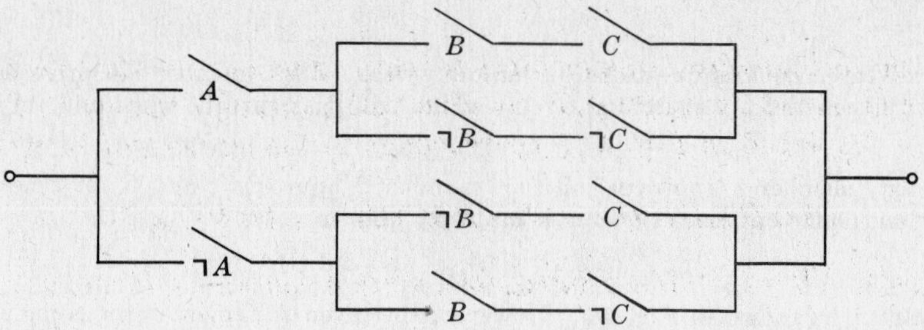

Fig. 4-13

4.3 BRIDGE CIRCUITS

Sometimes a series-parallel circuit can be replaced by an equivalent circuit which is not a series-parallel circuit.

Example 4.7.

A series-parallel circuit corresponding to the condition $[A \,\&\, (B \vee E)] \vee [C \,\&\, (\neg B \vee E \vee D)]$ is given in Fig. 4-14. This is equivalent to the circuit shown in Fig. 4-15. Clearly, the only paths through this circuit are $A \,\&\, B$, $A \,\&\, E \,\&\, D$, $A \,\&\, E \,\&\, \neg B$, $C \,\&\, E \,\&\, B$, $C \,\&\, D$, $C \,\&\, \neg B$. Hence, a condition for flow through this circuit is $(A \,\&\, B) \vee (A \,\&\, E \,\&\, D) \vee (A \,\&\, E \,\&\, \neg B) \vee (C \,\&\, E \,\&\, B) \vee (C \,\&\, D) \vee (C \,\&\, \neg B)$, which is logically equivalent to $[A \,\&\, (B \vee E)] \vee [C \,\&\, (\neg B \vee E \vee D)]$.

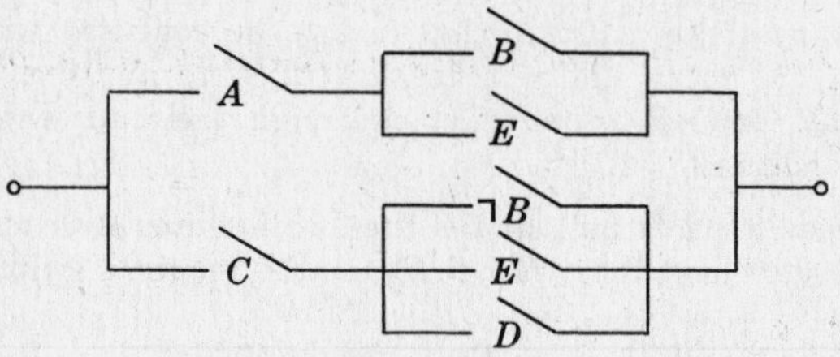

Fig. 4-14

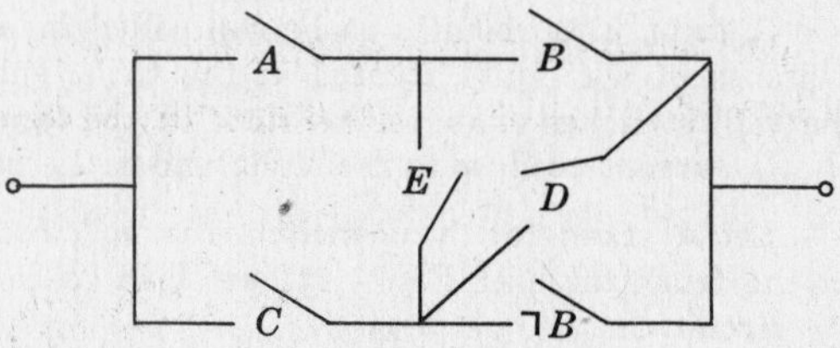

Fig. 4-15

The circuit of Fig. 4-15 is an example of a circuit which is not a series-parallel circuit. Such circuits are called *bridge circuits*. In Example 4.7, the bridge circuit had fewer switches (6) than the corresponding series-parallel circuit (7).

Another example of a bridge circuit is given in Fig. 4-16. A corresponding statement form is $(A \,\&\, [D \vee (C \,\&\, E)]) \vee [B \,\&\, (E \vee (C \,\&\, D))]$, whose series-parallel circuit is shown in Fig. 4-17.

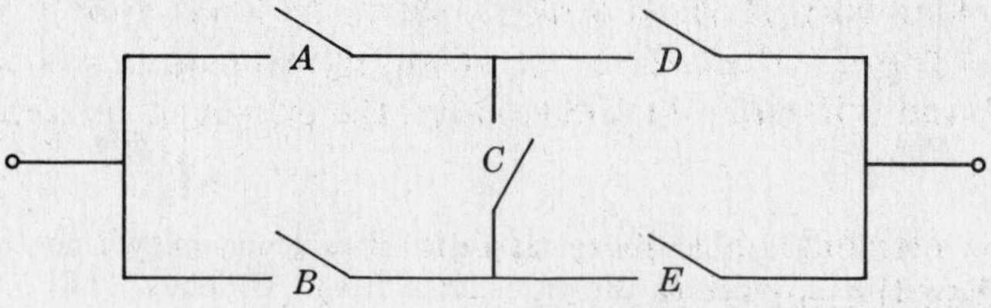

Fig. 4-16

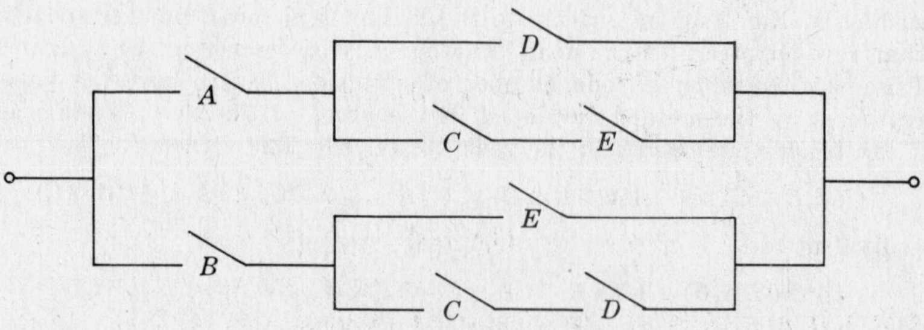

Fig. 4-17

Notice that any bridge circuit determines a truth function. A statement form for this truth function is obtained by finding all possible paths through the circuit. For example, the bridge circuit displayed in Fig. 4-18 corresponds to the statement form

$$(A \& B \& C \& D) \vee (A \& B \& \neg C) \vee (\neg B \& C \& D)$$

Notice that the path $\neg B \rightarrow C \rightarrow \neg C$ is impossible, since it contains a formula and its negation.

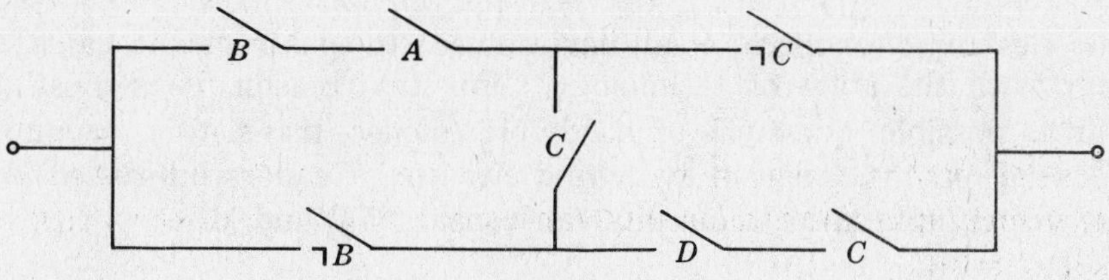

Fig. 4-18

4.4 LOGIC CIRCUITS

The processing of information is one of the most important roles of the modern digital computer. For this purpose, special devices are available.

An *and-gate* operates on two or more inputs $A_1, \ldots, A_n$ and produces their conjunction $A_1 \& A_2 \& \ldots \& A_n$. An and-gate is denoted (&).

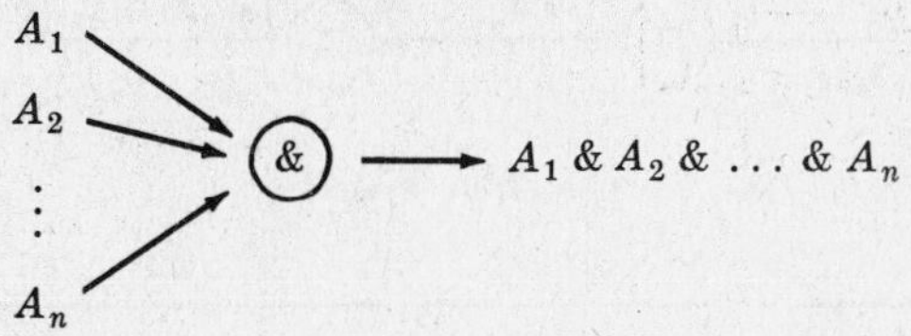

Fig. 4-19

More precisely, each input A_i has the form of a physical quantity (say, voltage level), of which we choose to distinguish two states, denoted 0 and 1. The state 1 occurs if A_i is true, and the state 0 if A_i is false. The output of the and-gate is likewise in two possible states, 0 and 1: it is 1 if and only if $A_1 \& A_2 \& \ldots \& A_n$ is true, and it is 0 if and only if $A_1 \& A_2 \& \ldots \& A_n$ is false. Often the state of an input or output is taken to be 1 if it is transmitting current and 0 if not. Arithmetically, the output of an and-gate is the *product* of the inputs.

Another common element of a logic circuit is an or-gate (∨). If the inputs are $A_1, \ldots, A_n$ $(n \geqq 2)$, then the output is $A_1 \vee A_2 \vee \cdots \vee A_n$.

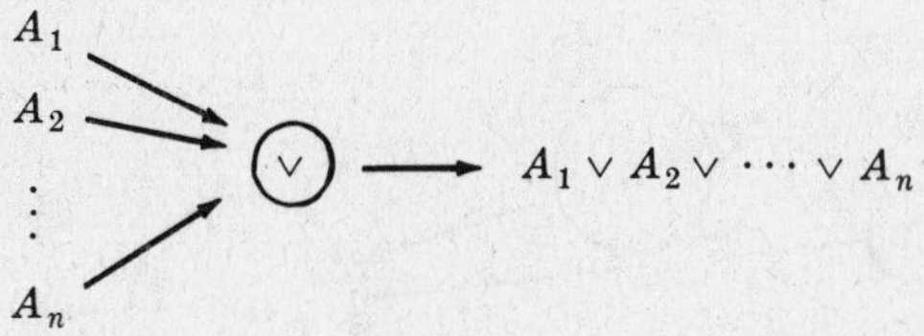

Fig. 4-20

Thus the output is 1 if and only if the output of at least one A_i is 1. Arithmetically, the output is the maximum of the inputs.

An *inverter* (¬) is a device which has one input A and produces as its output $\neg A$. Thus the output is 1 if the input is 0 and the output is 0 if the input is 1.

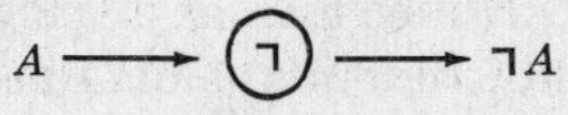

Fig. 4-21

A *logic circuit* is defined as a circuit constructed from various inputs by means of and-gates, or-gates, inverters, and possibly also other devices for performing truth-functional operations.

The actual electronic (or mechanical) devices used to construct and-gates, or-gates, and inverters vary with the state of technology. For this reason, it is most convenient to ignore (as far as possible) questions of hardware (diodes, transistors, vacuum tubes, etc.). This also holds for our treatment of switching circuits. Readers interested in the physical realization of switching and logic circuits can consult [53] and [13].

Example 4.8.

To construct a logic circuit producing the output $A_1 \leftrightarrow A_2$, notice that $A_1 \leftrightarrow A_2$ is logically equivalent to $(A_1 \& A_2) \vee (\neg A_1 \& \neg A_2)$ (Fig. 4-22) as well as to $(A_1 \& A_2) \vee \neg(A_1 \vee A_2)$ (Fig. 4-23). Clearly the second logic circuit is simpler.

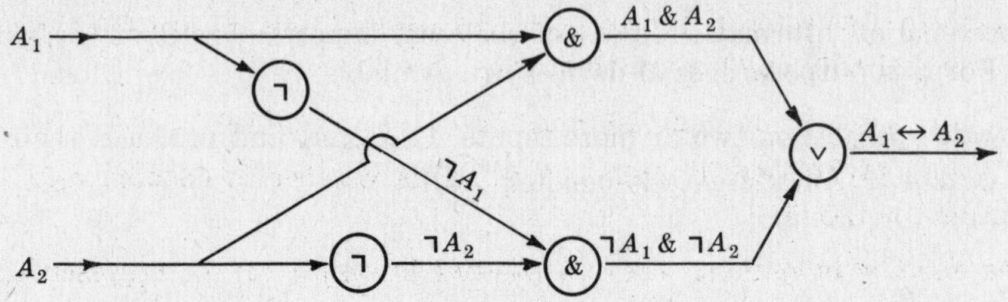

Fig. 4-22

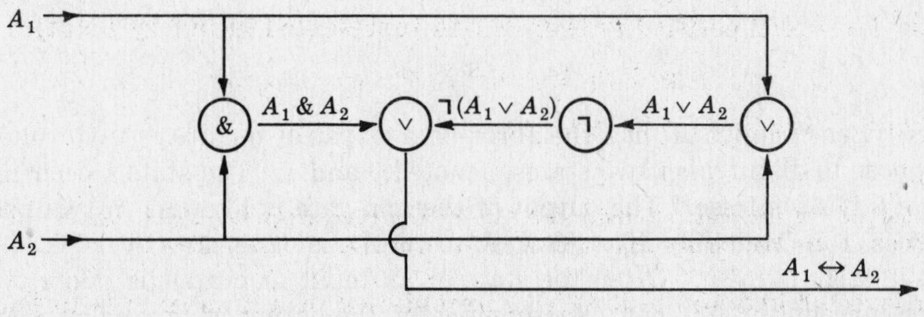

Fig. 4-23

Example 4.9.

Construct a logic circuit producing

$$(A_1 \& \neg A_2) \vee \neg A_1 \vee (A_2 \& A_3) \qquad (1)$$

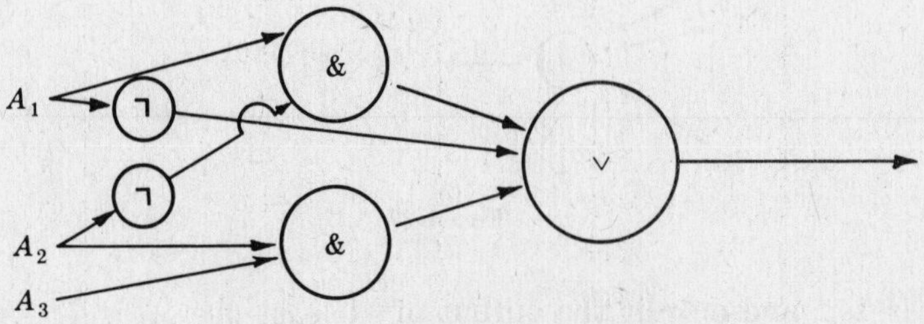

Fig. 4-24

Notice that, instead of a logic circuit, one could construct a series-parallel switching circuit through which current flows if and only if (*1*) is true (see Fig. 4-25).

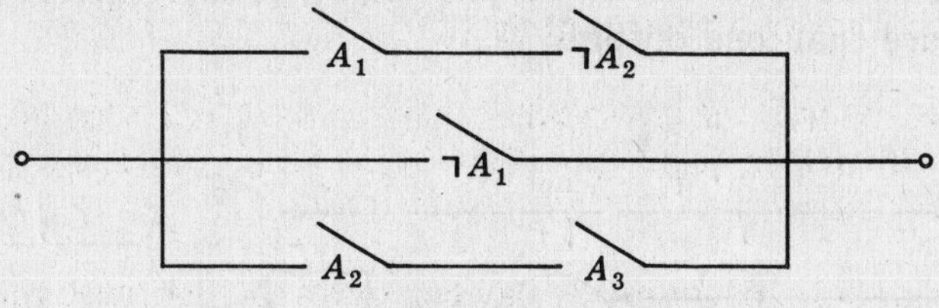

Fig. 4-25

Example 4.9 indicates that the same effect can be obtained by logic circuits as by series-parallel switching circuits. Indeed, connection in series corresponds to an and-gate, while connection in parallel corresponds to an or-gate.

4.5 THE BINARY NUMBER SYSTEM

We are accustomed to using the decimal number system. Thus 34,062 stands for the number $2 + 6 \cdot 10 + 0 \cdot 10^2 + 4 \cdot 10^3 + 3 \cdot 10^4$. In general, any positive integer can be represented in one and only one way in the form

$$a_0 + a_1 \cdot 10 + a_2 \cdot 10^2 + \cdots + a_k \cdot 10^k$$

where $0 \leq a_i \leq 9$ for $0 \leq i \leq k$ and $a_k > 0$. This number is denoted $a_k a_{k-1} \cdots a_2 a_1 a_0$ in standard decimal notation.

However, for any integer $r > 1$, every positive integer n can be represented uniquely in the form

$$a_0 + a_1 \cdot r + a_2 \cdot r^2 + \cdots + a_m \cdot r^m$$

where $0 \leq a_i \leq r - 1$ for $0 \leq i \leq m$ and $a_m > 0$. This can be proved by induction on n.

In particular, every positive integer can be represented in binary notation:

$$a_0 + a_1 \cdot 2 + a_2 \cdot 2^2 + \cdots + a_m \cdot 2^m$$

where $0 \leq a_i \leq 1$ for $0 \leq i \leq m$ and $a_m = 1$.

Example 4.10.

The number 23 (in decimal notation) has the binary representation 10111, i.e. $2^4 + 2^2 + 2 + 1$. The decimal number 101 has the binary representation 1100101, i.e. $2^6 + 2^5 + 2^2 + 1$.

A procedure for finding the binary representation of a number n is to find the highest power 2^m which is $\leq n$, subtract 2^m from n, then find the highest power 2^j which is $\leq n - 2^m$, etc.

Further examples:

Decimal Notation	Binary Notation	Decimal Notation	Binary Notation
1	1	11	1011
2	10	16	10000
3	11	35	100011
4	100	52	110100
5	101	117	1110101
6	110		
7	111		
8	1000		

4.6 MULTIPLE OUTPUT LOGIC CIRCUITS

Occurrence of the same logic circuit as part of other logic circuits suggests the use of logic circuits with more than one output.

Example 4.11.

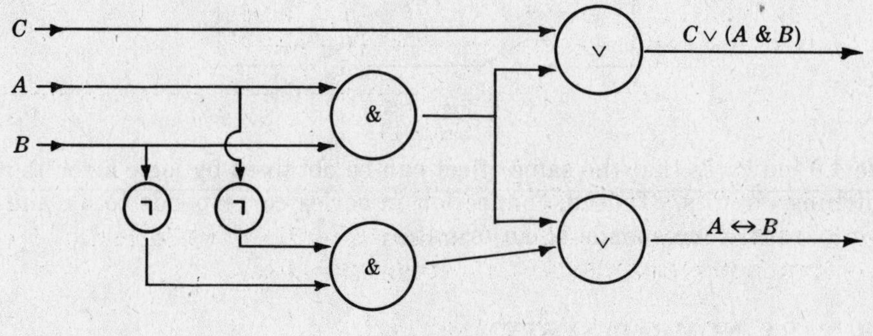

Fig. 4-26

Example 4.12.

Two numbers in binary notation are added in the same way as numbers in decimal notation.

Binary notation:	100101	Decimal notation:	37
	10111		23
	111100		60

If we just consider the addition of one digit numbers, 0 and 1, we have the following values for the *sum digit* s and the *carry digit* c.

A	B	s	c
1	1	0	1
0	1	1	0
1	0	1	0
0	0	0	0

Thus s corresponds to the exclusive-or (which we shall denote $A + B$), while c corresponds to the conjunction.

If we wished to construct a separate logic circuit for s we would obtain

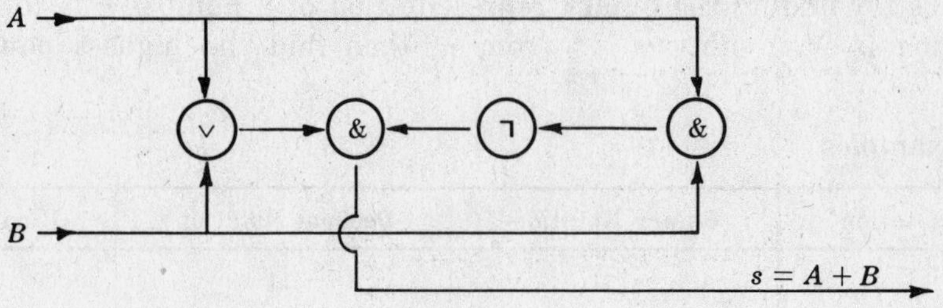

Fig. 4-27

Similarly we can construct a logic circuit for c:

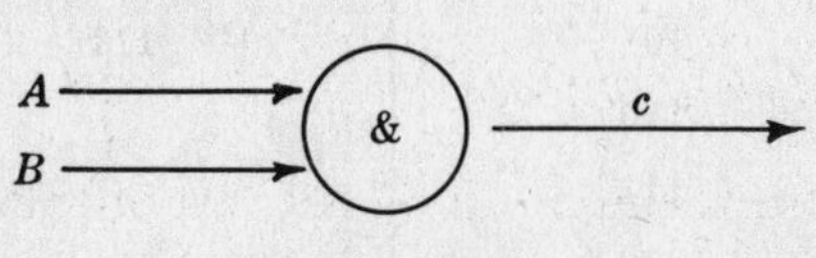

Fig. 4-28

However, we can combine these two circuits into a single multiple output circuit:

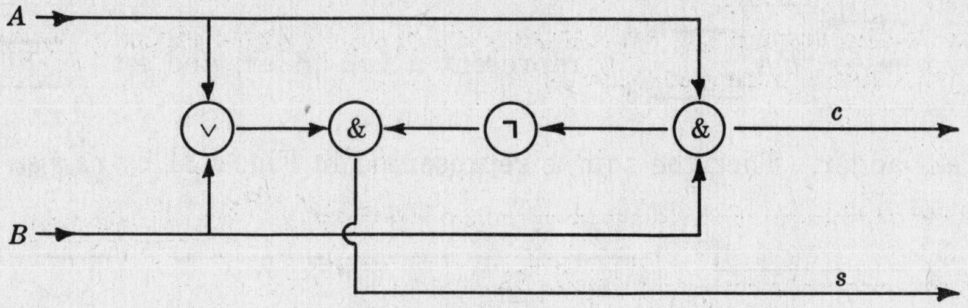

Fig. 4-29

The circuit in Fig. 4-29 is called a *half-adder*.

If we wish to add two single-digit numbers A and B†, while taking into account a carry-over C from a previous addition, we obtain the table

A	B	C	s	c
1	1	1	1	1
0	1	1	0	1
1	0	1	0	1
0	0	1	1	0
1	1	0	0	1
0	1	0	1	0
1	0	0	1	0
0	0	0	0	0

Thus s corresponds to the statement form

$$(C \mathbin{\&} \neg(A+B)) \vee ((A+B) \mathbin{\&} \neg C)$$

which is logically equivalent to $(A+B)+C$. The carry-over c corresponds to the statement form

$$(A \mathbin{\&} B) \vee (C \mathbin{\&} (A+B))$$

We can use the circuit constructed for $A+B$ in Fig. 4-27 to obtain the following diagram corresponding to the above statement.

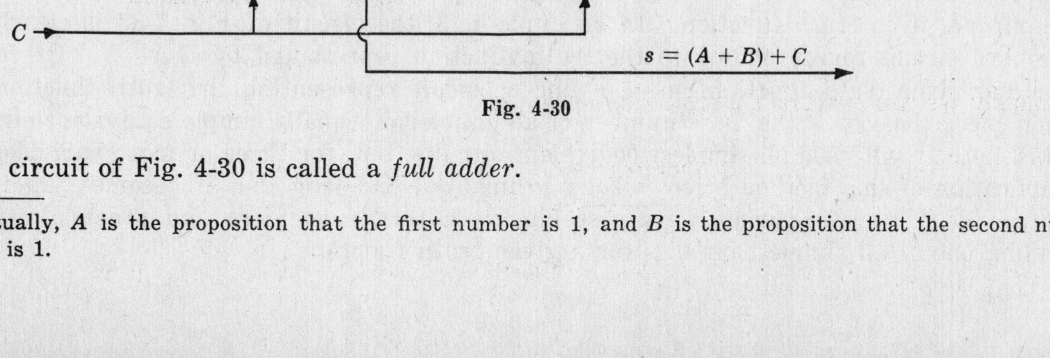

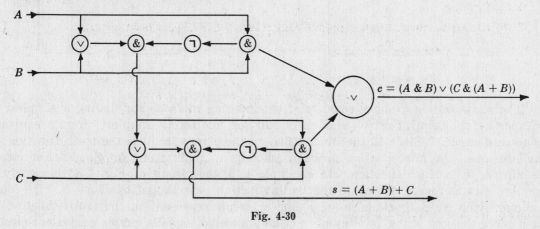

Fig. 4-30

The circuit of Fig. 4-30 is called a *full adder*.

†Actually, A is the proposition that the first number is 1, and B is the proposition that the second number is 1.

We can construct a circuit for adding two three-digit binary numbers $A_2A_1A_0$ and $B_2B_1B_0$. Let

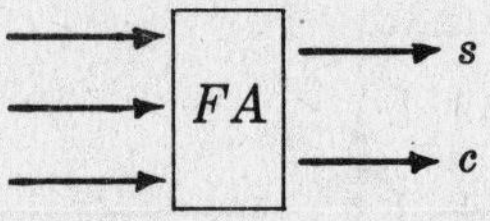

represent a full-adder, and let

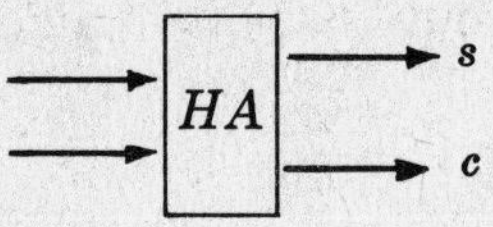

represent a half-adder. Then the sum is represented in Fig. 4-31 by $cs_2s_1s_0$.

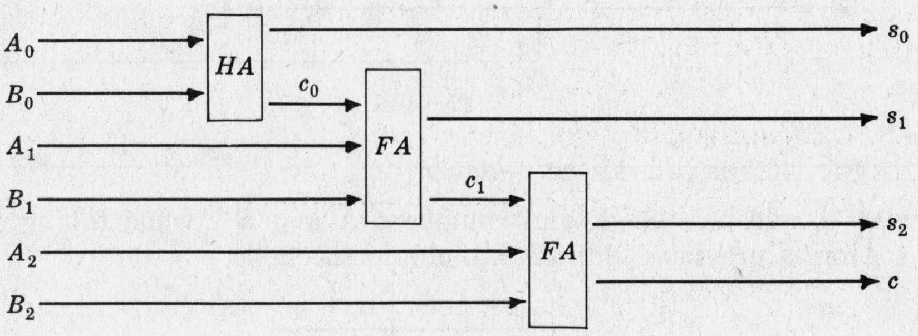

Fig. 4-31

4.7 MINIMIZATION

The cost of constructing and running a switching circuit or logic circuit depends upon the state of technology and therefore varies with time. However, at a given time, some circuits will be less expensive than other equivalent circuits.

Example 4.13.

The circuit (Fig. 4-32) corresponding to $\neg A \vee (B \& A)$ is more expensive than the equivalent circuit (Fig. 4-33) corresponding to $\neg A \vee B$, since the latter contains fewer occurrences of statement letters and fewer connectives. In general, decreasing the number of connectives (i.e. gates and inverters in a logic circuit) lowers the cost, other things being equal, and decreasing the number of occurrences of statement letters also lowers the cost, other things being equal. These two criteria usually are not the only measures of cost; the special hardware used for constructing circuits imposes other criteria.

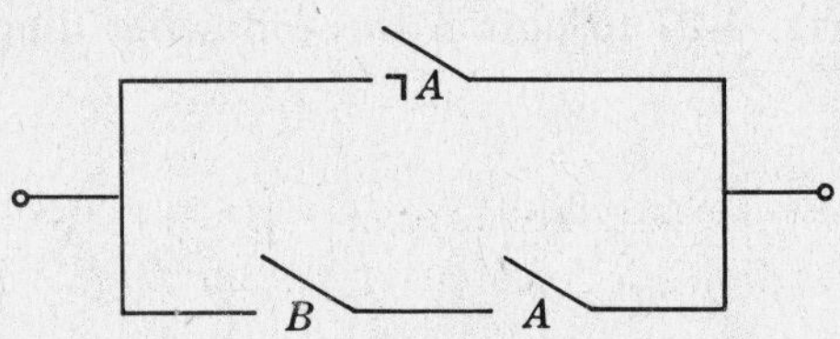

Fig. 4-32

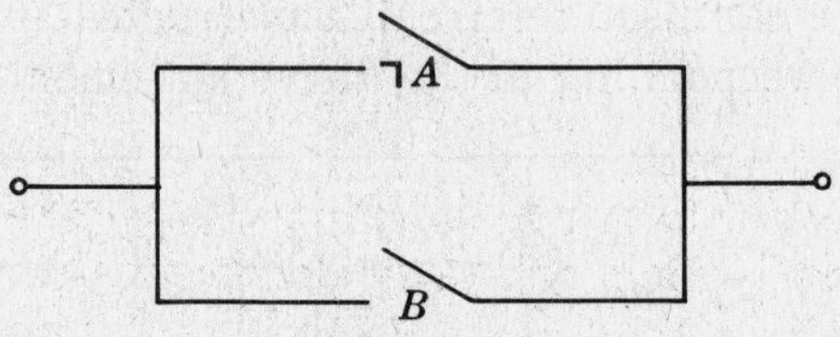

Fig. 4-33

The *minimization problem* consists of determining methods for finding a simplest (i.e. cheapest) circuit equivalent to a given circuit (or finding *all* simplest circuits equivalent to a given circuit). Since all that matters about the given circuit is the truth function that it determines, the minimization problem amounts to finding one or all simplest circuits defining a given truth function. In Example 4.13, the circuit of Fig. 4-33 is clearly the simplest circuit corresponding to the truth function represented by $\neg A \vee B$. Of course, for any given truth function, one can find a circuit representing the truth function and then check the cost of the *finite* number of all simpler or equally simple equivalent circuits. This method will yield all simplest equivalent circuits, but, for three or more variables, the application of this method often will be so involved and long that it becomes practically unfeasible. Therefore what we are seeking is a fast, convenient and practical way of finding one or all simplest circuits for a given truth function.

Let us assume that we are given a truth function. There are several forms of the minimization problem.

(I) Find the "simplest" disjunctive normal form (dnf) representing the given truth function.

(II) Find the "simplest" series-parallel switching circuit (or logic circuit) representing the given truth function.

(III) Find the "simplest" switching circuit (either series-parallel or bridge) representing the given truth function.

Example 4.14.

Consider the truth function given by the following table.

A	B	C	$g(A,B,C)$
T	T	T	T
F	T	T	F
T	F	T	T
F	F	T	F
T	T	F	T
F	T	F	F
T	F	F	F
F	F	F	F

A dnf for this function is $(A \,\&\, B \,\&\, C) \vee (A \,\&\, \neg B \,\&\, C) \vee (A \,\&\, B \,\&\, \neg C)$, which is logically equivalent to $(A \,\&\, B) \vee (A \,\&\, C)$. It is easy to check that this is the simplest dnf for g, thus solving Problem (I). The corresponding series-parallel circuit is given in Fig. 4-34. However, a logically equivalent statement form is $A \,\&\, (B \vee C)$, which has the simpler series-parallel circuit shown in Fig. 4-35. It is obvious that this circuit cannot be replaced by a simpler one. Thus, a minimal dnf solving Problem (I) need not be a solution of Problem (II), that is, a solution of Problem (II) need not be a dnf.

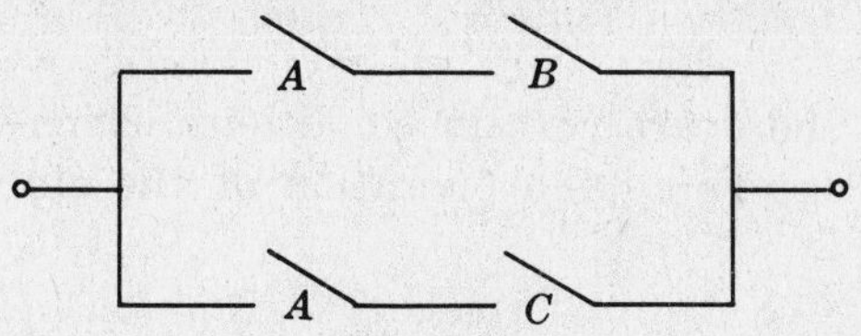

Fig. 4-34

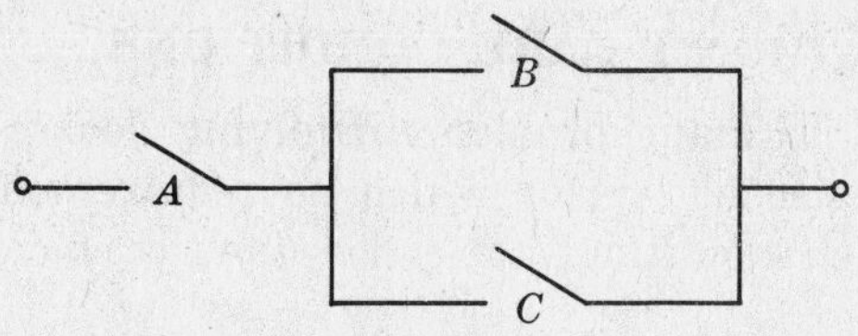

Fig. 4-35

The example shown in Fig. 4-16 demonstrates that there are bridge circuits which are simpler than any equivalent series-parallel circuit. Hence a solution of Problem (III) need not be a solution of Problem (II).

Remarks: (1) Solving minimization problem (I) requires consideration only of dnf's. The solution will provide what is called a two-stage and-or logic circuit. For example, the dnf $(A \,\&\, B \,\&\, \neg C) \vee (\neg A \,\&\, B) \vee (\neg B \,\&\, \neg C)$ corresponds to the circuit in Fig. 4-36.

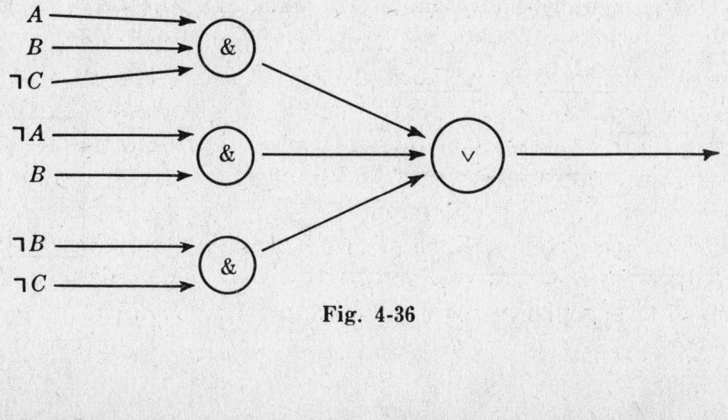

Fig. 4-36

Notice that we have not pictured inverters. This common convention stems from the fact that the presence or absence of a negation sign often results from an arbitrary decision as to which of two contradictory assertions is to be labeled by a letter, say A, rather than by $\neg A$. Since the number of negation signs often depends upon arbitrary decisions, it is advisable to consider negations of letters as initial inputs, on a par with letters, and not to count the number of inverters in computing the cost of the circuit.

(2) In solving minimization problem (II) for the case of logic circuits, we must consider arbitrary statement forms (not just dnf's). In this case, inverters are counted in computing the cost, since negations may be applied not only to letters but also to arbitrary statement forms. However, if we are only interested in switching circuits, consideration is restricted to statement forms in which negation is applied only to statement letters.

Notational Convention. In writing statement forms, it is often convenient to omit the conjunction sign &, and to write $\bar{\mathbf{A}}$ instead of $\neg\mathbf{A}$.

Example 4.15.

$$A\bar{B}C \vee \bar{A}B\bar{C} \quad \text{instead of} \quad (A \,\&\, \neg B \,\&\, C) \vee (\neg A \,\&\, B \,\&\, \neg C)$$

$$\bar{A}\bar{B} \vee \bar{A}B\bar{C}D \vee \bar{A}C \quad \text{instead of} \quad (\neg A \,\&\, \neg B) \vee (\neg A \,\&\, B \,\&\, \neg C \,\&\, D) \vee (\neg A \,\&\, C)$$

$$(\bar{A} \vee B)(A \vee \bar{B} \vee \bar{C}) \quad \text{instead of} \quad (\neg A \vee B) \,\&\, (A \vee \neg B \vee \neg C)$$

In Example 4.15 and in the sequel, we adopt the convention of omitting the parentheses around the disjuncts of a disjunctive normal form. Thus, we have written $A\bar{B}C \vee \bar{A}B\bar{C}$ instead of $(A\bar{B}C) \vee (\bar{A}B\bar{C})$, and $\bar{A}\bar{B} \vee \bar{A}B\bar{C}D \vee \bar{A}C$ instead of $(\bar{A}\bar{B}) \vee (\bar{A}B\bar{C}D) \vee (\bar{A}C)$. This alternative notation saves time and space, and is customary in work on circuits.

4.8 DON'T CARE CONDITIONS

In many problems involving design of circuits there are certain conditions which are impossible or for which no requirement is made concerning the operation of the circuit. Such conditions are called *don't care conditions.*

Example 4.16.

The switching circuit of Fig. 4-37 has a corresponding statement form $A(\bar{B} \vee C) \vee \bar{A}\,\bar{B}\,\bar{C}$. In the special case where A is "x is an even integer", B is "x is a perfect square", and C is "x is an integer divisible by 4", the three conditions $AB\bar{C}$, $\bar{A}\bar{B}C$, $\bar{A}BC$ are impossible. Hence there is no danger if we build a circuit which happens to allow current to flow if some of these impossible conditions occur. In particular, a circuit corresponding to the statement form

$$A(\bar{B} \vee C) \vee \bar{A}\bar{B}\bar{C} \vee AB\bar{C} \vee \bar{A}\bar{B}C$$

will accomplish the same task as the original circuit. But this statement form turns out to be logically equivalent to $A \vee \bar{B}$, which has the much simpler circuit of Fig. 4-38. (To derive the logical equivalence, notice that $\bar{A}\bar{B}\bar{C} \vee \bar{A}\bar{B}C$ is logically equivalent to $\bar{A}\bar{B}$, while $A(\bar{B} \vee C) \vee AB\bar{C}$ is logically equivalent to $A(\bar{B} \vee C \vee B\bar{C})$ and therefore to A. We are left with $A \vee \bar{A}\bar{B}$, which is logically equivalent to $A \vee \bar{B}$.)

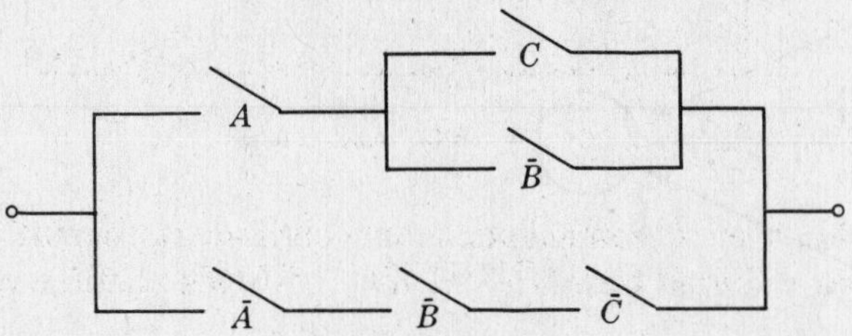

Fig. 4-37

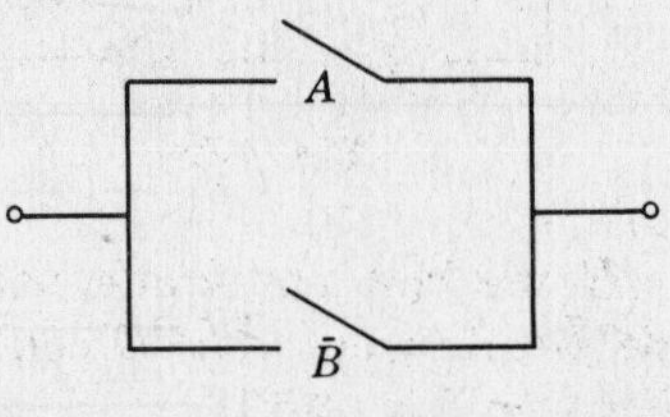

Fig. 4-38

Example 4.16 shows that addition of don't care conditions sometimes allows simplification of circuits. Later (Section 4.17), we shall learn a technique enabling us to choose those don't care conditions which lead to maximal simplification of the circuit.

Example 4.17.

The decimal digits 0 to 9 can be represented in binary notation as follows:

Decimal Notation	0	1	2	3	4	5	6	7	8	9
Binary Notation	0000	0001	0010	0011	0100	0101	0110	0111	1000	1001

Consider the sentences:

A The first (right-most) binary digit is 1

B The second binary digit is 1

C The third binary digit is 1

D The fourth binary digit is 1

Then

$\bar{A}\bar{B}\bar{C}\bar{D}$ corresponds to 0

$A\bar{B}\bar{C}\bar{D}$ corresponds to 1

$\bar{A}B\bar{C}\bar{D}$ corresponds to 2

....

$A\bar{B}\bar{C}D$ corresponds to 9

In terms of inputs $A, B, C, D,$ let us construct a switching circuit which passes current if and only if the number represented is 6, 7, or 8, i.e. under the condition $\bar{A}BC\bar{D} \vee ABC\bar{D} \vee \bar{A}\bar{B}\bar{C}D$. If the inputs A, B, C, D are such that they always represent a number between 0 and 9, then we can ignore the six possibilities 1010, 1011, ..., 1111 (i.e. the binary representations of 10 through 15). Hence the don't care conditions are $\bar{A}B\bar{C}D$, $AB\bar{C}D$, $\bar{A}\bar{B}CD$, $A\bar{B}CD$, $\bar{A}BCD$, $ABCD$. In particular, we can use

$$\bar{A}BC\bar{D} \vee ABC\bar{D} \vee \bar{A}\bar{B}\bar{C}D \vee ABCD \vee \bar{A}BCD \vee \bar{A}\bar{B}CD \vee \bar{A}B\bar{C}D$$

(Thus we are using four of the six don't care conditions.) This statement form is logically equivalent to $BC \vee \bar{A}D$. (This is left as an exercise. It can be done laboriously by a truth table, or much more easily using well-known logical equivalences from Chapter 1.) The circuit for $BC \vee \bar{A}D$ is given in Fig. 4-39.

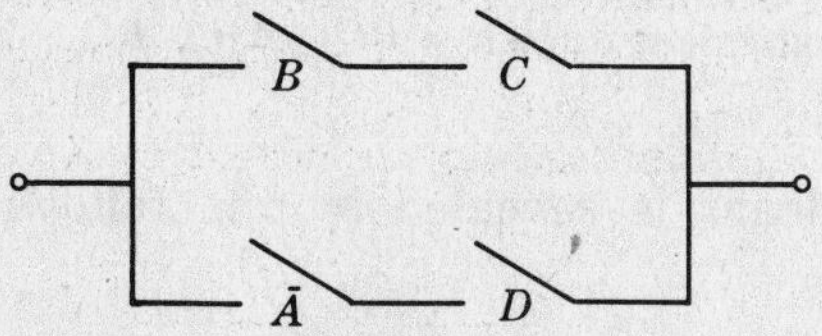

Fig. 4-39

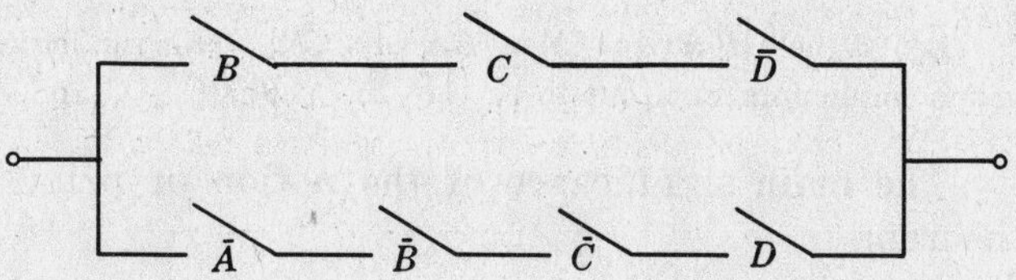

Fig. 4-40

If we had not made use of the don't care conditions, our original statement form $\bar{A}BC\bar{D} \vee ABC\bar{D} \vee \bar{A}\bar{B}\bar{C}D$ could have been reduced to $BC\bar{D} \vee \bar{A}\bar{B}\bar{C}D$, with the costlier circuit shown in Fig. 4-40.

4.9. MINIMAL DISJUNCTIVE NORMAL FORMS

Given any dnf Φ, say $A\bar{B}C \vee \bar{A}BD \vee ABC\bar{D} \vee A\bar{B}\bar{C}D$. Let l_Φ = the total number of literals (i.e. letters or negations of letters) in Φ, and d_Φ = the total number of disjuncts of Φ. In the example above, $l_\Phi = 14$ and $d_\Phi = 4$.

For dnf's Φ and Ψ, we say that Φ is *simpler* than Ψ if and only if $l_\Phi \leq l_\Psi$ and $d_\Phi \leq d_\Psi$ and at least one of these inequalities is strict (<).

This definition of *simpler* is most suitable in the case of logic circuits. If one is interested only in switching circuits, then the size of l_Φ alone would be a better measure of simplicity.

That we do not take into account the number of negation signs stems from the fact, already mentioned, that the number of such signs often depends only on arbitrary decisions as to which one of a proposition and its negation is to be represented by a statement letter.

A dnf Φ is said to be a *minimal dnf* for a statement form **A** if and only if Φ is logically equivalent to **A** and no other dnf simpler than Φ is logically equivalent to **A**. We shall now embark upon the task of describing various methods of finding the minimal dnf's for a given statement form.

We must emphasize again that we shall not distinguish between a fundamental conjunction and any other permutation of the literals in that conjunction. Thus we shall not distinguish between $A\bar{B}C$, $\bar{B}AC$, $\bar{B}CA$, $AC\bar{B}$, $CA\bar{B}$ and $C\bar{B}A$. Likewise we shall not distinguish between a given dnf and any other dnf obtained by permuting the disjuncts. Hence for our purposes, $AB \vee \bar{A}BC \vee A\bar{B}\bar{C}$ and $\bar{A}CB \vee BA \vee A\bar{C}\bar{B}$ are essentially the same.

4.10 PRIME IMPLICANTS

Let **A** be a statement form. A fundamental conjunction ψ is said to be a *prime implicant* of **A** if and only if ψ logically implies **A** but **A** is not logically implied by any other fundamental conjunction included in ψ. This is the same as saying that ψ logically implies **A** while any fundamental conjunction obtained by eliminating literals from ψ does not logically imply **A**. Clearly, a prime implicant of **A** is also a prime implicant of any statement form logically equivalent to **A**.

Example 4.18.

Let **A** be $AB \vee A\bar{B}C \vee \bar{A}\bar{B}C$. Then AC is a prime implicant of **A**. For, AC logically implies **A**, while A alone does not logically imply **A** and C alone does not logically imply **A**. Other prime implicants of **A** are AB and $\bar{B}C$. (Verification of these facts is left to the reader. We have no way of knowing at this point whether we have found all the prime implicants of **A**.)

Example 4.19.

Let **A** be $(\bar{B} \& (A \vee C)) \vee (B \& (\bar{A} \vee \bar{C}))$. This is logically equivalent to $A\bar{B} \vee \bar{B}C \vee \bar{A}B \vee B\bar{C}$. The prime implicants turn out to be $A\bar{B}$, $\bar{B}C$, $\bar{A}B$, $B\bar{C}$, $A\bar{C}$ and $\bar{A}C$.

The main significance of the notion of prime implicant is revealed by the following theorem.

Theorem 4.1. Any minimal dnf Φ for **A** is a disjunction of one or more prime implicants of **A**.

Proof. Let ψ be a disjunct of Φ. If ψ were not a prime implicant of **A**, then ψ would include a fundamental conjunction σ such that σ logically implies **A**. If $\Phi^\#$ is formed from

Φ by replacing ψ by σ, then clearly $\Phi^{\#}$ is logically equivalent to Φ. (For, on the one hand, ψ logically implies σ, and therefore Φ logically implies $\Phi^{\#}$. On the other hand, σ logically implies **A**, which is logically equivalent to Φ, and therefore $\Phi^{\#}$ logically implies Φ.) But $\Phi^{\#}$ is simpler than Φ, contradicting the assumption that Φ is a minimal dnf. ▶

We shall see shortly that a minimal dnf for **A** need not be a disjunction of *all* the prime implicants of **A**.

Remark: Every fundamental conjunction which logically implies a statement form **A** must include a prime implicant of **A**. For, if the fundamental conjunction ψ is not itself a prime implicant of **A**, then ψ must include a fundamental conjunction ψ_1 which logically implies **A**. If ψ_1 is not a prime implicant of **A**, then ψ_1 must include a fundamental conjunction ψ_2 which logically implies **A**, etc. This procedure must eventually stop, yielding a prime implicant of **A** which is included in ψ.

Definition. If ψ is a fundamental conjunction and Φ is a dnf, then we say that ψ is *superfluous* in $\psi \vee \Phi$ when and only when Φ is logically equivalent to $\psi \vee \Phi$.

If α is a literal, ψ is a fundamental conjunction, and Φ is a dnf, then we say that α is *superfluous* in $\alpha\psi \vee \Phi$ when and only when $\psi \vee \Phi$ is logically equivalent to $\alpha\psi \vee \Phi$.

Remarks: (1) ψ is superfluous in $\psi \vee \Phi$ if and only if ψ logically implies Φ. (2) α is superfluous in $\alpha\psi \vee \Phi$ if and only if ψ logically implies $\alpha \vee \Phi$.

We shall say that a dnf is *irredundant* if and only if it contains no superfluous disjuncts or literals.

Of course, we may obtain an irredundant equivalent of a given statement form by eliminating superfluous conjunctions and literals one by one. Clearly, an irredundant dnf is a disjunction of prime implicants, for, if one of its disjuncts were not a prime implicant, some literal of that disjunct would be superfluous (cf. Problem 4.10). In addition, any minimal dnf must be irredundant (for, elimination of superfluous disjuncts or literals would yield a simpler dnf).

Example 4.20.

Start with the dnf $AB \vee A\bar{B}C \vee \bar{A}\bar{B}C \vee B\bar{C}$.

(1) The first occurrence of $\bar{B}$ is superfluous. (For, AC implies $\bar{B} \vee AB \vee \bar{A}\bar{B}C \vee B\bar{C}$.) Thus we obtain $AB \vee AC \vee \bar{A}\bar{B}C \vee B\bar{C}$.

(2) AB is superfluous. (For, AB implies $AC \vee \bar{A}\bar{B}C \vee B\bar{C}$.) We now have $AC \vee \bar{A}\bar{B}C \vee B\bar{C}$.

(3) $\bar{A}$ is superfluous. (For, $\bar{B}C$ implies $AC \vee \bar{A} \vee B\bar{C}$.) This leaves us with $AC \vee \bar{B}C \vee B\bar{C}$, which turns out to be irredundant. (The reader can verify this without difficulty.)

Notice that AB is a prime implicant of the original dnf but that AB does not occur in the irredundant dnf that we have constructed. Thus *an irredundant dnf logically equivalent to a given statement form* **A** *need not contain all of the prime implicants of* **A**.

Example 4.21.

It is easy to verify that $A\bar{B} \vee \bar{A}B \vee B\bar{C} \vee \bar{B}C$ is irredundant. However, it is not a minimal dnf, since $A\bar{B} \vee \bar{A}C \vee B\bar{C}$ is a simpler logically equivalent dnf. Thus *an irredundant dnf need not be a minimal dnf*. Hence the very simple procedure of reducing to an irredundant dnf does not solve the problem of finding minimal dnf's.

4.11 THE QUINE-McCLUSKEY METHOD FOR FINDING ALL PRIME IMPLICANTS

If **A** is a statement form and ϕ_1 and ϕ_2 are fundamental conjunctions, then we say that ϕ_1 is a *completion* of ϕ_2 relative to **A** if and only if ϕ_1 includes ϕ_2 and the statement letters in ϕ_1 are precisely the letters occurring in **A**.

Example 4.22.

Let **A** be $A\bar{B}C \vee AD \vee \bar{A}B$, and let ϕ_1 be $A\bar{C}$. Then there are four completions of ϕ relative to **A**, namely: $AB\bar{C}D$, $AB\bar{C}\bar{D}$, $A\bar{B}\bar{C}\bar{D}$, $A\bar{B}\bar{C}D$.

Lemma 4.2. Let Φ be a full dnf (i.e. a dnf in which the letters contained in any one disjunct are precisely the letters in any other disjunct; see page 14). Let ϕ be a fundamental conjunction all of whose letters are in Φ. Then ϕ logically implies Φ if and only if all completions of ϕ relative to Φ are disjuncts of Φ.

Proof. (i) Assume ϕ logically implies Φ, but some completion ψ of ϕ relative to Φ is not a disjunct of Φ. Take the truth assignment corresponding to ψ (i.e. letters unnegated in ψ are T, while letters negated in ψ are F). Since ψ is a completion of ϕ, the assignment makes ϕ T, and, therefore, it also makes Φ T. But all disjuncts of Φ, being different from ψ in at least one letter, must be F. Hence Φ would also be F, not T.

(ii) Assume all completions of ϕ relative to Φ are disjuncts of Φ. Take any truth assignment making ϕ T. We must prove that Φ also is T. The truth assignment corresponds to some completion ψ of ϕ relative to Φ (a letter appears unnegated in ψ if it is T and negated if it is F). Then ψ is a disjunct of Φ. But, since ψ is T, so is Φ. ▸

Lemma 4.3. If **A** is not a tautology, no prime implicant ϕ of **A** contains any letters not in **A**.

Proof. Assume some letter, say B, is in ϕ but not in **A**. Let χ be the fundamental conjunction obtained from ϕ by removing the literal containing B. (Notice that ϕ is neither B nor $\bar{B}$. For, take a truth assignment making **A** false and choose the value of B so that ϕ is T. Then ϕ does not logically imply **A**.) χ also logically implies Φ. (For, given any truth assignment making χ T, extend it by making B true or false according as B or $\bar{B}$ occurs as a conjunct of ϕ. Then ϕ is T and therefore **A** is also T.) But this contradicts the assumption that ϕ is a prime implicant of **A**. ▸

Theorem 4.4. Let Φ be a non-tautologous full dnf, and let ϕ be some fundamental conjunction. Then ϕ is a prime implicant of Φ if and only if

(i) all letters of ϕ are also in Φ;

(ii) all completions of ϕ relative to Φ are disjuncts of Φ, but no other fundamental conjunction included in ϕ has this property.

Proof. Direct consequence of Lemmas 4.2-4.3. ▸

The Quine-McCluskey Method for Finding All Prime Implicants of a Non-Tautologous Full Dnf Φ: Let Φ be $\psi_1 \vee \cdots \vee \psi_k$.

(1) List $\psi_1, \ldots, \psi_k$.

(2) If two fundamental conjunctions ϕ and χ in the list are the same except that ϕ contains a certain letter unnegated while χ contains the same letter negated, add to the list the fundamental conjunction obtained by eliminating from ϕ the letter in which ϕ differs from χ. Place check marks next to ϕ and χ.

(3) Repeat the process indicated in (2) until it can no longer be applied. Fundamental conjunctions which already have been checked can be used again in applications of (2).

The unchecked fundamental conjunctions in the resulting list are the prime implicants of Φ. (This assertion will be justified after consideration of a few examples.)

Example 4.23.

Let Φ be $AB\bar{C} \vee A\bar{B}C \vee \bar{A}BC \vee \bar{A}\bar{B}C \vee \bar{A}\bar{B}\bar{C}$. Start with

$$\begin{array}{l} AB\bar{C} \\ A\bar{B}C \\ \bar{A}BC \\ \bar{A}\bar{B}C \\ \bar{A}\bar{B}\bar{C} \end{array}$$

Application of (2) yields

$$\begin{array}{ll} AB\bar{C} & \\ A\bar{B}C \;\checkmark & \bar{B}C \\ \bar{A}BC \;\checkmark & \bar{A}C \\ \bar{A}\bar{B}C \;\checkmark & \bar{A}\bar{B} \\ \bar{A}\bar{B}\bar{C} \;\checkmark & \end{array}$$

Notice that $A\bar{B}C$ and $\bar{A}\bar{B}C$ yield $\bar{B}C$; $\bar{A}BC$ and $\bar{A}\bar{B}C$ yield $\bar{A}C$; $\bar{A}\bar{B}C$ and $\bar{A}\bar{B}\bar{C}$ yield $\bar{A}\bar{B}$. Now (2) is no longer applicable. Hence the prime implicants are $AB\bar{C}, \bar{B}C, \bar{A}C, \bar{A}\bar{B}$.

Example 4.24.

Let Φ be

$$ABCD \vee AB\bar{C}D \vee A\bar{B}\bar{C}D \vee A\bar{B}CD \vee AB\bar{C}\bar{D} \vee \bar{A}B\bar{C}\bar{D} \vee \bar{A}BC\bar{D} \vee \bar{A}\bar{B}\bar{C}D$$

(1) List

$$\left\{ ABCD \right. \qquad \left\{ \begin{array}{l} AB\bar{C}D \\ A\bar{B}CD \end{array} \right. \qquad \left\{ \begin{array}{l} A\bar{B}\bar{C}D \\ AB\bar{C}\bar{D} \\ \bar{A}BC\bar{D} \end{array} \right. \qquad \left\{ \begin{array}{l} \bar{A}\bar{B}\bar{C}D \\ \bar{A}B\bar{C}\bar{D} \end{array} \right.$$

Notice that the disjuncts are listed in groups: first, those with no negations, then those with one negation, etc. Since process (2) is applicable only to a pair of fundamental conjunctions which differ by one in the number of negations, in seeking to apply (2) we need only compare fundamental conjunctions with those in the next group.

(2) Application of process (2) yields

$$\begin{array}{ll} ABCD \;\checkmark & \left\{ \begin{array}{l} ABD \\ ACD \end{array} \right. \\ \left\{ \begin{array}{l} AB\bar{C}D \;\checkmark \\ A\bar{B}CD \;\checkmark \end{array} \right. & \left\{ \begin{array}{l} A\bar{C}D \\ AB\bar{C} \\ A\bar{B}D \end{array} \right. \\ \left\{ \begin{array}{l} A\bar{B}\bar{C}D \;\checkmark \\ AB\bar{C}\bar{D} \;\checkmark \\ \bar{A}BC\bar{D} \;\checkmark \end{array} \right. & \left\{ \begin{array}{l} \bar{B}\bar{C}D \\ B\bar{C}\bar{D} \\ \bar{A}B\bar{D} \end{array} \right. \\ \left\{ \begin{array}{l} \bar{A}\bar{B}\bar{C}D \;\checkmark \\ \bar{A}B\bar{C}\bar{D} \;\checkmark \end{array} \right. & \end{array}$$

Further application of (2) produces

$ABCD$ ✓	ABD ✓	AD
	ACD ✓	
$AB\bar{C}D$ ✓		
$A\bar{B}CD$ ✓	$A\bar{C}D$ ✓	
$A\bar{B}\bar{C}D$ ✓	$AB\bar{C}$	
$AB\bar{C}\bar{D}$ ✓	$A\bar{B}D$ ✓	
$\bar{A}BC\bar{D}$ ✓		
	$\bar{B}\bar{C}D$	
$\bar{A}\bar{B}\bar{C}D$ ✓	$B\bar{C}\bar{D}$	
$\bar{A}B\bar{C}\bar{D}$ ✓	$\bar{A}B\bar{D}$	

Process (2) is no longer applicable. Hence the prime implicants are $AB\bar{C}$, $\bar{B}\bar{C}D$, $B\bar{C}\bar{D}$, $\bar{A}B\bar{D}$, AD.

Justification of the Quine-McCluskey Method: All fundamental conjunctions in the list logically imply Φ. (It is only necessary to observe that application of process (2) to fundamental conjunctions which logically imply Φ yields a fundamental conjunction which also logically implies Φ.) Every prime implicant ψ of Φ will appear unchecked. (For, by Theorem 4.4, all completions of ψ will be disjuncts in Φ. Hence ψ will eventually appear in the list after suitable applications of process (2) to these completions eliminate all the letters not in ψ. ψ itself will never be checked, since if it were it would not be a prime implicant of Φ.) On the other hand, no fundamental disjunction ψ which is not a prime implicant will remain unchecked. (For, ψ must include a fundamental conjunction ϕ which is a prime implicant of Φ. By Theorem 4.4, all completions of ϕ originally appear in the list. By suitable applications of process (2) to these completions, we obtain a fundamental disjunction χ differing from ψ in precisely one letter. Then application of process (2) to ψ and χ imposes a check on ψ.)

Limitation of the Quine-McCluskey Method: One must start with a full dnf. If we are given a dnf which is not full, we must expand it into a full dnf. This tedious and long process can be avoided by another procedure which we shall study later.

4.12 PRIME IMPLICANT TABLES

Once we have obtained all prime implicants of a given statement form Φ, we must find out which disjunctions of prime implicants are minimal dnf's.

Theorem 4.5. Let Φ be a non-tautologous full dnf, and let Ψ be a dnf. If Ψ is a minimal dnf for Φ, then each disjunct of Φ includes a disjunct of Ψ.

Proof. Assume not. Let ϕ be a disjunct of Φ which does not include any disjunct of Ψ. Hence each disjunct of Ψ differs in at least one literal from ϕ. But then the assignment of truth values making ϕ T makes Ψ F, and therefore makes Φ F. But ϕ is a disjunct of Φ; so Φ must be T, which is a contradiction. ▶

Our overall strategy can now be made clear. We choose a disjunction Ψ of prime implicants so that every disjunct of the full dnf Φ includes a disjunct of Ψ. (Clearly, Ψ is logically equivalent to Φ. Since Ψ is a disjunction of prime implicants of Φ, Ψ logically implies Φ. On the other hand, since each disjunct of Φ includes a disjunct of Ψ, Φ logically implies Ψ.) Among all such Ψ's we find the minimal ones. We shall indicate techniques for narrowing this choice to a relatively small number of Ψ's.

Construct a matrix, one row for each prime implicant of Φ and one column for each disjunct of Φ. Place a cross ($\times$) at the intersection of a row corresponding to a prime implicant ϕ and a column corresponding to a disjunct ψ of Φ such that ψ includes ϕ. This matrix is called the *prime implicant table* for Φ.

Example 4.25.

Let Φ be $A\bar{B}C \vee \bar{A}B\bar{C} \vee AB\bar{C} \vee ABC$. Using the Quine-McCluskey method, we obtain

$$\begin{array}{ll} ABC \ \checkmark & \\ \left\{\begin{array}{l} A\bar{B}C \ \checkmark \\ AB\bar{C} \ \checkmark \end{array}\right. & \begin{array}{l} AC \\ AB \\ B\bar{C} \end{array} \\ \bar{A}B\bar{C} \ \checkmark & \end{array}$$

Thus the prime implicants are $AC, AB, B\bar{C}$, and the prime implicant table is shown in Fig. 4-41.

	ABC	$A\bar{B}C$	$AB\bar{C}$	$\bar{A}B\bar{C}$
AC	$\times$	$\times$		
AB	$\times$		$\times$	
$B\bar{C}$			$\times$	$\times$

Fig. 4-41

We shall now describe various operations performed on prime implicant tables in order to obtain minimal dnf's.

Core Operation. Assume there is a disjunct ψ of Φ such that the column under ψ contains a single cross. Let ϕ be the prime implicant corresponding to this cross. ϕ belongs to what we shall call the *core* of Φ. By Theorem 4.5, ϕ must be a disjunct of every minimal dnf for Φ. We eliminate the row corresponding to ϕ as well as all columns containing a cross in the row corresponding to ϕ. (Since ϕ must be a disjunct of every minimal dnf, the condition of Theorem 4.5 is met for the disjuncts heading any such column.)

Example 4.25 (continued).

In Fig. 4-41, the columns under $A\bar{B}C$ and $\bar{A}B\bar{C}$ have a single cross each. Hence AC and $B\bar{C}$ belong to the core. But all the columns contain a cross in the rows corresponding to AC and $B\bar{C}$. Hence $AC \vee B\bar{C}$ is the unique minimal dnf for Φ.

Example 4.26.

Let Φ be

$$ABCD \vee AB\bar{C}D \vee AB\bar{C}\bar{D} \vee A\bar{B}CD \vee \bar{A}BCD \vee \bar{A}BC\bar{D} \vee \bar{A}B\bar{C}D \vee \bar{A}\bar{B}\bar{C}D$$

Find the prime implicants:

$$\begin{array}{lll} ABCD \ \checkmark & \left\{\begin{array}{l} ABD \ \checkmark \\ ACD \\ BCD \ \checkmark \end{array}\right. & BD \\ \left\{\begin{array}{l} AB\bar{C}D \ \checkmark \\ A\bar{B}CD \ \checkmark \\ \bar{A}BCD \ \checkmark \end{array}\right. & & \\ \left\{\begin{array}{l} AB\bar{C}\bar{D} \ \checkmark \\ \bar{A}BC\bar{D} \ \checkmark \\ \bar{A}B\bar{C}D \ \checkmark \end{array}\right. & \left\{\begin{array}{l} AB\bar{C} \\ B\bar{C}D \ \checkmark \\ \bar{A}BC \\ \bar{A}BD \ \checkmark \end{array}\right. & \\ \bar{A}\bar{B}\bar{C}D \ \checkmark & \bar{A}\bar{C}D & \end{array}$$

Hence the prime implicants are $BD, \bar{A}\bar{C}D, \bar{A}BC, AB\bar{C}, ACD$.

The prime implicant table is

	$ABCD$	$ABC\bar{D}$	$ABC\bar{D}$	$A\bar{B}CD$	$\bar{A}BCD$	$\bar{A}BC\bar{D}$	$\bar{A}B\bar{C}D$	$\bar{A}\bar{B}\bar{C}D$
BD	×	×			×		×	
$\bar{A}\bar{C}D$							⊠	⊗
$\bar{A}BC$					⊠	⊗		
$AB\bar{C}$		⊠	⊗					
ACD	⊠			⊗				

We draw circles around crosses which are the only ones in a given column. In this case there are such crosses in the columns under $AB\bar{C}\bar{D}$, $A\bar{B}CD$, $\bar{A}BC\bar{D}$, $\bar{A}\bar{B}\bar{C}D$. Hence $AB\bar{C}$, ACD, $\bar{A}BC$ and $\bar{A}\bar{C}D$ are in the core. We draw a square around each cross in any row in which there is an encircled cross. In Example 4.26 we then have a square or circle in every column. Thus every disjunct of Φ includes a prime implicant in the core. Hence $AB\bar{C} \vee ACD \vee \bar{A}BC \vee \bar{A}\bar{C}D$ is the unique minimal dnf for Φ.

The results of Examples 4.25 and 4.26 are exceptional. Sometimes a single application of the core operation is not sufficient. Wider coverage is afforded by adding the following two operations.

Dominant Column Operation. If a column β has a cross in every row in which a column α has a cross, then we can eliminate column β. (To satisfy the condition of Theorem 4.5, we have to use a prime implicant included in the fundamental conjunction heading column α. Then by assumption the same prime implicant is included in the fundamental conjunction heading column β.)

Dominated Row Operation. If the row corresponding to a prime implicant ψ_1 has a cross in every column in which the row corresponding to a prime implicant ψ_2 has a cross, and if the number of literals of ψ_1 is smaller than that of ψ_2, then we eliminate the row corresponding to ψ_2. (For, if a minimal dnf had ψ_2 as a disjunct, replacing ψ_2 by ψ_1 would lower the cost, contradicting the assumption that ψ_2 is minimal.)

Example 4.27.

Let Φ be

$$ABCD \vee A\bar{B}CD \vee \bar{A}BC\bar{D} \vee A\bar{B}\bar{C}\bar{D} \vee \bar{A}B\bar{C}\bar{D} \vee ABC\bar{D} \vee A\bar{B}C\bar{D} \vee A\bar{B}\bar{C}D \vee \bar{A}BCD \vee \bar{A}B\bar{C}D \vee \bar{A}\bar{B}\bar{C}\bar{D}$$

First we obtain the prime implicants:

$ABCD$ √	ACD √	AC
$A\bar{B}CD$ √	ABC √	BC
$ABC\bar{D}$ √	BCD √	$A\bar{B}$
$\bar{A}BCD$ √	$A\bar{B}C$ √	$\bar{A}B$
$\bar{A}BC\bar{D}$ √	$A\bar{B}D$ √	
$A\bar{B}C\bar{D}$ √	$BC\bar{D}$ √	
$A\bar{B}\bar{C}D$ √	$AC\bar{D}$ √	
$\bar{A}B\bar{C}D$ √	$\bar{A}BC$ √	
$A\bar{B}\bar{C}\bar{D}$ √	$\bar{A}BD$ √	
$\bar{A}B\bar{C}\bar{D}$ √	$\bar{A}B\bar{D}$ √	
$\bar{A}\bar{B}\bar{C}\bar{D}$ √	$A\bar{B}\bar{D}$ √	
	$A\bar{B}\bar{C}$ √	
	$\bar{A}B\bar{C}$ √	
	$\bar{B}\bar{C}\bar{D}$	
	$\bar{A}\bar{C}\bar{D}$	

Thus the prime implicants are $\bar{B}\bar{C}\bar{D}$, $\bar{A}\bar{C}\bar{D}$, AC, BC, $A\bar{B}$, $\bar{A}B$. The prime implicant table is

	$ABCD$	$A\bar{B}CD$	$\bar{A}BC\bar{D}$	$A\bar{B}\bar{C}\bar{D}$	$\bar{A}B\bar{C}\bar{D}$	$ABC\bar{D}$	$A\bar{B}C\bar{D}$	$A\bar{B}\bar{C}D$	$\bar{A}BCD$	$\bar{A}B\bar{C}D$	$\bar{A}\bar{B}\bar{C}\bar{D}$
$\bar{B}\bar{C}\bar{D}$				×							×
$\bar{A}\bar{C}\bar{D}$					×						×
AC	×	×				×	×				
BC	×		×			×			×		
$A\bar{B}$		⊠		⊠			⊠	⊗			
$\bar{A}B$			⊠		⊠				⊠	⊗	

The columns under $A\bar{B}\bar{C}D$ and $\bar{A}B\bar{C}D$ have unique crosses, which we circle. Thus $A\bar{B}$ and $\bar{A}B$ belong to the core. Put a square around each cross in the rows belonging to $A\bar{B}$ and $\bar{A}B$. Hence all the columns containing a circle or square can be eliminated. The new table is

	$ABCD$	$ABC\bar{D}$	$\bar{A}\bar{B}\bar{C}\bar{D}$
$\bar{B}\bar{C}\bar{D}$			×
$\bar{A}\bar{C}\bar{D}$			×
AC	×	×	
BC	×	×	

The columns under $ABCD$ and $ABC\bar{D}$ have crosses in the same row. Hence by a dominant column operation we may eliminate either column, say, the one under $ABC\bar{D}$. This leaves us with the new table:

		$ABCD$	$\bar{A}\bar{B}\bar{C}\bar{D}$
(1)	$\bar{B}\bar{C}\bar{D}$		×
(2)	$\bar{A}\bar{C}\bar{D}$		×
(3)	AC	×	
(4)	BC	×	

Fig. 4-42

None of our operations are applicable to this table. However, notice that in order that each column include some disjunct of the required minimal dnf, we must have ((1) or (2)) and ((3) or (4))†, i.e. $((1) \vee (2)) \,\&\, ((3) \vee (4))$. This is logically equivalent to

$$((1) \,\&\, (3)) \vee ((1) \,\&\, (4)) \vee ((2) \,\&\, (3)) \vee ((2) \,\&\, (4))$$

Thus there are four different ways of choosing the rest of the prime implicants, and we obtain the four dnf's:

$$A\bar{B} \vee \bar{A}B \vee \bar{B}\bar{C}\bar{D} \vee AC$$

$$A\bar{B} \vee \bar{A}B \vee \bar{B}\bar{C}\bar{D} \vee BC$$

$$A\bar{B} \vee \bar{A}B \vee \bar{A}\bar{C}\bar{D} \vee AC$$

$$A\bar{B} \vee \bar{A}B \vee \bar{A}\bar{C}\bar{D} \vee BC$$

These are the only possibilities for minimal dnf's. Since they all have the same cost, all four are minimal dnf's.

We shall call the method we have used for handling the table of Fig. 4-42 the *Boolean method.*

Example 4.28.

Let Φ be

$$ABCD \vee AB\bar{C}D \vee \bar{A}BCD \vee AB\bar{C}\bar{D} \vee A\bar{B}\bar{C}D$$
$$\vee \bar{A}BC\bar{D} \vee \bar{A}B\bar{C}D \vee A\bar{B}\bar{C}\bar{D} \vee \bar{A}\bar{B}C\bar{D} \vee \bar{A}\bar{B}\bar{C}\bar{D}$$

†By (1), we mean that row (1) appears as a disjunct in the required minimal dnf, by (2) we mean that row (2) appears as a disjunct, etc.

We obtain the prime implicants:

$ABCD$ ✓	ABD ✓	BD
$ABC\bar{D}$ ✓	BCD ✓	$A\bar{C}$
$\bar{A}BCD$ ✓	$AB\bar{C}$ ✓	
$AB\bar{C}\bar{D}$ ✓	$A\bar{C}D$ ✓	
$A\bar{B}\bar{C}D$ ✓	$B\bar{C}D$ ✓	
$\bar{A}BC\bar{D}$ ✓	$\bar{A}BC$	
$\bar{A}B\bar{C}D$ ✓	$\bar{A}BD$ ✓	
$A\bar{B}\bar{C}\bar{D}$ ✓	$A\bar{C}\bar{D}$ ✓	
$\bar{A}\bar{B}C\bar{D}$ ✓	$A\bar{B}\bar{C}$ ✓	
$\bar{A}\bar{B}\bar{C}\bar{D}$ ✓	$\bar{A}C\bar{D}$	
	$\bar{B}\bar{C}\bar{D}$	
	$\bar{A}\bar{B}\bar{D}$	

Thus the prime implicants are $\bar{A}BC$, $\bar{A}C\bar{D}$, $\bar{B}\bar{C}\bar{D}$, $\bar{A}\bar{B}\bar{D}$, BD, $A\bar{C}$. The prime implicant table is

	$ABCD$	$ABC\bar{D}$	$\bar{A}BCD$	$AB\bar{C}\bar{D}$	$A\bar{B}\bar{C}D$	$\bar{A}BC\bar{D}$	$\bar{A}B\bar{C}D$	$A\bar{B}\bar{C}\bar{D}$	$\bar{A}\bar{B}C\bar{D}$	$\bar{A}\bar{B}\bar{C}\bar{D}$
$\bar{A}BC$			×			×				
$\bar{A}C\bar{D}$						×			×	
$\bar{B}\bar{C}\bar{D}$								×		×
$\bar{A}\bar{B}\bar{D}$									×	×
BD	⊗	⊠	⊠				⊗			
$A\bar{C}$		⊠		⊗	⊗			⊠		

Circle the crosses which are unique in their columns. Hence BD and $A\bar{C}$ are in the core. Place squares around all crosses in the rows of BD and $A\bar{C}$. We then eliminate the rows of BD and $A\bar{C}$, and all columns containing squares or circles. The new table is

		$\bar{A}BC\bar{D}$	$\bar{A}\bar{B}C\bar{D}$	$\bar{A}\bar{B}\bar{C}\bar{D}$
(1)	$\bar{A}BC$	×		
(2)	$\bar{A}C\bar{D}$	×	×	
(3)	$\bar{B}\bar{C}\bar{D}$			×
(4)	$\bar{A}\bar{B}\bar{D}$		×	×

Fig. 4-43

Now we can apply the Boolean method used in Example 4.27. We obtain $[(1) \vee (2)] \,\&\, [(2) \vee (4)] \,\&\, [(3) \vee (4)]$, which is equivalent to

$$[(1) \,\&\, (2) \,\&\, (3)] \vee [(1) \,\&\, (2) \,\&\, (4)] \vee [(1) \,\&\, (4)] \vee [(1) \,\&\, (3) \,\&\, (4)] \vee [(2) \,\&\, (3)] \vee [(2) \,\&\, (4)] \vee [(2) \,\&\, (3) \,\&\, (4)]$$

But, taking into account the fact that, when (2) is used, (1) is not required, and that, when (4) is used, (3) is not required, we obtain

$$((2) \,\&\, (3)) \vee ((2) \,\&\, (4)) \vee ((1) \,\&\, (4))$$

Thus there are three possibilities for minimal dnf's:

$$BD \vee A\bar{C} \vee \bar{A}C\bar{D} \vee \bar{B}\bar{C}\bar{D}$$

$$BD \vee A\bar{C} \vee \bar{A}C\bar{D} \vee \bar{A}\bar{B}\bar{D}$$

$$BD \vee A\bar{C} \vee \bar{A}BC \vee \bar{A}\bar{B}\bar{D}$$

Since the costs are the same, all three are (the only) minimal dnf's for Φ.

Instead of the Boolean method, we may use the so-called *branching method* for handling the table of Fig. 4-43. We take a column with a minimal number of crosses. In our example, there are two crosses in each column; so we may choose any column, say the one under $\bar{A}BC\bar{D}$. To ensure that $\bar{A}BC\bar{D}$ includes a prime implicant of the sought-for dnf's, we may use either $\bar{A}BC$ or $\bar{A}C\bar{D}$. Hence we obtain two tables (in general, if there are n crosses in the column, we would obtain n tables) as follows: in the left-hand table (Fig. 4-44) we assume that $\bar{A}BC$ is taken as a disjunct and we eliminate the row containing $\bar{A}BC$ as well as every column containing a cross in that row. For the right-hand table (Fig. 4-45), we do the same with $\bar{A}C\bar{D}$.

	$\bar{A}\bar{B}C\bar{D}$	$\bar{A}\bar{B}\bar{C}\bar{D}$
$\bar{A}C\bar{D}$	×	
$\bar{B}\bar{C}\bar{D}$		×
$\bar{A}\bar{B}\bar{D}$	×	×

Fig. 4-44

	$\bar{A}\bar{B}\bar{C}\bar{D}$
$\bar{A}BC$	
$\bar{B}\bar{C}\bar{D}$	×
$\bar{A}\bar{B}\bar{D}$	×

Fig. 4-45

In the left-hand table, we can again apply the Boolean method or branching, but in this simple case it is obvious that only the choice of $\bar{A}\bar{B}\bar{D}$ will yield minimal cost. In the right-hand table, we can choose either $\bar{B}\bar{C}\bar{D}$ or $\bar{A}\bar{B}\bar{D}$. Thus we have the following possibilities:

From the left-hand table, $BD \vee A\bar{C} \vee \bar{A}BC \vee \bar{A}\bar{B}\bar{D}$.

From the right-hand table, $BD \vee A\bar{C} \vee \bar{A}C\bar{D} \vee \bar{B}\bar{C}\bar{D}$ and $BD \vee A\bar{C} \vee \bar{A}C\bar{D} \vee \bar{A}\bar{B}\bar{D}$.

This is identical with the result of the Boolean method.

4.13 MINIMIZING WITH DON'T CARE CONDITIONS

Let us assume that we must find a minimal dnf for a statement form Φ, assuming that the additional fundamental conjunctions $\psi_1, \ldots, \psi_k$ are don't care conditions. Then we can adapt the method used in the preceding section in the following manner. Find all prime implicants of $\Phi \vee \psi_1 \vee \cdots \vee \psi_k$. However, in constructing the prime implicant table, use columns only for the disjuncts of Φ, not for $\psi_1, \ldots, \psi_k$ (since we are concerned only that each disjunct of Φ include some prime implicant of the required minimal dnf's).

Example 4.29.

Let Φ be

$$AB\bar{C}D \vee A\bar{B}\bar{C}\bar{D} \vee \bar{A}BCD \vee \bar{A}BC\bar{D} \vee \bar{A}\bar{B}C\bar{D}$$

Let the don't care conditions be: $ABCD$, $AB\bar{C}\bar{D}$, $A\bar{B}\bar{C}D$, $\bar{A}B\bar{C}D$, $\bar{A}\bar{B}\bar{C}\bar{D}$. By the standard procedure (left as an exercise for the reader), we find that the prime implicants of $\Phi \vee ABCD \vee AB\bar{C}\bar{D} \vee A\bar{B}\bar{C}D \vee \bar{A}B\bar{C}D \vee \bar{A}\bar{B}\bar{C}\bar{D}$ are: BD, $A\bar{C}$, $\bar{A}BC$, $\bar{A}C\bar{D}$, $\bar{B}\bar{C}\bar{D}$, $\bar{A}\bar{B}\bar{D}$.

The prime implicant table is

	$AB\bar{C}D$	$A\bar{B}\bar{C}\bar{D}$	$\bar{A}BCD$	$\bar{A}BC\bar{D}$	$\bar{A}\bar{B}C\bar{D}$
BD	×		×		
$A\bar{C}$	×	×			
$\bar{A}BC$			×	×	
$\bar{A}C\bar{D}$				×	×
$\bar{B}\bar{C}\bar{D}$		×			
$\bar{A}\bar{B}\bar{D}$					×

Now $A\bar{C}$ has crosses wherever $\bar{B}\bar{C}\bar{D}$ has, and $A\bar{C}$ has fewer literals than $\bar{B}\bar{C}\bar{D}$. Hence by a dominant row operation we eliminate the row of $\bar{B}\bar{C}\bar{D}$, obtaining

	$AB\bar{C}D$	$A\bar{B}\bar{C}D$	$\bar{A}BCD$	$\bar{A}BC\bar{D}$	$\bar{A}\bar{B}C\bar{D}$
BD	×		×		
$A\bar{C}$	$\boxed{\times}$	$\otimes$			
$\bar{A}BC$			×	×	
$\bar{A}C\bar{D}$				×	×
$\bar{A}\bar{B}\bar{D}$					×

Now the column of $A\bar{B}\bar{C}D$ has a unique cross. Hence we place $A\bar{C}$ into what we call the *secondary core.* ($A\bar{C}$ must be a disjunct of every minimal dnf.) Then by the core operation we drop the row of $A\bar{C}$, together with the columns under $AB\bar{C}D$ and $A\bar{B}\bar{C}D$:

	$\bar{A}BCD$	$\bar{A}BC\bar{D}$	$\bar{A}\bar{B}C\bar{D}$
BD	×		
$\bar{A}BC$	×	×	
$\bar{A}C\bar{D}$		×	×
$\bar{A}\bar{B}\bar{D}$			×

Now we can apply either the Boolean method or the branching method. However, it is clear that the first and third rows yield the only minimal dnf. Hence the unique minimal dnf is $A\bar{C} \vee BD \vee \bar{A}C\bar{D}$.

4.14 THE CONSENSUS METHOD FOR FINDING PRIME IMPLICANTS

Given two fundamental conjunctions ψ_1 and ψ_2. If there is precisely one letter ρ which occurs negated in one of ψ_1 and ψ_2 and unnegated in the other, then the fundamental conjunction obtained from $\psi_1\psi_2$ by deleting ρ and $\bar{\rho}$ and omitting repetitions of any other literals is called the *consensus* of ψ_1 and ψ_2.

Example 4.30.

(i) The consensus of $A\bar{B}C$ and ABD is ACD.

(ii) The consensus of $\bar{A}\bar{B}$ and $A\bar{B}CD$ is $\bar{B}CD$.

(iii) There is no consensus of $\bar{A}\bar{B}C$ and ABD.

(iv) The consensus of A and $\bar{A}B$ is B.

(v) The consensus of $\bar{A}$ and AB is B.

Theorem 4.6. The consensus ϕ of ψ_1 and ψ_2 logically implies $\psi_1 \vee \psi_2$.

Proof. Consider any truth assignment making ϕ true. Let ρ be the letter occurring negated in ψ_1 and unnegated in ψ_2. If ρ is T, then ψ_2 is T. If ρ is F, then ψ_1 is T. In either case, $\psi_1 \vee \psi_2$ is T. ▶

Corollary 4.7. If ϕ is the consensus of ψ_1 and ψ_2, then $\psi_1 \vee \psi_2$ is logically equivalent to $\psi_1 \vee \psi_2 \vee \phi$.

Consider the following two operations transforming a dnf into a logically equivalent dnf.

(i) Eliminate any disjunct which includes another.

(ii) Add as a disjunct the consensus of two disjuncts, if that consensus neither is identical with nor includes some disjunct of the given dnf.

Given a dnf Φ. (If we are given an arbitrary statement form, first transform it into a logically equivalent dnf.) The *consensus method* consists of applying operations (i) and (ii) until these operations are no longer applicable. The result turns out to be the disjunction of all prime implicants of Φ.

Example 4.31.

Let Φ be $AB\bar{C} \vee A\bar{B}C\bar{D} \vee A\bar{B} \vee \bar{A}B\bar{C} \vee \bar{A}\bar{B}\bar{C}D$.

By (i), $AB\bar{C} \vee A\bar{B} \vee \bar{A}B\bar{C} \vee \bar{A}\bar{B}\bar{C}D$ ($A\bar{B}C\bar{D}$ includes $A\bar{B}$).

By (ii), $AB\bar{C} \vee A\bar{B} \vee \bar{A}B\bar{C} \vee \bar{A}\bar{B}\bar{C}D \vee A\bar{C}$ (Consensus of $AB\bar{C}$ and $A\bar{B}$).

By (i), $A\bar{B} \vee \bar{A}B\bar{C} \vee \bar{A}\bar{B}\bar{C}D \vee A\bar{C}$ ($AB\bar{C}$ includes $A\bar{C}$).

By (ii), $A\bar{B} \vee \bar{A}B\bar{C} \vee \bar{A}\bar{B}\bar{C}D \vee A\bar{C} \vee \bar{B}\bar{C}D$ (Consensus of $A\bar{B}$ and $\bar{A}\bar{B}\bar{C}D$).

By (i), $A\bar{B} \vee \bar{A}B\bar{C} \vee A\bar{C} \vee \bar{B}\bar{C}D$ ($\bar{A}\bar{B}\bar{C}D$ includes $\bar{B}\bar{C}D$).

By (ii), $A\bar{B} \vee \bar{A}B\bar{C} \vee A\bar{C} \vee \bar{B}\bar{C}D \vee B\bar{C}$ (Consensus of $\bar{A}B\bar{C}$ and $A\bar{C}$).

By (i), $A\bar{B} \vee A\bar{C} \vee \bar{B}\bar{C}D \vee B\bar{C}$ ($\bar{A}B\bar{C}$ includes $B\bar{C}$).

By (ii), $A\bar{B} \vee A\bar{C} \vee \bar{B}\bar{C}D \vee B\bar{C} \vee \bar{C}D$ (Consensus of $\bar{B}\bar{C}D$ and $B\bar{C}$).

By (i), $A\bar{B} \vee A\bar{C} \vee B\bar{C} \vee \bar{C}D$ ($\bar{B}\bar{C}D$ includes $\bar{C}D$).

Thus, the prime implicants are $A\bar{B}, A\bar{C}, B\bar{C}, \bar{C}D$.

Example 4.32.

Let Φ be $AB \vee A\bar{B}C\bar{D} \vee \bar{A}B\bar{C} \vee \bar{B}D$.

By (ii), $AB \vee A\bar{B}C\bar{D} \vee \bar{A}B\bar{C} \vee \bar{B}D \vee AC\bar{D}$ (Consensus of AB and $A\bar{B}C\bar{D}$).

By (i), $AB \vee \bar{A}B\bar{C} \vee \bar{B}D \vee AC\bar{D}$ ($A\bar{B}C\bar{D}$ includes $AC\bar{D}$).

By (ii), $AB \vee \bar{A}B\bar{C} \vee \bar{B}D \vee AC\bar{D} \vee B\bar{C}$ (Consensus of AB and $\bar{A}B\bar{C}$).

By (i), $AB \vee \bar{B}D \vee AC\bar{D} \vee B\bar{C}$ ($\bar{A}B\bar{C}$ includes $B\bar{C}$).

By (ii), $AB \vee \bar{B}D \vee AC\bar{D} \vee B\bar{C} \vee AD$ (Consensus of AB and $\bar{B}D$).

By (ii), $AB \vee \bar{B}D \vee AC\bar{D} \vee B\bar{C} \vee AD \vee A\bar{B}C$ (Consensus of $\bar{B}D$ and $AC\bar{D}$).

By (ii), $AB \vee \bar{B}D \vee AC\bar{D} \vee B\bar{C} \vee AD \vee A\bar{B}C \vee AC$ (Consensus of AB and $A\bar{B}C$).

By (i), $AB \vee \bar{B}D \vee B\bar{C} \vee AD \vee AC$ ($A\bar{B}C$ includes AC; $AC\bar{D}$ includes AC).

By (ii), $AB \vee \bar{B}D \vee B\bar{C} \vee AD \vee AC \vee \bar{C}D$ (Consensus of $\bar{B}D$ and $B\bar{C}$).

Hence the prime implicants are $AB, \bar{B}D, B\bar{C}, AD, AC, \bar{C}D$.

Example 4.33.

Let Φ be $ABCD \vee AB\bar{C}\bar{D} \vee \bar{A}B\bar{C}D \vee \bar{A}\bar{B}\bar{C} \vee A\bar{C}\bar{D}$.

By (i), $ABCD \vee \bar{A}B\bar{C}D \vee \bar{A}\bar{B}\bar{C} \vee A\bar{C}\bar{D}$ ($AB\bar{C}\bar{D}$ includes $A\bar{C}\bar{D}$).

By (ii), $ABCD \vee \bar{A}B\bar{C}D \vee \bar{A}\bar{B}\bar{C} \vee A\bar{C}\bar{D} \vee \bar{A}\bar{C}D$ (Consensus of $\bar{A}B\bar{C}D$ and $\bar{A}\bar{B}\bar{C}$).

By (i), $ABCD \vee \bar{A}\bar{B}\bar{C} \vee A\bar{C}\bar{D} \vee \bar{A}\bar{C}D$ ($\bar{A}B\bar{C}D$ includes $\bar{A}\bar{C}D$).

By (ii), $ABCD \vee \bar{A}\bar{B}\bar{C} \vee A\bar{C}\bar{D} \vee \bar{A}\bar{C}D \vee \bar{B}\bar{C}\bar{D}$ (Consensus of $\bar{A}\bar{B}\bar{C}$ and $A\bar{C}\bar{D}$).

Hence the prime implicants are $ABCD, \bar{A}\bar{B}\bar{C}, A\bar{C}\bar{D}, \bar{A}\bar{C}D, \bar{B}\bar{C}\bar{D}$.

Justification of the Consensus Method.

(1) *The process must come to an end.* Since there are only a finite number of dnf's using the letters of the given statement form Φ, we must show that there can be no cycles in the application of (i) and (ii). Once we drop a fundamental conjunction ϕ by (i), then ϕ can never reappear by virtue of (ii). For, in all future steps, there will always be a fundamental conjunction which is included in ϕ. Hence if there were a cycle, it would consist solely of applications of (ii). But (ii) increases the number of disjuncts.

(2) *Every prime implicant ϕ of Φ occurs as a disjunct in the dnf Ψ remaining at the end of the process.* Assume the contrary. Hence there must be a fundamental conjunction θ which has the maximum number of literals among all fundamental disjunctions τ such that: (*a*) τ includes ϕ; (*b*) τ includes no disjunct of Ψ; (*c*) the letters of τ occur in Φ. Notice that ϕ is such a fundamental conjunction τ. Clearly, by (*a*), θ logically implies Φ.

Also, θ cannot contain all the letters of Φ. (Otherwise, by (*b*), θ would logically imply the negation of each disjunct of Ψ, and therefore would logically imply $\neg\Phi$. But only contradictions logically imply both Φ and $\neg\Phi$, and no fundamental conjunction is a contradiction.) Let A be a letter of Φ not in θ. By the maximality of θ, $A\theta$ and $\bar{A}\theta$ must lack one of the properties (*a*)-(*c*). The only one they can lack is (*b*). Hence there are disjuncts ψ_1 and ψ_2 of Ψ such that $A\theta$ includes ψ_1 and $\bar{A}\theta$ includes ψ_2. By property (*b*) of θ, A must be a literal of ψ_1 and $\bar{A}$ must be a literal of ψ_2. Since $A\theta$ includes ψ_1 and $\bar{A}\theta$ includes ψ_2, ψ_1 and ψ_2 do not have any other literals which are negations of each other. Then the consensus μ of ψ_1 and ψ_2 is included in θ, and therefore, by (*b*), includes no disjunct of Ψ. Hence an application of (ii) can be made to ψ_1 and ψ_2, contradicting the assumption that the process has ended.

(3) Every disjunct ϕ of the dnf Ψ remaining at the end of the process must be a prime implicant of Φ. Otherwise ϕ would include some prime implicant ψ. By (2), ψ would be a disjunct of the final dnf, and operation (i) would still be applicable.

4.15 FINDING MINIMAL DNF's BY THE CONSENSUS METHOD

If we have obtained all the prime implicants of a statement form Φ by the consensus method, the problem remains to find the minimal dnf's. Of course, we could construct the full dnf for Φ and then apply the methods already described. However, constructing the full dnf for Φ sometimes would involve a long and tedious process, and it would be convenient to have ways of producing minimal dnf's without going through that process. One such method is to eliminate superfluous literals and disjuncts from the disjunction of the prime implicants, obtaining irredundant dnf's. Then one can compare the irredundant dnf's and pick out the minimal ones.†

Example 4.34.

We have already found (Example 4.33) that the dnf $ABCD \vee AB\bar{C}\bar{D} \vee \bar{A}B\bar{C}D \vee \bar{A}\bar{B}\bar{C} \vee A\bar{C}\bar{D}$ has as its prime implicants $ABCD$, $\bar{A}\bar{B}\bar{C}$, $A\bar{C}\bar{D}$, $\bar{A}\bar{C}D$, $\bar{B}\bar{C}\bar{D}$. Now we shall eliminate superfluous disjuncts from

$$ABCD \vee \bar{A}\bar{B}\bar{C} \vee A\bar{C}\bar{D} \vee \bar{A}\bar{C}D \vee \bar{B}\bar{C}\bar{D}$$

To determine whether a given disjunct ϕ is superfluous in a dnf $\phi \vee \Psi$ we check whether Ψ is logically equivalent to $\phi \vee \Psi$. This holds if and only if ϕ logically implies Ψ. But the latter holds if and only if the result is a tautology whenever, for each literal ρ in ϕ, we replace ρ in Ψ by T and the negation of ρ by F. Hence we construct the following table.

	$ABCD$	$\bar{A}\bar{B}\bar{C}$	$A\bar{C}\bar{D}$	$\bar{A}\bar{C}D$	$\bar{B}\bar{C}\bar{D}$
$ABCD$		F	F	F	F
$\bar{A}\bar{B}\bar{C}$	F		F	D	$\bar{D}$
$A\bar{C}\bar{D}$	F	F		F	$\bar{B}$
$\bar{A}\bar{C}D$	F	$\bar{B}$	F		F
$\bar{B}\bar{C}\bar{D}$	F	$\bar{A}$	A	F	

In the row corresponding to a fundamental conjunction ϕ we calculate what each of the other disjuncts must be when ϕ is T. Then we check to see whether the disjunction of the results in that row is a tautology. In the table above, this reveals that $\bar{A}\bar{B}\bar{C}$ and $\bar{B}\bar{C}\bar{D}$ are the superfluous disjuncts. All the other disjuncts must occur in every minimal dnf. Thus we are reduced to

$$ABCD \vee A\bar{C}\bar{D} \vee \bar{A}\bar{C}D$$

†The method we shall outline is due to M. J. Ghazala [26].

To eliminate superfluous literals, remember that a literal ρ is superfluous in $\rho\psi \vee \Psi$ if and only if ψ logically implies $\rho \vee \Psi$. A quick check shows that none of the literals is superfluous. Hence we have a unique irredundant dnf, which must be the only minimal dnf.

Example 4.35.

Consider the statement form Φ:

$$\bar{B}C \vee B\bar{C} \vee BD \vee CD \vee AD$$

Since the consensus method yields no additional disjuncts, this is already the disjunction of all the prime implicants of Φ. For finding superfluous disjuncts, we construct a table as in the preceding example.

	ϕ_1 $\bar{B}C$	ϕ_2 $B\bar{C}$	ϕ_3 BD	ϕ_4 CD	ϕ_5 AD	
ϕ_1 $\bar{B}C$		F	F	D	AD	
ϕ_2 $B\bar{C}$	F		D	F	AD	
ϕ_3 BD	F	$\bar{C}$		C	A	superfluous
ϕ_4 CD	$\bar{B}$	F	B		A	superfluous
ϕ_5 AD	$\bar{B}C$	$B\bar{C}$	B	C		

Since the disjunctions of the terms in the first, second and fifth rows (respectively) are not tautologies, $\bar{B}C$, $B\bar{C}$ and AD are not superfluous and occur in all minimal dnf's for Φ. Let σ_i mean that ϕ_i occurs as a disjunct in a given dnf for Φ. Hence from the third row, if $\overline{\sigma_3}$ then $\sigma_2\sigma_4$, for, if ϕ_3 does not occur in the given dnf for Φ, then both ϕ_2 and ϕ_4 must occur. (Otherwise, when ϕ_3 is T, then Φ would not necessarily be T, contradicting the fact that ϕ_3 logically implies Φ.) Thus $\sigma_3 \vee (\sigma_2\sigma_4)$ is true. Similarly, from the fourth row, $\sigma_4 \vee (\sigma_1\sigma_3)$ is true. Hence we must have

$$\sigma_1\sigma_2\sigma_5(\sigma_3 \vee \sigma_2\sigma_4)(\sigma_4 \vee \sigma_1\sigma_3)$$

which is equivalent to

$$\sigma_1\sigma_2\sigma_4\sigma_5 \vee \sigma_1\sigma_2\sigma_3\sigma_5$$

Hence the two irredundant dnf's are $\bar{B}C \vee B\bar{C} \vee CD \vee AD$ and $\bar{B}C \vee B\bar{C} \vee BD \vee AD$. Since these are of equal cost, they are both minimal dnf's. Notice that in this example we could have guessed this immediately from the third and fourth rows.

Example 4.36.

Let the prime implicants of a statement form be $\bar{D}E$, $CD\bar{E}$, $\bar{A}CD$, $\bar{A}CE$, $A\bar{B}D$, $A\bar{B}E$, $\bar{B}CD$, $\bar{B}CE$. (Observe that the consensus method is no longer applicable to the disjunction Φ of these fundamental conjunctions, and therefore the latter are all the prime implicants of Φ.)

	ϕ_1 $\bar{D}E$	ϕ_2 $CD\bar{E}$	ϕ_3 $\bar{A}CD$	ϕ_4 $\bar{A}CE$	ϕ_5 $A\bar{B}D$	ϕ_6 $A\bar{B}E$	ϕ_7 $\bar{B}CD$	ϕ_8 $\bar{B}CE$	
$\bar{D}E$		F	F	$\bar{A}C$	F	$A\bar{B}$	F	$\bar{B}C$	
$CD\bar{E}$	F		$\bar{A}$	F	$A\bar{B}$	F	$\bar{B}$	F	
$\bar{A}CD$	F	$\bar{E}$		E	F	F	$\bar{B}$	$\bar{B}E$	superfluous
$\bar{A}CE$	$\bar{D}$	F	D		F	F	$\bar{B}D$	$\bar{B}$	superfluous
$A\bar{B}D$	F	$C\bar{E}$	F	F		E	C	CE	
$A\bar{B}E$	$\bar{D}$	F	F	F	D		CD	C	superfluous
$\bar{B}CD$	F	$\bar{E}$	$\bar{A}$	$\bar{A}E$	A	AE		E	superfluous
$\bar{B}CE$	$\bar{D}$	F	$\bar{A}D$	$\bar{A}$	AD	A	D		superfluous

Rows 3, 4, 6, 7, 8 show that $\bar{A}CD$, $\bar{A}CE$, $A\bar{B}E$, $\bar{B}CD$, $\bar{B}CE$ are superfluous. Thus σ_1, σ_2, σ_5 are true. From the third row, $\sigma_3 \vee \sigma_2\sigma_4$ is true. From the fourth row, $\sigma_4 \vee \sigma_1\sigma_3$ is true. From the sixth row, $\sigma_6 \vee \sigma_1\sigma_5$ is true. From the seventh row,

$$\sigma_7 \vee \sigma_2\sigma_8 \vee \sigma_2\sigma_4\sigma_6 \vee \sigma_2\sigma_3\sigma_6 \vee \sigma_2\sigma_4\sigma_5 \vee \sigma_3\sigma_5$$

is true. This has been obtained by finding those subsets of the entries in the seventh row, the disjunction of which is a tautology and such that no proper subset of this subset has the same property. This process can be carried out by constructing a table for the entries in the seventh row similar to the one constructed above. From the eighth row,

$$\sigma_8 \vee \sigma_1\sigma_7 \vee \sigma_1\sigma_3\sigma_5 \vee \sigma_1\sigma_3\sigma_6 \vee \sigma_1\sigma_4\sigma_5 \vee \sigma_4\sigma_6$$

is true. Hence we have

$$\sigma_1\sigma_2\sigma_5(\sigma_3 \vee \sigma_2\sigma_4)(\sigma_4 \vee \sigma_1\sigma_3)(\sigma_6 \vee \sigma_1\sigma_5)$$

$$(\sigma_7 \vee \sigma_2\sigma_8 \vee \sigma_2\sigma_4\sigma_6 \vee \sigma_2\sigma_3\sigma_6 \vee \sigma_2\sigma_4\sigma_5 \vee \sigma_3\sigma_5)(\sigma_8 \vee \sigma_1\sigma_7 \vee \sigma_1\sigma_3\sigma_5 \vee \sigma_1\sigma_3\sigma_6 \vee \sigma_1\sigma_4\sigma_5 \vee \sigma_4\sigma_6)$$

This is equivalent to

$$\sigma_1\sigma_2\sigma_3\sigma_5 \vee \sigma_1\sigma_2\sigma_4\sigma_5$$

(When multiplying out, we see that these are two of the disjuncts in the expansion, and, since all the other disjuncts include one of these, all other disjuncts may be dropped.)

Thus the irredundant dnf's are

$$\bar{D}E \vee CD\bar{E} \vee \bar{A}CD \vee \bar{A}\bar{B}\bar{D} \quad \text{and} \quad \bar{D}E \vee CD\bar{E} \vee \bar{A}CE \vee \bar{A}\bar{B}D$$

Notice that none of the literals is superfluous. Since the costs of these dnf's are equal, both are minimal dnf's.

4.16 KARNAUGH MAPS

There is a pictorial method for obtaining minimal dnf's which is convenient for problems involving at most six statement letters.†

Let us start with the case of two statement letters. In this case the Karnaugh map is based upon Fig. 4-46. Each square represents the fundamental conjunction whose conjuncts are the literals standing at the head of the row and column determining the square. To represent a full dnf Φ we place a check in each square corresponding to a disjunct of Φ.

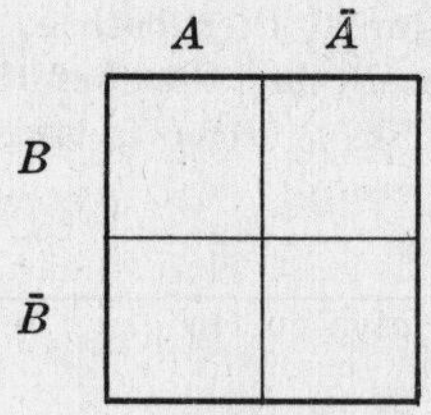

Fig. 4-46

Example 4.37.

$AB \vee \bar{A}B$ is represented in the Karnaugh map of Fig. 4-47 and $A\bar{B} \vee \bar{A}B \vee \bar{A}\bar{B}$ in the Karnaugh map of Fig. 4-48.

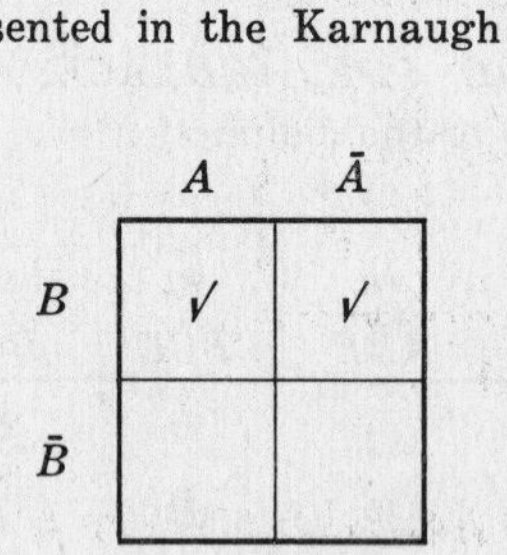

Fig. 4-47

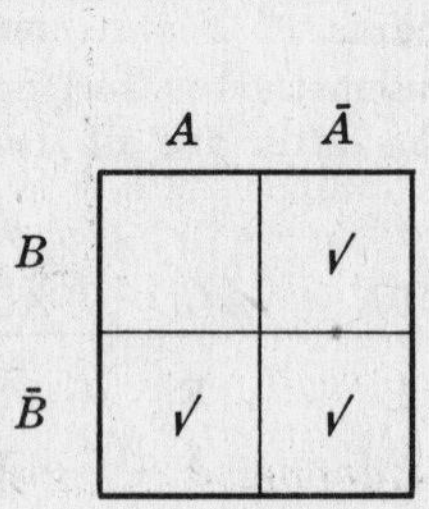

Fig. 4-48

By *adjacent* squares we mean squares which have a side in common. Clearly, a single square represents a fundamental conjunction consisting of two literals, while two adjacent squares differ in one statement letter and therefore represent a single literal. Thus in Fig. 4-47 we have $AB \vee \bar{A}B$, which is logically equivalent to B. In Fig. 4-49 we notice two pairs of checked adjacent

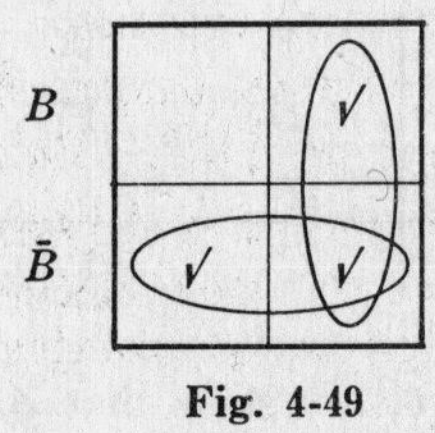

Fig. 4-49

†See Karnaugh [42]. Another pictorial method, somewhat less graphic than Karnaugh's, has been given by Veitch [93].

squares. We place loops around the checks in each pair. Then the corresponding dnf is logically equivalent to $\bar{B} \vee \bar{A}$. ($\bar{B}$ corresponds to the horizontal loop, and $\bar{A}$ to the vertical loop.) Clearly, $\bar{B} \vee \bar{A}$ is minimal.

Now let us turn to the case of three statement letters (Fig. 4-50). Each square represents the conjunction of the fundamental conjunctions heading the column and row intersecting in that square.

	AB	$A\bar{B}$	$\bar{A}\bar{B}$	$\bar{A}B$
C				
$\bar{C}$				

Fig. 4-50

Example 4.38.

$A\bar{B}C \vee \bar{A}\bar{B}C \vee AB\bar{C}$ is represented in Fig. 4-51.

	AB	$A\bar{B}$	$\bar{A}\bar{B}$	$\bar{A}B$
C		✓	✓	
$\bar{C}$	✓			

Fig. 4-51

Example 4.39.

$ABC \vee A\bar{B}C \vee \bar{A}BC \vee AB\bar{C} \vee \bar{A}B\bar{C}$ is represented in Fig. 4-52.

	AB	$A\bar{B}$	$\bar{A}\bar{B}$	$\bar{A}B$
C	✓	✓		✓
$\bar{C}$	✓			✓

Fig. 4-52

In Fig. 4-50, by *adjacent* squares we mean squares which differ in precisely one literal. Thus two squares which have a side in common are adjacent. (Observe that we have used the labeling AB, $A\bar{B}$, $\bar{A}\bar{B}$, $\bar{A}B$ so that as we move from one square to an adjoining one, only one literal changes.) In addition, in the first row the left-most square ABC is adjacent to the right-most square $\bar{A}BC$; and in the second row, $AB\bar{C}$ is adjacent to $\bar{A}B\bar{C}$. This amounts to an identification of the left-most vertical line with the right-most vertical line. Pictorially we can imagine the left-most vertical line glued to the right-most vertical line so as to form a cylinder (Fig. 4-53).

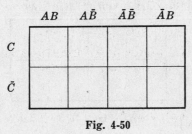

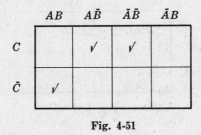

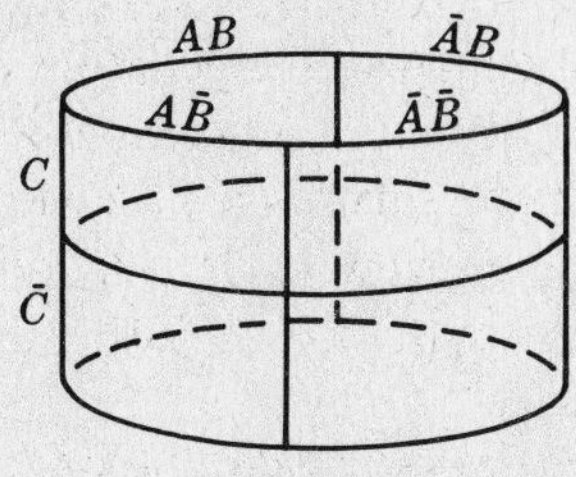

Fig. 4-53

On the cylinder, *adjacent* squares are adjacent in the usual geometric sense.

In Fig. 4-50, a single square represents a fundamental conjunction of three literals, while two adjacent squares differ in one literal and therefore represent a fundamental conjunction of two literals.

Example 4.40.

Fig. 4-54 shows $A\bar{B}C \vee A\bar{B}\bar{C}$ which is logically equivalent to $A\bar{B}$.

	AB	$A\bar{B}$	$\bar{A}\bar{B}$	$\bar{A}B$
C		✓		
$\bar{C}$		✓		

Fig. 4-54

Example 4.41.

$ABC \vee \bar{A}BC$ is represented in Fig. 4-55 and is logically equivalent to BC.

	AB	$A\bar{B}$	$\bar{A}\bar{B}$	$\bar{A}B$
C	✓			✓
$\bar{C}$				

Fig. 4-55

Furthermore, four squares forming a square array or arranged in one row represent a single literal.

Example 4.42.

	AB	$A\bar{B}$	$\bar{A}\bar{B}$	$\bar{A}B$
C		✓	✓	
$\bar{C}$		✓	✓	

Fig. 4-56

Fig. 4-56 exhibits $A\bar{B}C \vee \bar{A}\bar{B}C \vee A\bar{B}\bar{C} \vee \bar{A}\bar{B}\bar{C}$, which is logically equivalent to $\bar{B}$.

Example 4.43.

	AB	$A\bar{B}$	$\bar{A}\bar{B}$	$\bar{A}B$
C	✓	✓		
$\bar{C}$	✓	✓		

Fig. 4-57

In Fig. 4-57 we see $ABC \vee A\bar{B}C \vee AB\bar{C} \vee A\bar{B}\bar{C}$, which is logically equivalent to A.

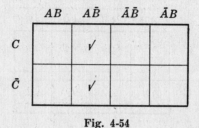

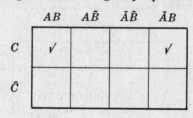

Example 4.44.

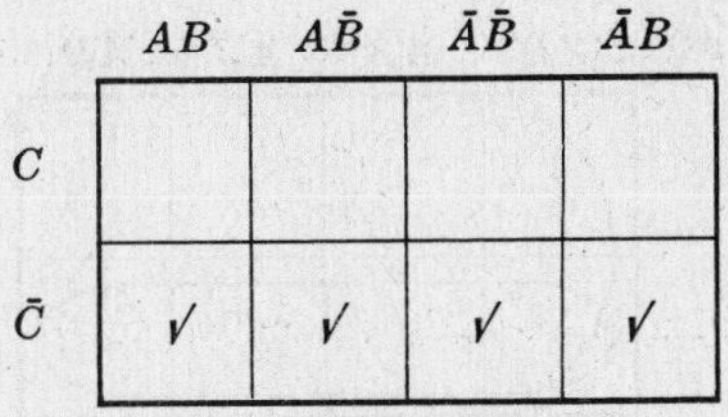

Fig. 4-58

In Fig. 4-58 we have $AB\bar{C} \vee A\bar{B}\bar{C} \vee \bar{A}\bar{B}\bar{C} \vee \bar{A}B\bar{C}$, which is logically equivalent to $\bar{C}$.

Example 4.45.

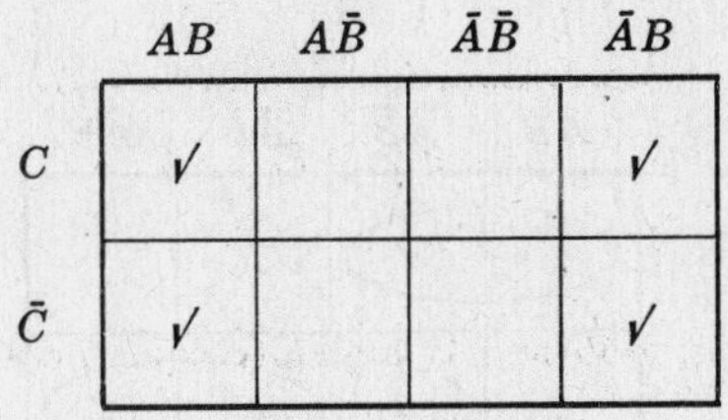

Fig. 4-59

Fig. 4-59 represents $ABC \vee \bar{A}BC \vee AB\bar{C} \vee \bar{A}B\bar{C}$, which is logically equivalent to B.

Notice that if we picture Fig. 4-59 on a cylinder, the four checks form a square array.

The technique for minimization is straightforward. We draw loops around single checks, or pairs of adjacent checks, or groups of four checks (forming a square array or arranged along a row), in such a way that every check belongs to at least one loop. We try to make maximal use of groups of four checks or two checks so as to minimize the number of disjuncts and literals.

Example 4.46.

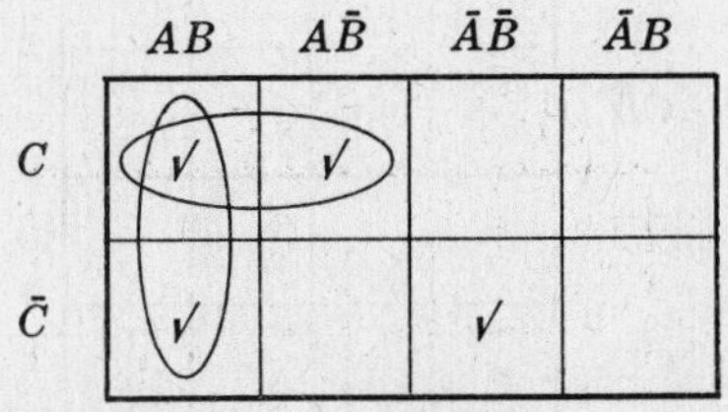

Fig. 4-60

The Karnaugh map of Fig. 4-60 represents $ABC \vee A\bar{B}C \vee AB\bar{C} \vee \bar{A}\bar{B}\bar{C}$. The unique minimal dnf is $AB \vee AC \vee \bar{A}\bar{B}\bar{C}$. AB corresponds to the vertical loop, and AC to the horizontal loop.

Example 4.47.

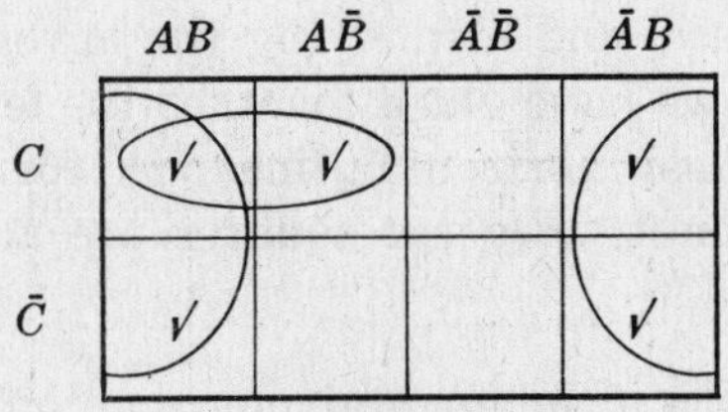

Fig. 4-61

Fig. 4-61 represents $ABC \vee A\bar{B}C \vee \bar{A}BC \vee AB\bar{C} \vee \bar{A}B\bar{C}$. There is a unique minimal dnf: $B \vee AC$. B corresponds to the four checks ABC, $AB\bar{C}$, $\bar{A}BC$, $AB\bar{C}$, while AC corresponds to the horizontal loop covering ABC and $A\bar{B}C$.

Example 4.48.

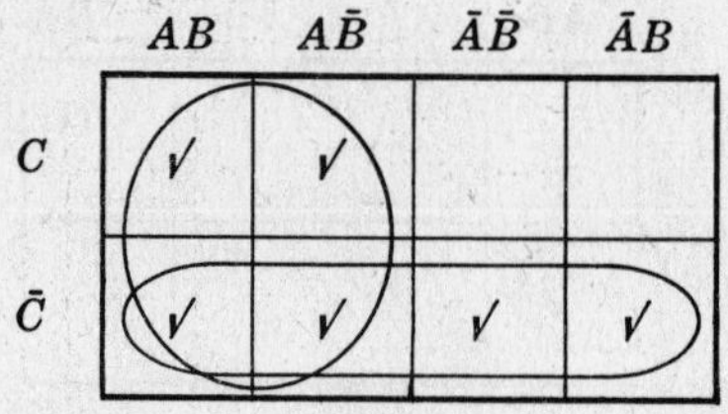

Fig. 4-62

The unique minimal dnf is $A \vee \bar{C}$.

Example 4.49.

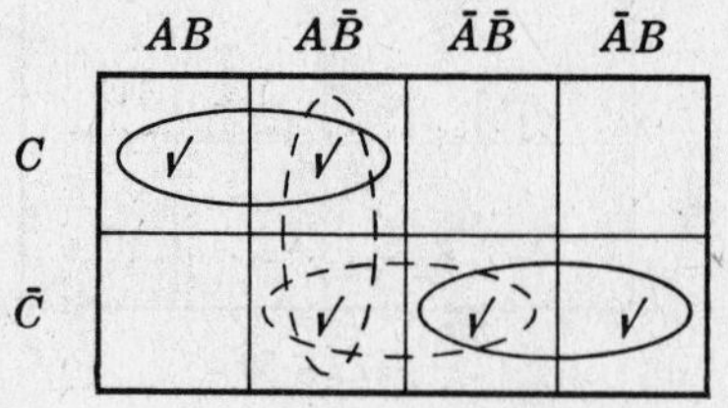

Fig. 4-63

In this case, $A\bar{B}\bar{C}$ can be combined with either $A\bar{B}C$ or $\bar{A}\bar{B}\bar{C}$. Hence we have two minimal dnf's:

$$AC \vee \bar{A}\bar{C} \vee A\bar{B}, \quad AC \vee \bar{A}\bar{C} \vee \bar{B}\bar{C}$$

Let us consider now Karnaugh maps for four statement letters (Fig. 4-64).

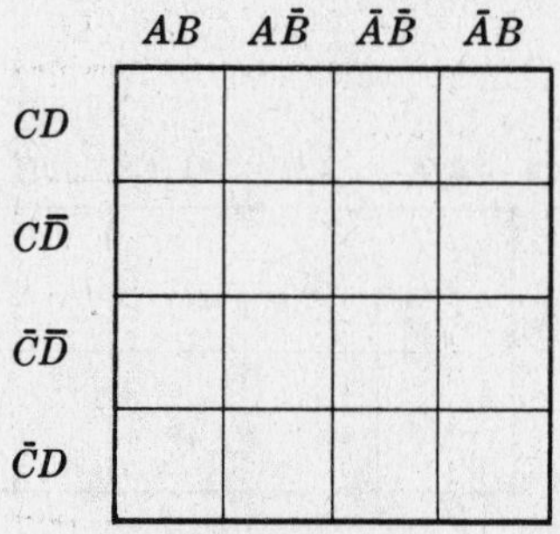

Fig. 4-64

Again, *adjacent* squares are those which differ in exactly one literal. In particular, $A\bar{B}CD$ and $A\bar{B}\bar{C}D$ are adjacent, as are $AB\bar{C}D$ and $\bar{A}B\bar{C}D$. This amounts to identifying the left-most and right-most vertical lines, and identifying the lowest and highest horizontal lines. Pictorially we can imagine that we have glued together the left-most and right-most vertical lines, and the lowest and highest horizontal lines, to form a doughnut-shaped surface (called a *torus*). On this doughnut, *adjacent* squares are adjacent in the usual geometric sense.

A single square represents a fundamental conjunction of four literals. A pair of adjacent squares represents a fundamental conjunction of three literals. Four squares, in a square array or along a single row or along a single column, represent a fundamental conjunction of two literals. Finally, eight squares arranged in two *adjacent* columns or in two *adjacent* rows represent a single literal.

Example 4.50.

	AB	$A\bar{B}$	$\bar{A}\bar{B}$	$\bar{A}B$
CD		✓		
$C\bar{D}$		✓		
$\bar{C}\bar{D}$		✓		
$\bar{C}D$		✓		

Fig. 4-65

The Karnaugh map in Fig. 4-65 represents $A\bar{B}$.

Example 4.51.

	AB	$A\bar{B}$	$\bar{A}\bar{B}$	$\bar{A}B$
CD	✓	✓		
$C\bar{D}$				
$\bar{C}\bar{D}$				
$\bar{C}D$	✓	✓		

Fig. 4-66

Fig. 4-66 represents AD.

Example 4.52.

	AB	$A\bar{B}$	$\bar{A}\bar{B}$	$\bar{A}B$
CD	✓			✓
$C\bar{D}$				
$\bar{C}\bar{D}$				
$\bar{C}D$	✓			✓

Fig. 4-67

Fig. 4-67 is the Karnaugh map for BD.

Example 4.53.

	AB	$A\bar{B}$	$\bar{A}\bar{B}$	$\bar{A}B$
CD		✓	✓	
$C\bar{D}$		✓	✓	
$\bar{C}\bar{D}$		✓	✓	
$\bar{C}D$		✓	✓	

Fig. 4-68

Fig. 4-68 represents $\bar{B}$.

Example 4.54.

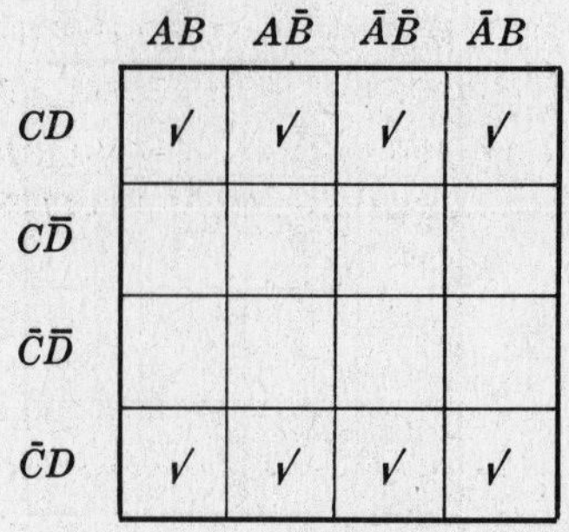

Fig. 4-69

Fig. 4-69 represents D.

Minimization techniques for four statement letters are similar to those for three.

Example 4.55.

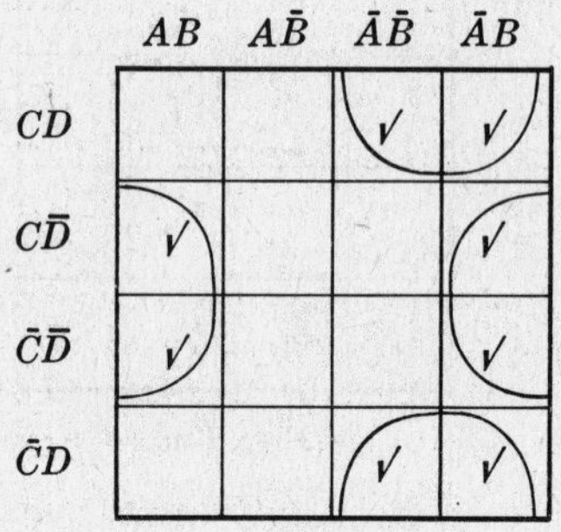

Fig. 4-70

The unique minimal dnf is $\bar{A}D \vee B\bar{D}$. Observe that the four squares in the column under $\bar{A}B$ are not joined by a loop, since the corresponding fundamental conjunction $\bar{A}B$ would be superfluous.

Example 4.56.

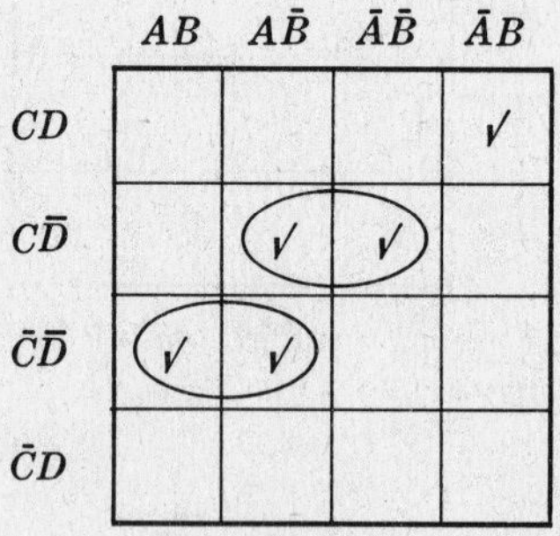

Fig. 4-71

The check in $\bar{A}BCD$ cannot be combined with any other. Hence $\bar{A}BCD$ must be in any minimal dnf. The check in $\bar{A}\bar{B}C\bar{D}$ can be combined only with the check in $A\bar{B}C\bar{D}$. Hence $\bar{B}C\bar{D}$ must be a disjunct of any minimal dnf. Similarly, the check in $AB\bar{C}\bar{D}$ can be combined only with the check in $A\bar{B}\bar{C}\bar{D}$. Hence $A\bar{C}\bar{D}$ is a disjunct of any minimal dnf. Now the checks in $A\bar{B}C\bar{D}$ and $A\bar{B}\bar{C}\bar{D}$ already have been covered. Thus the unique minimal dnf is

$$\bar{A}BCD \vee \bar{B}C\bar{D} \vee A\bar{C}\bar{D}$$

Examples 4.55-4.56 illustrate the method to be used. For each checked square, determine whether there is a *unique* largest combination of checked squares containing it. If so, put a loop around that combination. To avoid superfluous disjuncts, first handle each checked square whose unique largest combination consists only of itself; then, among the remaining uncovered checks, handle those whose unique largest combination consists of two checks; among the still uncovered checks, handle those whose unique largest combination consists of four checks, etc. For any remaining checked square, determine all the possible largest combinations containing them, and, among the corresponding dnf's,† find the minimal ones.

Example 4.57.

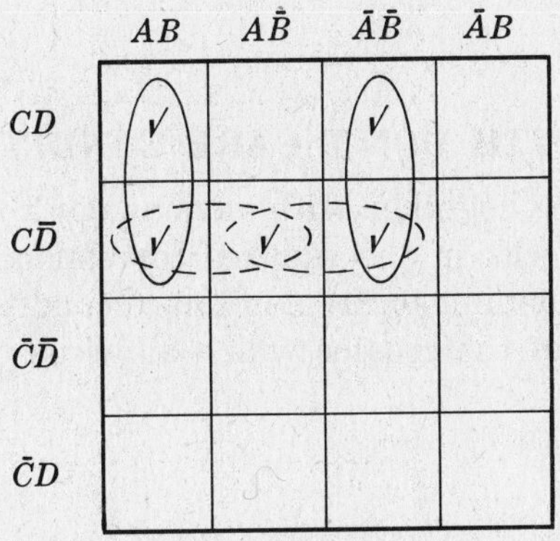

Fig. 4-72

Considering $ABCD$, we see that ABC must be a disjunct (covering $ABCD$ and $ABC\bar{D}$). Looking at $\bar{A}\bar{B}CD$, we note that $\bar{A}\bar{B}C$ must be a disjunct (covering $\bar{A}\bar{B}CD$ and $\bar{A}\bar{B}C\bar{D}$). None of the other three checks belongs to a unique largest combination. The only uncovered check is $A\bar{B}C\bar{D}$, which can be combined either with $ABC\bar{D}$ or with $\bar{A}\bar{B}C\bar{D}$. Hence we obtain two minimal dnf's:

$$ABC \vee \bar{A}\bar{B}C \vee AC\bar{D} \quad \text{and} \quad ABC \vee \bar{A}\bar{B}C \vee \bar{B}C\bar{D}$$

In the case of five statement letters, we can use a three-dimensional Karnaugh map (Fig. 4-73).

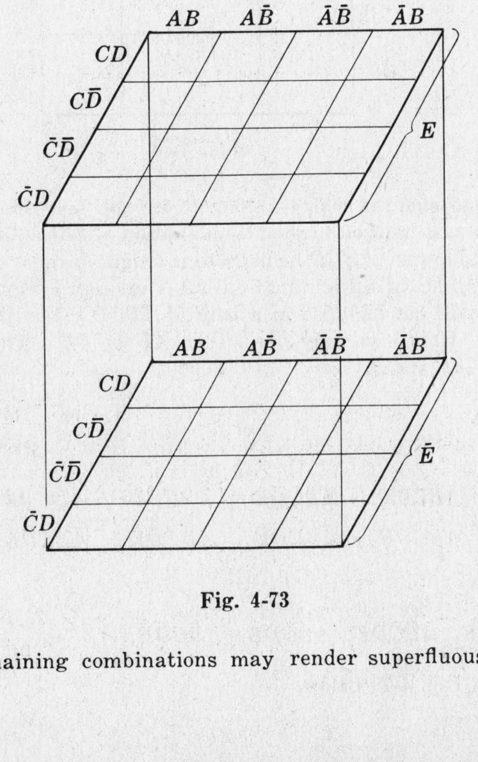

Fig. 4-73

†Some choices among the remaining combinations may render superfluous some of the disjuncts already obtained (cf. Problem 4.22).

The usual definition of *adjacent* squares implies that corresponding squares in the two planes (e.g. $A\bar{B}C\bar{D}E$ and $A\bar{B}C\bar{D}\bar{E}$) are adjacent. Combinations of sixteen squares are possible and yield a fundamental conjunction consisting of a single literal. Combinations of eight squares yield fundamental conjunctions of two literals, etc.

For six statement letters one could use four planes, but in that case, and even more so for larger numbers of statement letters, the geometric picture often is too complex to permit easy construction of minimal dnf's.

4.17 KARNAUGH MAPS WITH DON'T CARE CONDITIONS

If we are given a full dnf Φ, together with various don't care conditions, we construct a Karnaugh map by placing checks in the squares corresponding to the disjuncts of Φ and crosses in the squares corresponding to the don't care conditions. In constructing minimal dnf's, we are free to use any of the crosses which allow us to form larger combinations of squares.

Example 4.58.

Let Φ be

$$ABCD \vee A\bar{B}CD \vee A\bar{B}C\bar{D} \vee \bar{A}BC\bar{D} \vee A\bar{B}\bar{C}D$$

and assume that $\bar{A}BCD$, $AB\bar{C}D$, $\bar{A}B\bar{C}D$, $A\bar{B}\bar{C}\bar{D}$, $\bar{A}\bar{B}\bar{C}\bar{D}$ are don't care conditions. The Karnaugh map is shown in Fig. 4-74.

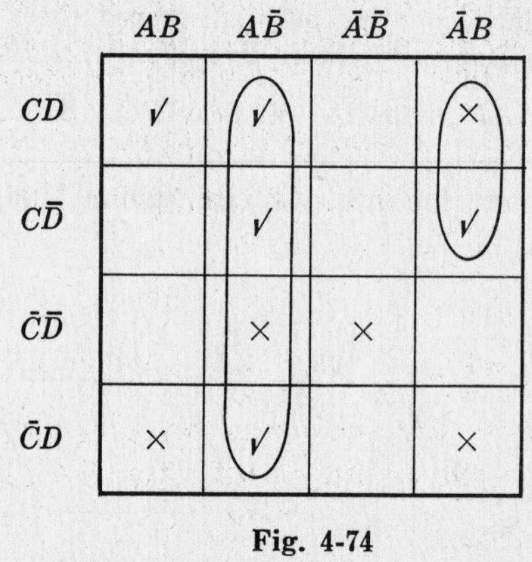

Fig. 4-74

First we handle the checked squares which belong to unique largest combinations (possibly including crosses). Thus $\bar{A}BC\bar{D}$ belongs to a unique largest combination: $\{\bar{A}BC\bar{D}, \bar{A}BCD\}$. Hence $\bar{A}BC$ must be a disjunct of all minimal dnf's. Likewise, $A\bar{B}C\bar{D}$ belongs to a unique largest combination (the second column), and therefore $A\bar{B}$ must be a disjunct of all minimal dnf's. The other checks do not belong to unique largest combinations. The only check still not included in a loop is $ABCD$. For the latter, there are two possible combinations of four squares. Hence we may use either AD or BD. Thus there are two minimal dnf's: $\bar{A}BC \vee A\bar{B} \vee AD$ and $\bar{A}BC \vee A\bar{B} \vee BD$.

Example 4.59.

Let Φ be

$$ABCDE \vee \bar{A}\bar{B}CDE \vee A\bar{B}C\bar{D}E \vee \bar{A}\bar{B}C\bar{D}E \vee \bar{A}\bar{B}\bar{C}\bar{D}E \vee \bar{A}B\bar{C}DE$$
$$\vee ABCD\bar{E} \vee \bar{A}BCD\bar{E} \vee A\bar{B}\bar{C}\bar{D}\bar{E} \vee \bar{A}\bar{B}\bar{C}\bar{D}\bar{E} \vee AB\bar{C}D\bar{E}$$

Let the don't care conditions be

$$\bar{A}BCDE,\ A\bar{B}\bar{C}\bar{D}E,\ AB\bar{C}DE,\ A\bar{B}C\bar{D}\bar{E},\ \bar{A}\bar{B}C\bar{D}\bar{E},\ \bar{A}B\bar{C}D\bar{E}$$

The Karnaugh map is shown in Fig. 4-75 below.

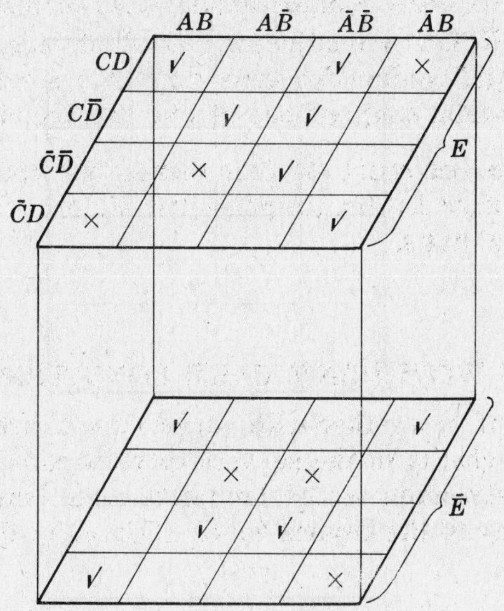

Fig. 4-75

We seek the checks belonging to unique largest combinations. First, $ABCDE$ belongs to such a combination (the four corners of both planes). Hence BD is a disjunct of all minimal dnf's. We get the same result from the checks for $\bar{A}B\bar{C}DE$, $ABCD\bar{E}$, $\bar{A}BCD\bar{E}$, $AB\bar{C}D\bar{E}$. The checks in the middle squares of both planes belong to a unique 8-square combination: $\{A\bar{B}C\bar{D}E, \bar{A}\bar{B}C\bar{D}E, A\bar{B}\bar{C}\bar{D}E, \bar{A}\bar{B}\bar{C}\bar{D}E, A\bar{B}C\bar{D}\bar{E}, \bar{A}\bar{B}C\bar{D}\bar{E}, A\bar{B}\bar{C}\bar{D}\bar{E}, \bar{A}\bar{B}\bar{C}\bar{D}\bar{E}\}$, yielding the disjunct $\bar{B}\bar{D}$. The only check still unaccounted for is $\bar{A}\bar{B}CDE$. This belongs to two 2-square combinations. Hence we must have either $\bar{A}\bar{B}CE$ or $\bar{A}CDE$. Thus there are two minimal dnf's:

$$BD \vee \bar{B}\bar{D} \vee \bar{A}\bar{B}CE$$

$$BD \vee \bar{B}\bar{D} \vee \bar{A}CDE$$

4.18 MINIMAL DNF's OR CNF's

Given a statement form Φ, we can obtain the minimal dnf's for Φ and we also can obtain the minimal dnf's for $\neg\Phi$. But the minimal dnf's for $\neg\Phi$ yield minimal cnf's (conjunctive normal forms) for Φ.

Example 4.60.

Recall that a cnf is a conjunction of one or more disjunctions of one or more literals. The cnf

$$(A \vee \bar{B} \vee C) \;\&\; (\bar{A} \vee B) \;\&\; (B \vee \bar{C})$$

has as its negation

$$\bar{A}B\bar{C} \vee A\bar{B} \vee \bar{B}C$$

Thus by comparing the costs of the minimal dnf's and minimal cnf's for Φ, we can obtain those statement forms which are minimal among all dnf's or cnf's for Φ.

Example 4.61.

Let Φ be

$$\bar{A}\bar{B}C\bar{D} \vee \bar{A}\bar{B}CD \vee \bar{A}B\bar{C}D \vee \bar{A}BCD \vee A\bar{B}\bar{C}\bar{D} \vee A\bar{B}C\bar{D} \vee AB\bar{C}\bar{D}$$

If we examine the Karnaugh map for Φ (Fig. 4-76, below), we find that there are three minimal dnf's:

$$\bar{A}BD \vee A\bar{C}\bar{D} \vee \bar{A}CD \vee \bar{B}C\bar{D}$$

$$\bar{A}BD \vee A\bar{C}\bar{D} \vee A\bar{B}\bar{D} \vee \bar{A}\bar{B}C$$

$$\bar{A}BD \vee A\bar{C}\bar{D} \vee \bar{A}\bar{B}C \vee \bar{B}C\bar{D}$$

	AB	$A\bar{B}$	$\bar{A}\bar{B}$	$\bar{A}B$
CD			√	√
$C\bar{D}$		√	√	
$\bar{C}\bar{D}$	√	√		
$\bar{C}D$				√

Fig. 4-76

On the other hand, $\neg\Phi$ has the Karnaugh map (Fig. 4-77) obtained by putting checks in the empty squares and erasing the checks already present in Fig. 4-76.

	AB	$A\bar{B}$	$\bar{A}\bar{B}$	$\bar{A}B$
CD	√	√		
$C\bar{D}$	√			√
$\bar{C}\bar{D}$			√	√
$\bar{C}D$	√	√	√	

Fig. 4-77

From Fig. 4-77, we see that the minimal dnf's for $\neg\Phi$ are

$$AD \vee ABC \vee \bar{A}B\bar{D} \vee \bar{A}\bar{B}\bar{C}$$
$$AD \vee BC\bar{D} \vee \bar{A}B\bar{D} \vee \bar{A}\bar{B}\bar{C}$$
$$AD \vee BC\bar{D} \vee \bar{A}\bar{C}\bar{D} \vee \bar{A}\bar{B}\bar{C}$$
$$AD \vee BC\bar{D} \vee \bar{A}\bar{C}\bar{D} \vee \bar{B}\bar{C}D$$

Thus the minimal cnf's for Φ are

$$(\bar{A} \vee \bar{D}) \,\&\, (\bar{A} \vee \bar{B} \vee \bar{C}) \,\&\, (A \vee \bar{B} \vee D) \,\&\, (A \vee B \vee C)$$
$$(\bar{A} \vee \bar{D}) \,\&\, (\bar{B} \vee \bar{C} \vee D) \,\&\, (A \vee \bar{B} \vee D) \,\&\, (A \vee B \vee C)$$
$$(\bar{A} \vee \bar{D}) \,\&\, (\bar{B} \vee \bar{C} \vee D) \,\&\, (A \vee C \vee D) \,\&\, (A \vee B \vee C)$$
$$(\bar{A} \vee \bar{D}) \,\&\, (\bar{B} \vee \bar{C} \vee D) \,\&\, (A \vee C \vee D) \,\&\, (B \vee C \vee \bar{D})$$

Since these are cheaper than the minimal dnf's for Φ, these are minimal among all dnf's or cnf's for Φ.

The procedure in the above example for finding the minimal statement forms among all dnf's or cnf's for Φ does not provide a general method for finding minimal statement forms for Φ (i.e. minimal series-parallel switching circuits, or minimal logic circuits). For example, $(A \,\&\, B) \vee (C \,\&\, (D \vee E))$ is minimal, but it is neither a dnf nor a cnf.

Final Remarks: (1) We have indicated methods for finding minimal dnf's (or minimal dnf's or cnf's). This constitutes a solution of Problem (I) on page 81, although possibly not the best solution (cf. Remark (2)). No reasonably good general solutions for Problems (II) or (III), page 81, are known. (2) The methods we have given for finding minimal dnf's require us to find all prime implicants. (This is not true of Karnaugh maps, but these are useful only for statement forms involving at most five or six statement letters.) However, there are certain cases in which the number of prime implicants is so large that our methods are not practical.† There is need then for a method for finding minimal dnf's which does not use the set of *all* prime implicants, but no general method of this kind is available.

Solved Problems

SWITCHING CIRCUITS. SIMPLIFICATION

4.1. Replace the series-parallel circuit of Fig. 4-78 by a simpler bridge circuit.

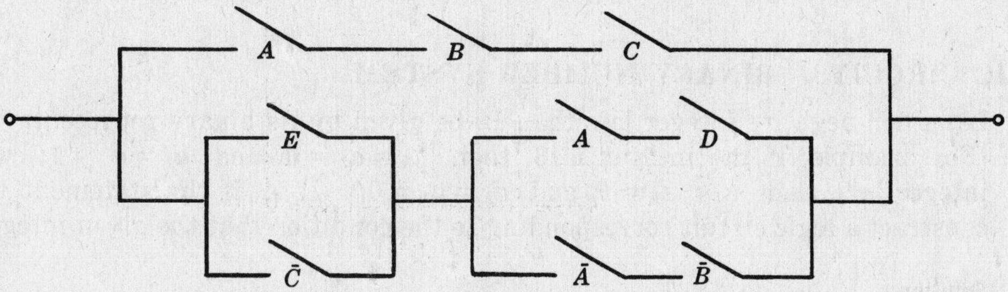

Fig. 4-78

Solution:

Consider the bridge circuit shown in Fig. 4-79. The paths through this circuit are $A \& B \& C$, $A \& D \& E$, $A \& D \& \bar{C}$, $\bar{A} \& \bar{B} \& \bar{C}$, $\bar{A} \& \bar{B} \& E$. Hence a condition for passage of current is

$$(A \& B \& C) \vee (A \& D \& E) \vee (A \& D \& \bar{C}) \vee (\bar{A} \& \bar{B} \& \bar{C}) \vee (\bar{A} \& \bar{B} \& E)$$

which is logically equivalent to $[A \& B \& C] \vee [(E \vee \bar{C}) \& ((A \& D) \vee (\bar{A} \& \bar{B}))]$. But this is a condition for passage of current through the given circuit.

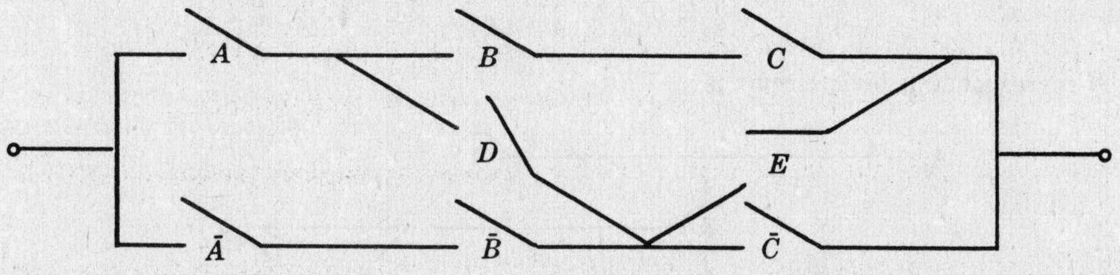

Fig. 4-79

†Fridshal [24] states that, for nine statement letters, the full dnf whose fundamental conjunctions are those with 1, 3, 4, 5, 6, or 8 negated literals has 1698 prime implicants. The full dnf whose fundamental conjunctions are those with 0, 1, 5, 6, or 7 negated literals has 765 prime implicants, and its negation has the same number of prime implicants.

4.2. A committee consists of the chairman, president, secretary, and treasurer. A motion passes if and only if it receives a majority vote or the vote of the chairman plus one other member. Each member presses a button to indicate approval of a motion. Design a switching circuit controlled by the buttons which passes current if and only if a motion is approved.

Solution:

Let C, P, S, T stand for "The chairman approves", "The president approves", etc. Then the condition for approval is

$$[C \,\&\, (P \vee S \vee T)] \;\vee\; (P \,\&\, S \,\&\, T)$$

which has the corresponding series-parallel circuit shown in Fig. 4-80.

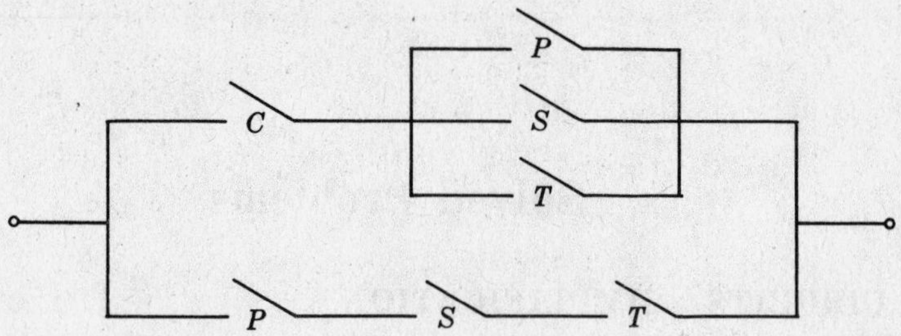

Fig. 4-80

LOGIC CIRCUITS. BINARY NUMBER SYSTEM

4.3. Let a non-negative integer less than 10 be given by its binary representation $a_3a_2a_1a_0$. (For example, if the integer is 3, then $a_3 = a_2 = 0$ and $a_1 = a_0 = 1$; while if the integer is 9, then $a_3 = a_0 = 1$ and $a_2 = a_1 = 0$.) If A_i is the statement that a_i is 1, construct a logic circuit corresponding to the condition that the given integer is prime.

Solution:

The prime integers are

Decimal Notation	2	3	5	7
Binary Notation	0010	0011	0101	0111

A corresponding statement form is

$$\neg A_3 \,\&\, [(\neg A_2 \,\&\, A_1 \,\&\, \neg A_0) \vee (\neg A_2 \,\&\, A_1 \,\&\, A_0) \vee (A_2 \,\&\, \neg A_1 \,\&\, A_0) \vee (A_2 \,\&\, A_1 \,\&\, A_0)]$$

which is logically equivalent to

$$\neg A_3 \,\&\, ([A_0 \,\&\, (A_1 \vee A_2)] \vee [\neg A_0 \,\&\, A_1 \,\&\, \neg A_2])$$

A corresponding logic circuit is

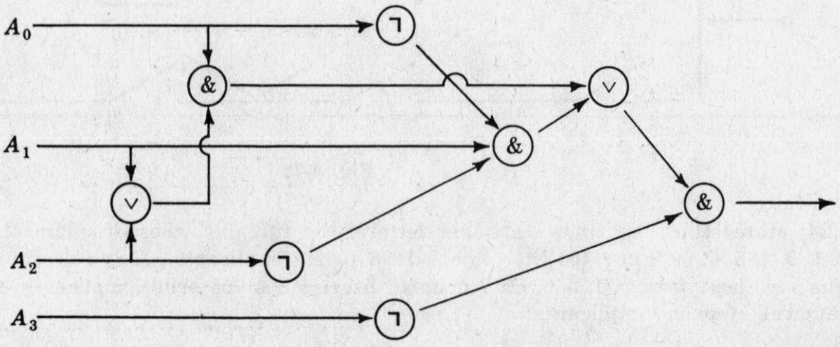

4.4. Justify the following algorithm for translating a number x from decimal into binary notation.

Use two columns. Place x at the top of the left column (in our example, $x = 43$). Divide x by 2, putting the remainder r_0 in the right hand column and the quotient q_0 in the left hand column below the given number. Repeat this process with q_0, etc. Stop when we get a quotient 0. The resulting binary number is to be found by reading the right hand column from the bottom up (in our example, 101011).

43	1
21	1
10	0
5	1
2	0
1	1
0	

Solution:

$$x = 2q_0 + r_0, \qquad r_0 = 0 \text{ or } 1$$
$$q_0 = 2q_1 + r_1, \qquad r_1 = 0 \text{ or } 1$$
$$\cdots\cdots\cdots\cdots \qquad \cdots\cdots\cdots$$
$$q_{k-2} = 2q_{k-1} + r_{k-1}, \qquad r_{k-1} = 0 \text{ or } 1$$
$$q_{k-1} = 2 \cdot 1 + r_k, \qquad r_k = 0 \text{ or } 1$$
$$1 = 2 \cdot 0 + 1$$

Then

$$\begin{aligned} x &= 2(2q_1 + r_1) + r_0 = 4q_1 + 2r_1 + r_0 = 4(2q_2 + r_2) + 2r_1 + r_0 \\ &= 8q_2 + 4r_2 + 2r_1 + r_0 = 16q_3 + 8r_3 + 4r_2 + 2r_1 + r_0 = \cdots \\ &= 2^{k+1} + 2^k r_k + \cdots + 2^3 r_3 + 2^2 r_2 + 2r_1 + r_0 \end{aligned}$$

Thus the binary expansion of x is $1 r_k r_{k-1} \cdots r_2 r_1 r_0$.

4.5. Construct a logic circuit for adding 1 to a four-digit binary number $a_3 a_2 a_1 a_0$.

Solution:

Let A_i stand for "a_i is 1". Let $b_4 b_3 b_2 b_1 b_0$ be the result of adding 1, and let B_i stand for "b_i is 1". Then

$B_0 = \neg A_0$ and the carry $C_0 = A_0$;

$B_1 = (A_1 \,\&\, \neg C_0) \vee (\neg A_1 \,\&\, C_0) = A_1 + C_0$ and the carry $C_1 = A_1 \,\&\, C_0 = A_1 \,\&\, A_0$;

$B_2 = (A_2 \,\&\, \neg C_1) \vee (\neg A_2 \,\&\, C_1) = A_2 + C_1$ and the carry $C_2 = A_2 \,\&\, C_1 = A_2 \,\&\, A_1 \,\&\, A_0$;

$B_3 = (A_3 \,\&\, \neg C_2) \vee (\neg A_3 \,\&\, C_2) = A_3 + C_2$ and the carry $C_3 = A_3 \,\&\, C_2 = A_3 \,\&\, A_2 \,\&\, A_1 \,\&\, A_0 = B_4$.

If we use $\oplus$ to designate the circuit of Fig. 4-27, we obtain

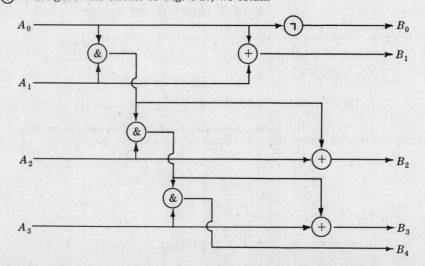

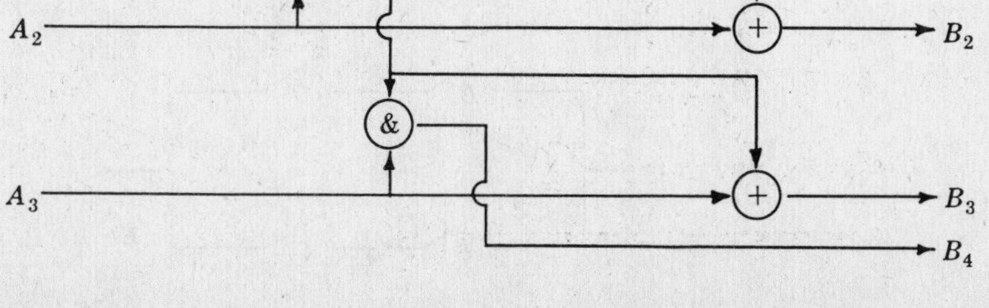

4.6. Describe a method for reducing the operation of subtraction of binary numbers to the use of addition only.

Solution:

Assume given two numbers x and y in binary notation. Let us assume that they are at most n-digit numbers.

Example: $x = 11010$

$y = \ \ 1101 \quad (n = 5)$

Change all digits of y to their opposites; in our example, 10010. Add this new number z to x.

$$\begin{array}{r} 11010 \\ +\ 10010 \\ \hline 101100 \end{array}$$

Add 1 to the result: 101101. Omit the leading 1: 1101. This is $x - y$.

What we did in obtaining z from y was to form $(2^n - 1) - y = \underbrace{111\ldots1}_{n \text{ digits}} - y$. Adding this to x yielded $x + [(2^n - 1) - y] = (x - y) + (2^n - 1)$. Addition of 1 then gave $(x - y) + 2^n$, and omission of the leading 1 finally reduced to $x - y$.

Another example: $x = 101111$ and $y = 110100$. Then $z = 001011$.

$$\begin{array}{r} 101111 \\ +\ 001011 \\ \hline 111010 \\ 1 \\ \hline 111011. \end{array} \quad \textit{Answer}\text{: } 11011.$$

The purpose of reducing subtraction to other operations (addition, adding 1, etc.) is to facilitate its implementation by logic circuits.

What does the process described above yield when y is greater than x?

4.7. Assume that a number between 0 and 9 is given as a four-digit binary number. Employing the notation of Example 4.17, make use of don't care conditions in order to construct a simple switching circuit (or logic circuit) for the condition that the given number is a prime.

Solution:

The condition for being a prime is

$$\bar{A}B\bar{C}\bar{D} \vee AB\bar{C}\bar{D} \vee A\bar{B}C\bar{D} \vee ABC\bar{D}$$

The don't care conditions correspond to the numbers 10 through 15:

$$\bar{A}B\bar{C}D,\ AB\bar{C}D,\ \bar{A}\bar{B}CD,\ A\bar{B}CD,\ \bar{A}BCD,\ ABCD$$

Of these six conditions, we select four and obtain the dnf:

$$\bar{A}B\bar{C}\bar{D} \vee AB\bar{C}\bar{D} \vee A\bar{B}C\bar{D} \vee ABC\bar{D} \vee ABCD \vee AB\bar{C}D \vee \bar{A}B\bar{C}D \vee A\bar{B}CD$$

(A method for choosing the proper don't care conditions is presented in Section 4.17.) This statement form is logically equivalent to $B\bar{C} \vee AC$ (exercise for the reader), which has the switching circuit

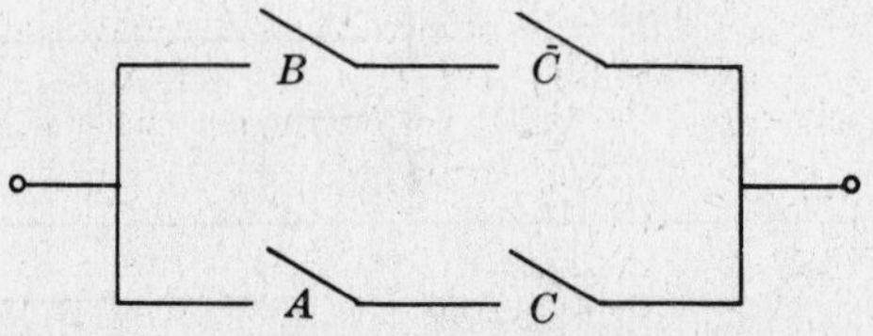

MINIMAL DISJUNCTIVE NORMAL FORMS. PRIME IMPLICANTS

4.8. To test whether a fundamental conjunction ψ is superfluous in $\psi \vee \Phi$, show that it suffices to replace all occurrences in Φ of unnegated letters of ψ by T and occurrences of negated letters of ψ by F, and then observe whether the result is a tautology.

Example: To see if $A\bar{C}$ is superfluous in $AB \vee A\bar{C} \vee \bar{B}\bar{C}$, we obtain T$B \vee \bar{B}$T, i.e. $B \vee \bar{B}$, which is a tautology. (Note that any T in a fundamental conjunction may be dropped, and any fundamental conjunction containing an F as a conjunct also may be omitted.)

Solution:

We must see whether ψ logically implies Φ, i.e. whenever ψ is T, then Φ also must be T. When ψ is T, all the unnegated letters of ψ are T and all the negated letters are F. When this truth assignment is made in Φ, we observe whether the result is always T, i.e. whether the result is a tautology.

4.9. If ψ is a fundamental conjunction and α is a literal, to test whether α is superfluous in $\alpha\psi \vee \Phi$, show that it suffices to replace all occurrences in $\alpha \vee \Phi$ of unnegated letters of ψ by T and occurrences of negated letters by F, and then to observe whether the result is a tautology.

Example: Is $\bar{B}$ superfluous in $AB \vee A\bar{B} \vee \bar{A}BC$? We obtain T$B \vee \bar{B} \vee$ FBC, i.e. $B \vee \bar{B}$, which is a tautology.

Solution:

We must see whether ψ logically implies $\alpha \vee \Phi$; i.e. whenever ψ is T, then $\alpha \vee \Phi$ also must be T. But, if ψ is T, the unnegated letters of ψ are T and the negated ones are F. Then the result of the indicated substitution must always be T, i.e. must be a tautology.

4.10. Prove that if one of the disjuncts of a dnf Φ is not a prime implicant of Φ, then a literal of that disjunct is superfluous. Hence an irredundant dnf Φ must be a disjunction of prime implicants of Φ.

Solution:

Let Φ be $\psi \vee \Psi$, where ψ is not a prime implicant of Φ. This means that there is a fundamental conjunction θ which is a proper part of ψ and such that θ logically implies $\psi \vee \Psi$. Let α be any literal of ψ which is not a literal of θ, and let $\hat{\psi}$ be obtained from ψ by deleting α. Thus θ is included in $\hat{\psi}$. Hence $\hat{\psi}$ logically implies θ, and therefore $\hat{\psi}$ logically implies $\psi \vee \Psi$. From this we may conclude that $\hat{\psi}$ logically implies $\alpha \vee \Psi$. Thus α is superfluous in $\alpha\hat{\psi} \vee \Psi$, i.e. in Φ.

4.11. Find an irredundant dnf logically equivalent to

$$ABC \vee ACD \vee A\bar{B}D \vee \bar{A}B\bar{C} \vee B\bar{C}D$$

Solution:

(1) ACD is superfluous (since $B \vee \bar{B}$ is a tautology). This leaves

$$ABC \vee A\bar{B}D \vee \bar{A}B\bar{C} \vee B\bar{C}D$$

(2) $\bar{B}$ is superfluous (since $BC \vee \bar{B} \vee B\bar{C}$ is a tautology). This leaves

$$ABC \vee AD \vee \bar{A}B\bar{C} \vee B\bar{C}D$$

(3) $B\bar{C}D$ is superfluous (since $A \vee \bar{A}$ is a tautology). This leaves

$$ABC \vee AD \vee \bar{A}B\bar{C}$$

which is irredundant.

4.12. Find all prime implicants of $((A \vee \bar{B}) \rightarrow C) \vee \bar{A}\bar{B}\bar{C}$ by the Quine-McCluskey method.

Solution:

First we must expand the given statement form into a full dnf:

$$\neg(A \vee \bar{B}) \vee C \vee \bar{A}\bar{B}\bar{C}$$

$$\bar{A}B \vee C \vee \bar{A}\bar{B}\bar{C}$$

$$\bar{A}BC \vee \bar{A}B\bar{C} \vee ABC \vee A\bar{B}C \vee \bar{A}\bar{B}C \vee \bar{A}\bar{B}\bar{C}$$

Then

ABC ✓	$\{BC$ ✓	C
	AC ✓	$\bar{A}$
$\{\bar{A}BC$ ✓		
$A\bar{B}C$ ✓	$\{\bar{A}B$ ✓	
$\{\bar{A}B\bar{C}$ ✓	$\bar{A}C$ ✓	
$\bar{A}\bar{B}C$ ✓	$\bar{B}C$ ✓	
$\bar{A}\bar{B}\bar{C}$ ✓	$\{\bar{A}\bar{C}$ ✓	
	$\bar{A}\bar{B}$ ✓	

Thus there are two prime implicants: C, $\bar{A}$.

4.13. Find all minimal dnf's logically equivalent to

$$ABCDE \vee A\bar{B}C\bar{D}E \vee AB\bar{C}DE \vee A\bar{B}CDE \vee \bar{A}BC\bar{D}E \vee ABC\bar{D}\bar{E}$$
$$\vee\ \bar{A}B\bar{C}D\bar{E} \vee \bar{A}\bar{B}\bar{C}D\bar{E} \vee \bar{A}\bar{B}\bar{C}\bar{D}E \vee \bar{A}\bar{B}C\bar{D}E$$

using the Quine-McCluskey method and prime implicant tables.

Solution:

First we find the prime implicants.

$ABCDE$ ✓	$\{ABDE$
$\{AB\bar{C}DE$ ✓	$ACDE$
$A\bar{B}CDE$ ✓	$A\bar{B}CE$
$\{A\bar{B}C\bar{D}E$ ✓	$\{\bar{B}C\bar{D}E$
$\bar{A}BC\bar{D}E$ ✓	$\bar{A}C\bar{D}E$
$ABC\bar{D}\bar{E}$	$\{\bar{A}\bar{C}D\bar{E}$
$\{\bar{A}B\bar{C}D\bar{E}$ ✓	$\bar{A}\bar{B}\bar{D}E$
$\bar{A}\bar{B}C\bar{D}E$ ✓	
$\{\bar{A}\bar{B}\bar{C}D\bar{E}$ ✓	
$\bar{A}\bar{B}\bar{C}\bar{D}E$ ✓	

The prime implicants are $ABC\bar{D}\bar{E}$, $ABDE$, $ACDE$, $A\bar{B}CE$, $\bar{A}C\bar{D}E$, $\bar{B}C\bar{D}E$, $\bar{A}\bar{C}D\bar{E}$, $\bar{A}\bar{B}\bar{D}E$. The prime implicant table is

	$ABCDE$	$A\bar{B}C\bar{D}E$	$AB\bar{C}DE$	$A\bar{B}CDE$	$\bar{A}BC\bar{D}E$	$ABC\bar{D}\bar{E}$	$\bar{A}B\bar{C}D\bar{E}$	$\bar{A}\bar{B}\bar{C}D\bar{E}$	$\bar{A}\bar{B}\bar{C}\bar{D}E$	$\bar{A}\bar{B}C\bar{D}E$
$ABC\bar{D}\bar{E}$						⊗				
$ABDE$	⊠		⊗							
$ACDE$	×			×						
$A\bar{B}CE$		×		×						
$\bar{A}C\bar{D}E$					⊗					⊠
$\bar{B}C\bar{D}E$		×								×
$\bar{A}\bar{C}D\bar{E}$							⊗	⊗		
$\bar{A}\bar{B}\bar{D}E$									⊗	⊠

Hence the core consists of $ABC\bar{D}\bar{E}$, $ABDE$, $\bar{A}C\bar{D}E$, $\bar{A}\bar{C}D\bar{E}$, $\bar{A}\bar{B}\bar{D}E$. Thus we obtain the new table:

	$A\bar{B}C\bar{D}E$	$A\bar{B}CDE$
$ACDE$		×
$A\bar{B}CE$	×	×
$\bar{B}C\bar{D}E$	×	

It is clear from this table that a minimal dnf is obtained only if we choose $A\bar{B}CE$ (since all other ways of covering both columns would require two disjuncts, each having four literals). Hence there is a unique minimal dnf:

$$ABC\bar{D}\bar{E} \vee ABDE \vee \bar{A}C\bar{D}E \vee \bar{A}\bar{C}D\bar{E} \vee \bar{A}\bar{B}\bar{D}E \vee A\bar{B}CE$$

4.14. Find the minimal dnf's for the dnf

$$ABCDE \vee AB\bar{C}DE \vee ABCD\bar{E} \vee A\bar{B}\bar{C}DE \vee \bar{A}B\bar{C}DE$$
$$\vee\ \bar{A}\bar{B}C\bar{D}E \vee \bar{A}B\bar{C}D\bar{E} \vee \bar{A}\bar{B}C\bar{D}\bar{E}$$

with the don't care conditions $A\bar{B}CDE$, $A\bar{B}C\bar{D}E$, $\bar{A}BC\bar{D}\bar{E}$, $\bar{A}\bar{B}\bar{C}D\bar{E}$.

Solution:

First find the prime implicants:

$ABCDE$ √	$ABDE$ √	ADE
$AB\bar{C}DE$ √	$ACDE$ √	
$A\bar{B}CDE$ √	$ABCD$	
$ABCD\bar{E}$ √		
	$A\bar{C}DE$ √	
$A\bar{B}\bar{C}DE$ √	$B\bar{C}DE$	
$\bar{A}B\bar{C}DE$ √	$A\bar{B}DE$ √	
$A\bar{B}C\bar{D}E$ √	$A\bar{B}CE$	
$\bar{A}\bar{B}C\bar{D}E$ √	$\bar{A}B\bar{C}D$	
$\bar{A}BC\bar{D}\bar{E}$ √	$\bar{B}C\bar{D}E$	
$\bar{A}B\bar{C}D\bar{E}$ √		
	$\bar{A}\bar{B}C\bar{D}$	
$\bar{A}\bar{B}C\bar{D}\bar{E}$ √	$\bar{A}C\bar{D}\bar{E}$	
$\bar{A}\bar{B}\bar{C}D\bar{E}$ √	$\bar{A}\bar{C}D\bar{E}$	

Therefore the prime implicants are $ABCD$, $B\bar{C}DE$, $A\bar{B}CE$, $\bar{A}B\bar{C}D$, $\bar{B}C\bar{D}E$, $\bar{A}\bar{B}C\bar{D}$, $\bar{A}C\bar{D}\bar{E}$, $\bar{A}\bar{C}D\bar{E}$, ADE. We obtain the prime implicant table:

	$ABCDE$	$AB\bar{C}DE$	$ABCD\bar{E}$	$A\bar{B}\bar{C}DE$	$\bar{A}B\bar{C}DE$	$\bar{A}\bar{B}C\bar{D}E$	$\bar{A}B\bar{C}D\bar{E}$	$\bar{A}\bar{B}C\bar{D}\bar{E}$
$ABCD$	[×]		(×)					
$B\bar{C}DE$		×			×			
$A\bar{B}CE$								
$\bar{A}B\bar{C}D$					×		×	
$\bar{B}C\bar{D}E$						×		
$\bar{A}\bar{B}C\bar{D}$						×		×
$\bar{A}C\bar{D}\bar{E}$								×
$\bar{A}\bar{C}D\bar{E}$							×	
ADE	[×]	[×]		(×)				

Hence $ABCD$ and ADE belong to the core. We obtain the following new table. (Notice that row $A\bar{B}CE$ has been dropped, since it is empty.)

	$\bar{A}B\bar{C}DE$	$\bar{A}\bar{B}C\bar{D}E$	$\bar{A}B\bar{C}D\bar{E}$	$\bar{A}\bar{B}C\bar{D}\bar{E}$
$B\bar{C}DE$	×			
$\bar{A}B\bar{C}D$	×		×	
$\bar{B}C\bar{D}E$		×		
$\bar{A}\bar{B}C\bar{D}$		×		×
$\bar{A}C\bar{D}\bar{E}$				×
$\bar{A}\bar{C}D\bar{E}$			×	

We may now use the Boolean or branching method. However, in this case, it is obvious that the minimal dnf is obtained using $\bar{A}B\bar{C}D$ and $\bar{A}\bar{B}C\bar{D}$ (since any other way of covering all the columns would require more than two disjuncts). Hence there is a unique minimal dnf:

$$ABCD \vee ADE \vee \bar{A}B\bar{C}D \vee \bar{A}\bar{B}C\bar{D}$$

4.15. Find all solutions of Problem 4.7, using the Quine-McCluskey method and a prime implicant table.

Solution:

The given dnf is $\bar{A}B\bar{C}\bar{D} \vee AB\bar{C}\bar{D} \vee A\bar{B}C\bar{D} \vee ABC\bar{D}$. The don't care conditions are: $\bar{A}B\bar{C}D$, $AB\bar{C}D$, $\bar{A}\bar{B}CD$, $A\bar{B}CD$, $\bar{A}BCD$, $ABCD$. First we find the prime implicants.

$ABCD$ ✓	$\{ABC$ ✓	$\{AB$
	ABD ✓	AC
$\{ABC\bar{D}$ ✓	ACD ✓	BD
$AB\bar{C}D$ ✓	BCD ✓ $\}$	$CD\}$
$A\bar{B}CD$ ✓		
$\bar{A}BCD$ ✓ $\}$	$\{AB\bar{D}$ ✓	$B\bar{C}$
	$AC\bar{D}$ ✓	
$\{AB\bar{C}\bar{D}$ ✓	$AB\bar{C}$ ✓	
$A\bar{B}C\bar{D}$ ✓	$B\bar{C}D$ ✓	
$\bar{A}B\bar{C}D$ ✓	$A\bar{B}C$ ✓	
$\bar{A}\bar{B}CD$ ✓ $\}$	$\bar{B}CD$ ✓	
	$\bar{A}BD$ ✓	
$\bar{A}B\bar{C}\bar{D}$ ✓	$\bar{A}CD$ ✓ $\}$	
	$\{B\bar{C}\bar{D}$ ✓	
	$\bar{A}B\bar{C}$ ✓ $\}$	

Hence the prime implicants are AB, AC, BD, CD, $B\bar{C}$; and we obtain the following prime implicant table

	$\bar{A}B\bar{C}\bar{D}$	$AB\bar{C}\bar{D}$	$A\bar{B}C\bar{D}$	$ABC\bar{D}$
AB		×		×
AC			$\otimes$	$\boxtimes$
BD				
CD				
$B\bar{C}$	$\otimes$	$\boxtimes$		

Thus, AC and $B\bar{C}$ are in the core. Every disjunct of the given dnf includes a prime implicant in the core. Hence $AC \vee B\bar{C}$ is the unique minimal dnf.

4.16. Using the Quine-McCluskey method and prime implicant tables, find all minimal dnf's for

$$AB\bar{C}DEG \vee A\bar{B}\bar{C}DEG \vee ABCD\bar{E}G \vee \bar{A}BCD\bar{E}G \vee AB\bar{C}D\bar{E}G \vee A\bar{B}\bar{C}D\bar{E}G$$
$$\vee AB\bar{C}DE\bar{G} \vee A\bar{B}\bar{C}DE\bar{G} \vee AB\bar{C}\bar{D}E\bar{G} \vee A\bar{B}\bar{C}\bar{D}E\bar{G} \vee AB\bar{C}D\bar{E}\bar{G} \vee \bar{A}B\bar{C}D\bar{E}\bar{G}$$

with the don't care conditions

$$\bar{A}B\bar{C}DEG,\ \bar{A}\bar{B}\bar{C}DEG,\ A\bar{B}CD\bar{E}G,\ \bar{A}\bar{B}CD\bar{E}G,\ \bar{A}B\bar{C}D\bar{E}G$$
$$ABC\bar{D}E\bar{G},\ A\bar{B}C\bar{D}E\bar{G},\ A\bar{B}\bar{C}D\bar{E}\bar{G},\ \bar{A}\bar{B}\bar{C}D\bar{E}\bar{G}$$

Solution:

The prime implicants turn out to be

$$B\bar{C}DG,\ \bar{C}DEG,\ CD\bar{E}G,\ BD\bar{E}G,\ AD\bar{E}G,\ B\bar{C}D\bar{E},\ A\bar{C}E\bar{G},\ A\bar{D}E\bar{G},\ \bar{C}D\bar{E}\bar{G},\ A\bar{C}D$$

(Verification is left as an exercise for the reader.) We then draw up the prime implicant Table I.

Table I

	$AB\bar{C}DEG$	$A\bar{B}\bar{C}DEG$	$ABCD\bar{E}G$	$\bar{A}BCD\bar{E}G$	$AB\bar{C}D\bar{E}G$	$A\bar{B}\bar{C}D\bar{E}G$
$B\bar{C}DG$	×				×	
$\bar{C}DEG$	×	×				
$CD\bar{E}G$			×	×		
$BD\bar{E}G$			×	×	×	
$AD\bar{E}G$			×		×	×
$B\bar{C}D\bar{E}$					×	
$A\bar{C}E\bar{G}$						
$A\bar{D}E\bar{G}$						
$\bar{C}D\bar{E}\bar{G}$						
$A\bar{C}D$	×	×			×	×

	$AB\bar{C}DE\bar{G}$	$A\bar{B}\bar{C}DE\bar{G}$	$AB\bar{C}\bar{D}E\bar{G}$	$A\bar{B}\bar{C}\bar{D}E\bar{G}$	$AB\bar{C}D\bar{E}\bar{G}$	$\bar{A}B\bar{C}D\bar{E}\bar{G}$
$B\bar{C}DG$						
$\bar{C}DEG$						
$CD\bar{E}G$						
$BD\bar{E}G$						
$AD\bar{E}G$						
$B\bar{C}D\bar{E}$					×	×
$A\bar{C}E\bar{G}$	×	×	×	×		
$A\bar{D}E\bar{G}$			×	×		
$\bar{C}D\bar{E}\bar{G}$					×	×
$A\bar{C}D$	×	×			×	

No applications of the core operation are possible. However, we can eliminate the following columns by dominant column operations:

$AB\bar{C}DEG$ (since it dominates $A\bar{B}\bar{C}DEG$),

$ABCD\bar{E}G$ (since it dominates $\bar{A}BCD\bar{E}G$),

$AB\bar{C}D\bar{E}G$ (since it dominates $A\bar{B}\bar{C}D\bar{E}G$),

$AB\bar{C}DE\bar{G}$ (since it dominates $A\bar{B}\bar{C}DE\bar{G}$),

$AB\bar{C}\bar{D}E\bar{G}$ (since it dominates $A\bar{B}\bar{C}\bar{D}E\bar{G}$),

$AB\bar{C}D\bar{E}\bar{G}$ (since it dominates $\bar{A}B\bar{C}D\bar{E}\bar{G}$).

Thus, we obtain Table II.

Table II

	$A\bar{B}\bar{C}DEG$	$\bar{A}BCD\bar{E}G$	$A\bar{B}\bar{C}D\bar{E}G$	$A\bar{B}\bar{C}DE\bar{G}$	$A\bar{B}\bar{C}\bar{D}E\bar{G}$	$\bar{A}B\bar{C}D\bar{E}\bar{G}$
$B\bar{C}DG$						
$\bar{C}DEG$	×					
$CD\bar{E}G$		×				
$BD\bar{E}G$		×				
$AD\bar{E}G$			×			
$B\bar{C}D\bar{E}$						×
$A\bar{C}E\bar{G}$				×	×	
$A\bar{D}E\bar{G}$					×	
$\bar{C}D\bar{E}\bar{G}$						×
$A\bar{C}D$	×		×	×		

In Table II we can apply the dominated row operation to eliminate the rows of $\bar{C}DEG$ and $AD\bar{E}G$ (both dominated by $A\bar{C}D$). We can also drop the first row, since it is empty. Thus we obtain Table III.

Table III

	$A\bar{B}\bar{C}DEG$	$\bar{A}BCD\bar{E}G$	$A\bar{B}\bar{C}D\bar{E}G$	$A\bar{B}\bar{C}DE\bar{G}$	$A\bar{B}\bar{C}\bar{D}E\bar{G}$	$\bar{A}B\bar{C}D\bar{E}\bar{G}$
$CD\bar{E}G$		×				
$BD\bar{E}G$		×				
$B\bar{C}D\bar{E}$						×
$A\bar{C}E\bar{G}$				×	×	
$A\bar{D}E\bar{G}$					×	
$\bar{C}D\bar{E}\bar{G}$						×
$A\bar{C}D$	⊗		⊗	⊠		

In Table III, the first and third columns have unique entries. Hence $A\bar{C}D$ belongs to the secondary core. We then can drop the last row and the first, third and fourth columns, obtaining Table IV.

Table IV

	$\bar{A}BCD\bar{E}G$	$A\bar{B}\bar{C}\bar{D}E\bar{G}$	$\bar{A}B\bar{C}D\bar{E}\bar{G}$
$CD\bar{E}G$	×		
$BD\bar{E}G$	×		
$B\bar{C}D\bar{E}$			×
$A\bar{C}E\bar{G}$		×	
$A\bar{D}E\bar{G}$		×	
$\bar{C}D\bar{E}\bar{G}$			×

Clearly, application of the Boolean method to Table IV yields eight different minimal dnf's (by choosing either the first or second row, either the third or sixth row, and either the fourth or fifth row):

$$A\bar{C}D \vee CD\bar{E}G \vee A\bar{C}E\bar{G} \vee B\bar{C}D\bar{E}$$
$$A\bar{C}D \vee CD\bar{E}G \vee A\bar{C}E\bar{G} \vee \bar{C}D\bar{E}\bar{G}$$
$$A\bar{C}D \vee CD\bar{E}G \vee A\bar{D}E\bar{G} \vee B\bar{C}D\bar{E}$$
$$A\bar{C}D \vee CD\bar{E}G \vee A\bar{D}E\bar{G} \vee \bar{C}D\bar{E}\bar{G}$$
$$A\bar{C}D \vee BD\bar{E}G \vee A\bar{C}E\bar{G} \vee B\bar{C}D\bar{E}$$
$$A\bar{C}D \vee BD\bar{E}G \vee A\bar{C}E\bar{G} \vee \bar{C}D\bar{E}\bar{G}$$
$$A\bar{C}D \vee BD\bar{E}G \vee A\bar{D}E\bar{G} \vee B\bar{C}D\bar{E}$$
$$A\bar{C}D \vee BD\bar{E}G \vee A\bar{D}E\bar{G} \vee \bar{C}D\bar{E}\bar{G}$$

Observe that $A\bar{C}D$ must be present because it is in the secondary core.

4.17. Using the consensus method, find all minimal dnf's for the dnf of Problem 4.11.

Solution:

(1) $ABC \vee ACD \vee A\bar{B}D \vee \bar{A}B\bar{C} \vee B\bar{C}D$

(2) $ABC \vee ACD \vee A\bar{B}D \vee \bar{A}B\bar{C} \vee B\bar{C}D \vee ABD$ (Consensus of ABC and $B\bar{C}D$)

(3) $ABC \vee ACD \vee A\bar{B}D \vee \bar{A}B\bar{C} \vee B\bar{C}D \vee ABD \vee A\bar{C}D$ (Consensus of $A\bar{B}D$ and $B\bar{C}D$)

(4) $ABC \vee ACD \vee A\bar{B}D \vee \bar{A}B\bar{C} \vee B\bar{C}D \vee ABD \vee A\bar{C}D \vee AD$ (Consensus of ACD and $A\bar{C}D$)

(5) $ABC \vee \bar{A}B\bar{C} \vee B\bar{C}D \vee AD$ (since ACD, $A\bar{B}D$, ABD, $A\bar{C}D$ all include AD)

Operations (i) and (ii), page 94, are no longer applicable. Hence ABC, $\bar{A}B\bar{C}$, $B\bar{C}D$, AD are the prime implicants.

Let us draw the following table:

	ϕ_1 ABC	ϕ_2 $\bar{A}B\bar{C}$	ϕ_3 $B\bar{C}D$	ϕ_4 AD	
ABC		F	F	D	
$\bar{A}B\bar{C}$	F		D	F	
$B\bar{C}D$	F	$\bar{A}$		A	superfluous
AD	BC	F	$B\bar{C}$		

From the third row, $\sigma_3 \vee \sigma_2\sigma_4$, and we obtain $\sigma_1\sigma_2(\sigma_3 \vee \sigma_2\sigma_4)\sigma_4$, which is equivalent to $\sigma_1\sigma_2\sigma_4$. Hence the only irredundant dnf is

$$ABC \vee \bar{A}B\bar{C} \vee AD$$

Therefore this is the only minimal dnf.

4.18. Apply the consensus method to find all minimal dnf's for

$$ABC \vee B\bar{D} \vee A\bar{C}D \vee \bar{A}\bar{B}\bar{C}$$

Solution:

(1) $ABC \vee B\bar{D} \vee A\bar{C}D \vee \bar{A}\bar{B}\bar{C}$

(2) $ABC \vee B\bar{D} \vee A\bar{C}D \vee \bar{A}\bar{B}\bar{C} \vee ABD$ (Consensus of ABC and $A\bar{C}D$)

(3) $ABC \vee B\bar{D} \vee A\bar{C}D \vee \bar{A}\bar{B}\bar{C} \vee ABD \vee AB\bar{C}$ (Consensus of $B\bar{D}$ and $A\bar{C}D$)

(4) $ABC \vee B\bar{D} \vee A\bar{C}D \vee \bar{A}\bar{B}\bar{C} \vee ABD \vee AB\bar{C} \vee \bar{A}\bar{C}\bar{D}$ (Consensus of $B\bar{D}$ and $\bar{A}\bar{B}\bar{C}$)

(5) $ABC \vee B\bar{D} \vee A\bar{C}D \vee \bar{A}\bar{B}\bar{C} \vee ABD \vee AB\bar{C} \vee \bar{A}\bar{C}\bar{D} \vee AB$ (Consensus of $B\bar{D}$ and ABD)

(6) $B\bar{D} \vee A\bar{C}D \vee \bar{A}\bar{B}\bar{C} \vee \bar{A}\bar{C}\bar{D} \vee AB$ (ABC, ABD, $AB\bar{C}$ all include AB)

(7) $B\bar{D} \vee A\bar{C}D \vee \bar{A}\bar{B}\bar{C} \vee \bar{A}\bar{C}\bar{D} \vee AB \vee \bar{B}\bar{C}D$ (Consensus of $A\bar{C}D$ and $\bar{A}\bar{B}\bar{C}$)

Operations (i) and (ii), page 94, are no longer applicable. Hence the prime implicants are $B\bar{D}$, $A\bar{C}D$, $\bar{A}\bar{B}\bar{C}$, $\bar{A}\bar{C}\bar{D}$, AB, $\bar{B}\bar{C}D$.

Now we construct the following table:

	ϕ_1 $B\bar{D}$	ϕ_2 $A\bar{C}D$	ϕ_3 $\bar{A}\bar{B}\bar{C}$	ϕ_4 $\bar{A}\bar{C}\bar{D}$	ϕ_5 AB	ϕ_6 $\bar{B}\bar{C}D$	
$B\bar{D}$		F	F	$\bar{A}\bar{C}$	A	F	
$A\bar{C}D$	F		F	F	B	$\bar{B}$	superfluous
$\bar{A}\bar{B}\bar{C}$	F	F		$\bar{D}$	F	D	superfluous
$\bar{A}\bar{C}\bar{D}$	B	F	$\bar{B}$		F	F	superfluous
AB	$\bar{D}$	$\bar{C}D$	F	F		F	
$\bar{B}\bar{C}D$	F	A	$\bar{A}$	F	F		superfluous

Hence we have the result

$$\sigma_1(\sigma_2 \vee \sigma_5\sigma_6)(\sigma_3 \vee \sigma_4\sigma_6)(\sigma_4 \vee \sigma_1\sigma_3)\sigma_5(\sigma_6 \vee \sigma_2\sigma_3)$$

which is equivalent to

$$\sigma_1\sigma_3\sigma_5\sigma_6 \vee \sigma_1\sigma_2\sigma_3\sigma_5 \vee \sigma_1\sigma_4\sigma_5\sigma_6$$

Therefore there are three irredundant dnf's:

$$B\bar{D} \vee \bar{A}\bar{B}\bar{C} \vee AB \vee \bar{B}\bar{C}D$$

$$B\bar{D} \vee \bar{A}\bar{B}\bar{C} \vee AB \vee A\bar{C}D$$

$$B\bar{D} \vee \bar{A}\bar{C}\bar{D} \vee AB \vee \bar{B}\bar{C}D$$

Since they are of equal cost, all three are minimal dnf's.

KARNAUGH MAPS

4.19. Using a Karnaugh map, find all minimal dnf's for

$$\bar{A}\bar{B}CD \vee ABC\bar{D} \vee A\bar{B}C\bar{D} \vee \bar{A}\bar{B}C\bar{D} \vee A\bar{B}\bar{C}\bar{D} \vee \bar{A}\bar{B}\bar{C}\bar{D} \vee \bar{A}B\bar{C}\bar{D} \vee A\bar{B}\bar{C}D$$

Solution:

Draw the Karnaugh map:

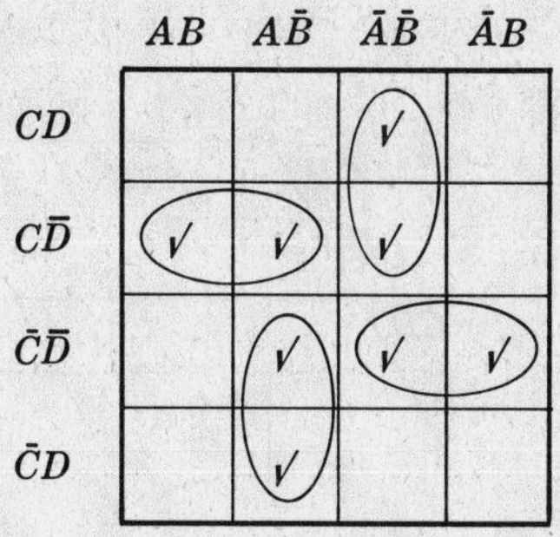

Handling the checks whose unique largest combination consists of two squares, we obtain the four loops indicated in the diagram. Since all checks are covered, the unique minimal dnf is

$$\bar{A}\bar{B}C \vee AC\bar{D} \vee A\bar{B}\bar{C} \vee \bar{A}\bar{C}\bar{D}$$

Notice that, although each of the four checks in the middle belongs to a unique largest combination of four squares, that combination is not required, since all the checks in it already have been covered.

4.20. Use a Karnaugh map to find all minimal dnf's for the dnf of Problem 4.18:

$$ABC \vee B\bar{D} \vee A\bar{C}D \vee \bar{A}\bar{B}\bar{C}$$

Solution:

To use a Karnaugh map we need not expand the given dnf into a full dnf. It suffices to place a check in every square containing one of the disjuncts. (For example, ABC generates the two checks in $ABCD$ and $ABC\bar{D}$; $B\bar{D}$ generates four checks, etc.)

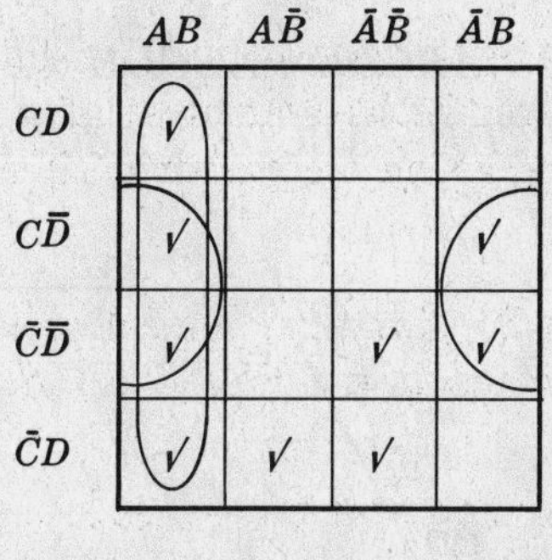

There are no isolated checks and no checks with a unique largest 2-square combination. However, there are checks belonging to unique largest 4-square combinations. The first column gives AB, and the other 4-square combination yields $B\bar{D}$. The remaining three checks produce three minimal dnf's:

$AB \vee B\bar{D} \vee A\bar{C}D \vee \bar{A}\bar{B}\bar{C}$ ($A\bar{B}\bar{C}D$ combines with $AB\bar{C}D$, and $\bar{A}\bar{B}\bar{C}\bar{D}$ with $\bar{A}\bar{B}\bar{C}D$)

$AB \vee B\bar{D} \vee \bar{B}\bar{C}D \vee \bar{A}\bar{B}\bar{C}$ ($A\bar{B}\bar{C}D$ combines with $\bar{A}\bar{B}\bar{C}D$, and $\bar{A}\bar{B}\bar{C}\bar{D}$ with $\bar{A}\bar{B}\bar{C}D$)

$AB \vee B\bar{D} \vee \bar{B}\bar{C}D \vee \bar{A}\bar{C}\bar{D}$ ($A\bar{B}\bar{C}D$ combines with $\bar{A}\bar{B}\bar{C}D$, and $\bar{A}\bar{B}\bar{C}\bar{D}$ with $\bar{A}B\bar{C}\bar{D}$)

This verifies the solution of Problem 4.18.

4.21. By means of a Karnaugh map, find all minimal dnf's for

$$ABCDE \vee A\bar{B}CDE \vee \bar{A}\bar{B}CDE \vee \bar{A}BC\bar{D}E \vee A\bar{B}\bar{C}\bar{D}E$$
$$\vee \bar{A}\bar{B}\bar{C}\bar{D}E \vee A\bar{B}\bar{C}DE \vee \bar{A}\bar{B}\bar{C}DE \vee A\bar{B}CD\bar{E} \vee \bar{A}\bar{B}CD\bar{E}$$
$$\vee A\bar{B}\bar{C}\bar{D}\bar{E} \vee \bar{A}\bar{B}\bar{C}\bar{D}\bar{E} \vee A\bar{B}\bar{C}D\bar{E} \vee \bar{A}\bar{B}\bar{C}D\bar{E}$$

Solution:

The Karnaugh map is

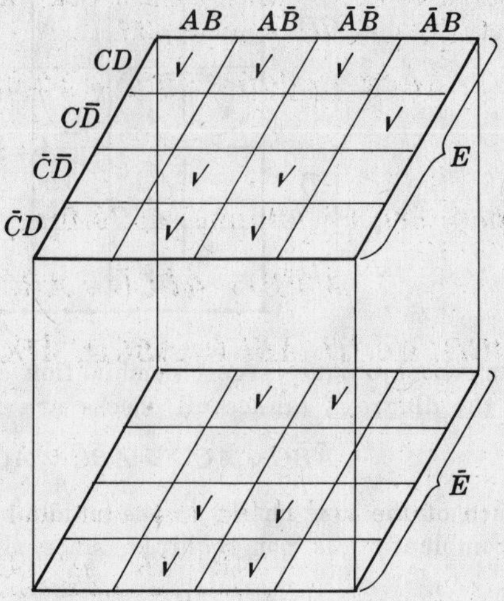

There is one isolated check, yielding the disjunct $\bar{A}BC\bar{D}E$. There is one check ($ABCDE$) belonging to a unique 2-square combination, which yields the disjunct $ACDE$. There are two checks ($A\bar{B}\bar{C}\bar{D}E$ and $\bar{A}\bar{B}\bar{C}\bar{D}E$) belonging to a unique 8-square combination, which yields $\bar{B}\bar{C}$. Another check ($\bar{A}\bar{B}CD\bar{E}$) belongs to a unique 8-square combination, yielding $\bar{B}D$. Since all the checks are covered, we have obtained a unique minimal dnf:

$$\bar{A}BC\bar{D}E \vee ACDE \vee \bar{B}\bar{C} \vee \bar{B}D$$

4.22. Use a Karnaugh map to find all minimal dnf's for

$$\bar{A}\bar{B}CDE \vee \bar{A}BCDE \vee A\bar{B}C\bar{D}E \vee \bar{A}\bar{B}C\bar{D}E \vee A\bar{B}\bar{C}\bar{D}E$$
$$\vee \bar{A}\bar{B}\bar{C}\bar{D}E \vee A\bar{B}\bar{C}DE \vee A\bar{B}C\bar{D}\bar{E} \vee \bar{A}BCD\bar{E} \vee \bar{A}\bar{B}\bar{C}\bar{D}\bar{E}$$

Solution:

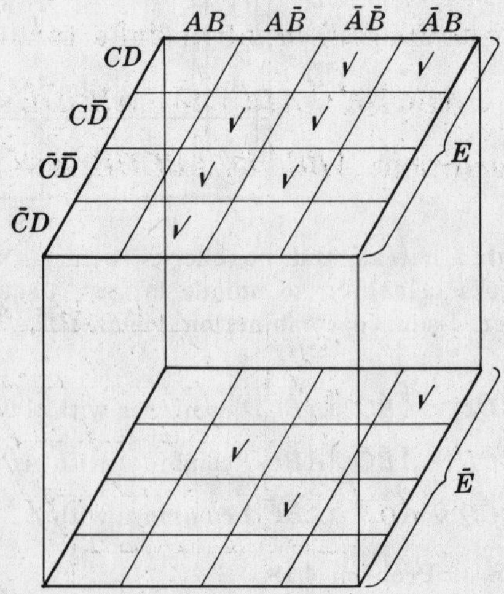

There are four checks belonging to unique largest 2-square combinations ($A\bar{B}\bar{C}DE$ with $A\bar{B}\bar{C}\bar{D}E$, $A\bar{B}C\bar{D}\bar{E}$ with $A\bar{B}C\bar{D}E$, $\bar{A}\bar{B}\bar{C}\bar{D}\bar{E}$ with $\bar{A}\bar{B}\bar{C}\bar{D}E$, and $\bar{A}BCD\bar{E}$ with $\bar{A}BCDE$), yielding the disjuncts $A\bar{B}\bar{C}E$, $A\bar{B}C\bar{D}$, $\bar{A}\bar{B}\bar{C}\bar{D}$, and $\bar{A}BCD$. There is a unique largest 4-square combination containing $\bar{A}\bar{B}C\bar{D}E$, and this yields the disjunct $\bar{B}\bar{D}E$. Now there is one check still uncovered: $\bar{A}\bar{B}CDE$. There are *apparently* two equally simple ways of covering this check:

(1) $\bar{A}\bar{B}CDE$ with $\bar{A}BCDE$, yielding the disjunct $\bar{A}CDE$.

(2) $\bar{A}\bar{B}CDE$ with $\bar{A}\bar{B}C\bar{D}E$, yielding the disjunct $\bar{A}\bar{B}CE$.

However, notice that in case (2) the 4-square combination $\{A\bar{B}C\bar{D}E,\ \bar{A}\bar{B}C\bar{D}E,\ A\bar{B}\bar{C}\bar{D}E,\ \bar{A}\bar{B}\bar{C}\bar{D}E\}$ has been covered by four different 2-square combinations, and the disjunct $\bar{B}\bar{D}E$ becomes superfluous. Thus there is actually only one minimal dnf:

$$A\bar{B}\bar{C}E \vee A\bar{B}C\bar{D} \vee \bar{A}\bar{B}\bar{C}\bar{D} \vee \bar{A}BCD \vee \bar{A}\bar{B}CE$$

4.23. Using a Karnaugh map, find all minimal dnf's for

$$A\bar{B}C\bar{D} \vee \bar{A}\bar{B}C\bar{D} \vee A\bar{B}\bar{C}\bar{D}$$

assuming $A\bar{B}CD$, $\bar{A}\bar{B}CD$, $A\bar{B}\bar{C}D$, $\bar{A}\bar{B}\bar{C}D$, $\bar{A}BCD$, $\bar{A}BC\bar{D}$, $AB\bar{C}\bar{D}$, $\bar{A}B\bar{C}\bar{D}$ are don't care conditions.

Solution:

We use checks for the disjuncts of the given dnf, and crosses for the don't care conditions.

	AB	$A\bar{B}$	$\bar{A}\bar{B}$	$\bar{A}B$
CD		×	×	×
$C\bar{D}$		✓	✓	×
$\bar{C}\bar{D}$	×	✓		×
$\bar{C}D$		×	×	

The check in $A\bar{B}\bar{C}\bar{D}$ is in a unique largest 4-square combination, yielding the disjunct $A\bar{B}$. The only uncovered check is in $\bar{A}\bar{B}C\bar{D}$. This belongs to two 4-square combinations. Hence there are two minimal dnf's: $A\bar{B} \vee \bar{A}C$ and $A\bar{B} \vee \bar{B}C$.

4.24. Use a Karnaugh map to find all minimal dnf's for

$$\bar{A}\bar{B}CD\bar{E} \vee ABCDE \vee A\bar{B}\bar{C}DE \vee A\bar{B}C\bar{D}\bar{E} \vee \bar{A}\bar{B}\bar{C}D\bar{E} \vee \bar{A}\bar{B}C\bar{D}\bar{E}$$

with the don't care conditions $A\bar{B}C\bar{D}E$, $\bar{A}\bar{B}CDE$, $\bar{A}B\bar{C}\bar{D}\bar{E}$, $A\bar{B}\bar{C}\bar{D}E$, $A\bar{B}\bar{C}\bar{D}\bar{E}$, $\bar{A}\bar{B}\bar{C}\bar{D}\bar{E}$.

Solution:

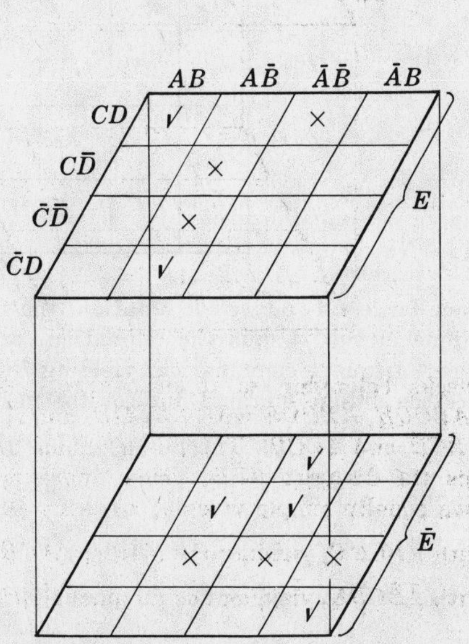

There is an isolated check at $ABCDE$. Hence this must be a disjunct. The checks at $A\bar{B}\bar{C}DE$ and $\bar{A}B\bar{C}D\bar{E}$ belong to unique largest 2-square combinations, producing the disjuncts $A\bar{B}\bar{C}E$ and $\bar{A}B\bar{C}\bar{E}$. The check in $\bar{A}\bar{B}C\bar{D}\bar{E}$ belongs to a unique largest 4-square combination, yielding the disjunct $\bar{B}\bar{D}\bar{E}$. The only uncovered check is $\bar{A}\bar{B}CD\bar{E}$. This belongs to two 2-square combinations, yielding either $\bar{A}\bar{B}C\bar{E}$ or $\bar{A}\bar{B}CD$. Hence there are two minimal dnf's:

$$ABCDE \vee A\bar{B}\bar{C}E \vee \bar{A}B\bar{C}\bar{E} \vee \bar{B}\bar{D}\bar{E} \vee \bar{A}\bar{B}C\bar{E}$$

$$ABCDE \vee A\bar{B}\bar{C}E \vee \bar{A}B\bar{C}\bar{E} \vee \bar{B}\bar{D}\bar{E} \vee \bar{A}\bar{B}CD$$

4.25. Find all minimal dnf's or cnf's for

$$\bar{A}\bar{B}CD \vee \bar{A}\bar{B}C\bar{D} \vee \bar{A}B\bar{C}D \vee A\bar{B}\bar{C}D \vee A\bar{B}\bar{C}\bar{D} \vee \bar{A}\bar{B}\bar{C}D \vee \bar{A}B\bar{C}\bar{D}$$

Solution:

Draw the following Karnaugh map.

	AB	$A\bar{B}$	$\bar{A}\bar{B}$	$\bar{A}B$
CD			✓	
$C\bar{D}$			✓	
$\bar{C}\bar{D}$		✓		✓
$\bar{C}D$		✓	✓	✓

The checks in $A\bar{B}\bar{C}\bar{D}$, $\bar{A}B\bar{C}\bar{D}$ and $\bar{A}\bar{B}C\bar{D}$ belong to unique largest 2-square combinations, yielding the disjuncts $A\bar{B}\bar{C}$, $\bar{A}B\bar{C}$ and $\bar{A}\bar{B}C$. The remaining check in $\bar{A}\bar{B}\bar{C}D$ belongs to three 2-square combinations. Hence there are three minimal dnf's:

$$A\bar{B}\bar{C} \vee \bar{A}B\bar{C} \vee \bar{A}\bar{B}C \vee \bar{B}\bar{C}D$$

$$A\bar{B}\bar{C} \vee \bar{A}B\bar{C} \vee \bar{A}\bar{B}C \vee \bar{A}\bar{B}D$$

$$A\bar{B}\bar{C} \vee \bar{A}B\bar{C} \vee \bar{A}\bar{B}C \vee \bar{A}\bar{C}D$$

To find the minimal cnf's, we draw the Karnaugh map for the negation:

	AB	$A\bar{B}$	$\bar{A}\bar{B}$	$\bar{A}B$
CD	✓	✓		✓
$C\bar{D}$	✓	✓		✓
$\bar{C}\bar{D}$	✓		✓	
$\bar{C}D$	✓			

$AB\bar{C}D$ belongs to a unique largest 4-square combination (the first column), yielding the disjunct AB. $A\bar{B}CD$ belongs to a unique largest 4-square combination, producing the disjunct AC, and $\bar{A}BCD$ belongs to a unique largest 4-square combination, yielding BC. The only uncovered check is $\bar{A}\bar{B}\bar{C}\bar{D}$, which is isolated. Hence the unique minimal dnf for the negation is

$$AB \vee AC \vee BC \vee \bar{A}\bar{B}\bar{C}\bar{D}$$

Therefore the unique minimal cnf is

$$(\bar{A} \vee \bar{B})(\bar{A} \vee \bar{C})(\bar{B} \vee \bar{C})(A \vee B \vee C \vee D)$$

This is simpler than any of the minimal dnf's. Hence it is the minimal dnf or cnf.

4.26. Find all minimal dnf's or cnf's for

$$ACDE \vee ABC\bar{D}E \vee A\bar{B}C\bar{D}\bar{E} \vee \bar{A}\bar{B}\bar{C}\bar{D} \vee \bar{A}\bar{B}\bar{C}D\bar{E} \vee \bar{A}B\bar{C}DE$$

with the don't care conditions $ACD\bar{E}$, $ABC\bar{E}$, $\bar{B}C\bar{D}E$, $A\bar{B}\bar{D}E$, $\bar{A}BC\bar{D}\bar{E}$, $AB\bar{C}\bar{E}$, $\bar{A}B\bar{C}D\bar{E}$.

Solution:

First we draw the Karnaugh map.

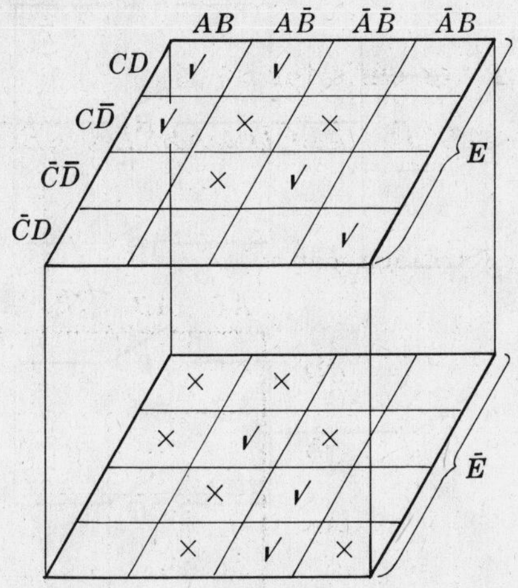

The check in $\bar{A}B\bar{C}DE$ is in a unique largest 2-square combination, yielding the disjunct $\bar{A}B\bar{C}D$. The check in $\bar{A}\bar{B}\bar{C}D\bar{E}$ is in a unique largest 4-square combination, yielding the disjunct $\bar{B}\bar{C}\bar{E}$. The other checks are covered by two 8-square combinations, yielding $\bar{B}\bar{D}$ and AC. Hence there is a unique minimal dnf:

$$\bar{A}B\bar{C}D \vee \bar{B}\bar{C}\bar{E} \vee \bar{B}\bar{D} \vee AC \tag{1}$$

The Karnaugh map for the negation is:

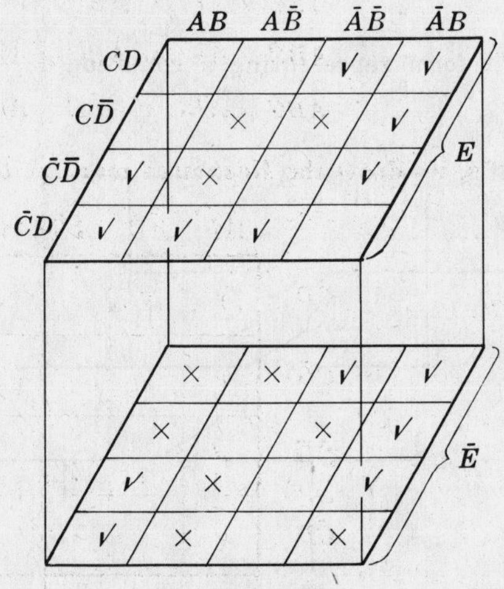

The check in $\bar{A}B\bar{C}\bar{D}\bar{E}$ belongs to a unique largest 8-square combination, yielding $B\bar{E}$. The checks in $AB\bar{C}DE$ and $\bar{A}BCDE$ belong to unique 8-square combinations, yielding $A\bar{C}$ and $\bar{A}C$, respectively. The only remaining uncovered check in $\bar{A}\bar{B}\bar{C}DE$ belongs to two 2-square combinations, yielding either $\bar{A}BDE$ or $\bar{B}\bar{C}DE$. Hence there are two minimal dnf's for the negation:

$$\bar{A}BDE \vee B\bar{E} \vee A\bar{C} \vee \bar{A}C \quad \text{and} \quad \bar{B}\bar{C}DE \vee B\bar{E} \vee A\bar{C} \vee \bar{A}C$$

These yield two minimal cnf's for the given dnf:

$$(A \vee \bar{B} \vee \bar{D} \vee \bar{E})\ (\bar{B} \vee E)\ (\bar{A} \vee C)\ (A \vee \bar{C}) \quad \text{and} \quad (B \vee C \vee \bar{D} \vee \bar{E})\ (\bar{B} \vee E)\ (\bar{A} \vee C)\ (A \vee \bar{C})$$

These cnf's have lower cost than the dnf (*1*), and, therefore, are the minimal dnf's or cnf's.

Supplementary Problems

SWITCHING CIRCUITS. SIMPLIFICATION. BRIDGE CIRCUITS

4.27. Write down a statement form representing a condition for flow of current through each of the following series-parallel circuits.

(a)

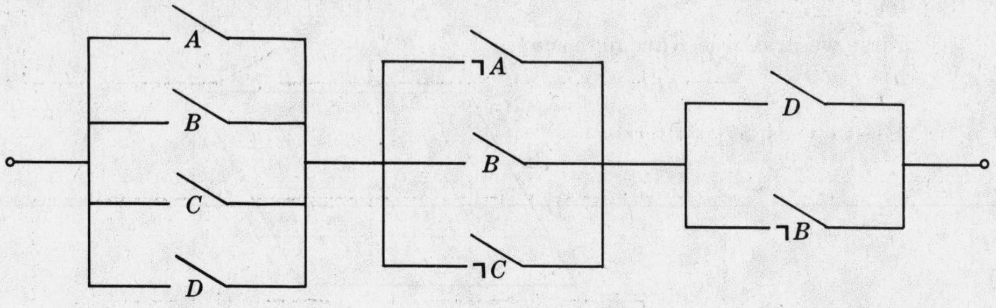

(b)

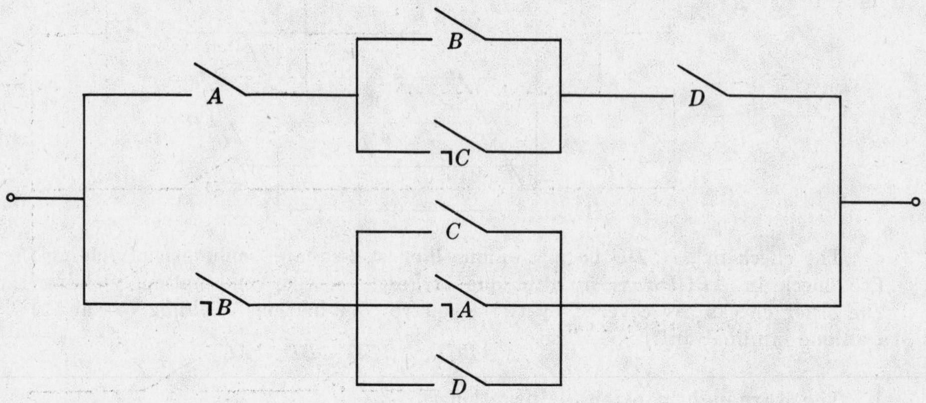

4.28. Write down a statement form representing a condition for flow of current through each of the following bridge circuits.

(a)

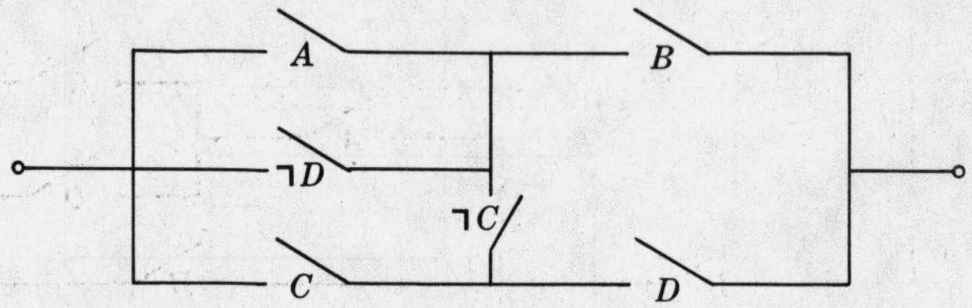

(b)

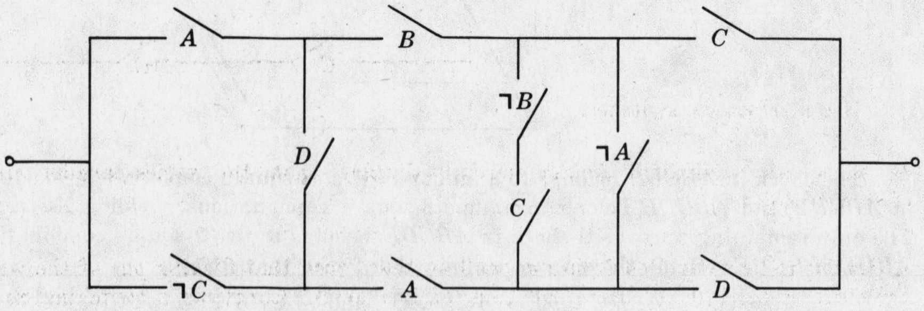

4.29. Draw a switching circuit having the following corresponding statement forms.

(a) $[(B \vee A) \,\&\, (\neg B \vee C)] \vee (C \,\&\, \neg A)$

(b) $(A \,\&\, \neg B \,\&\, (C \vee D)) \vee (\neg A \,\&\, (B \vee C))$

4.30. Replace the following series-parallel switching circuits by simpler bridge circuits.

(*a*)

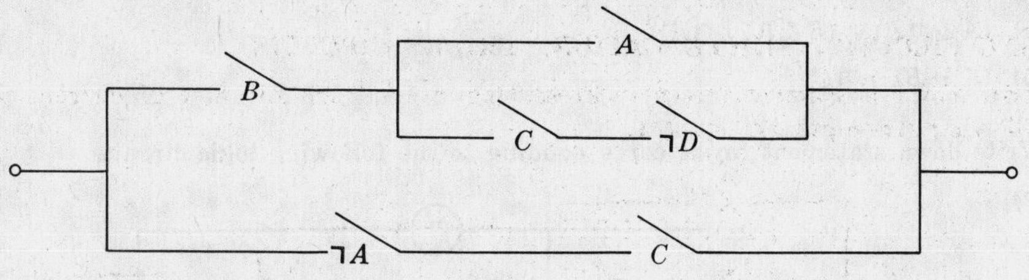

Use at most five switches.

(*b*)

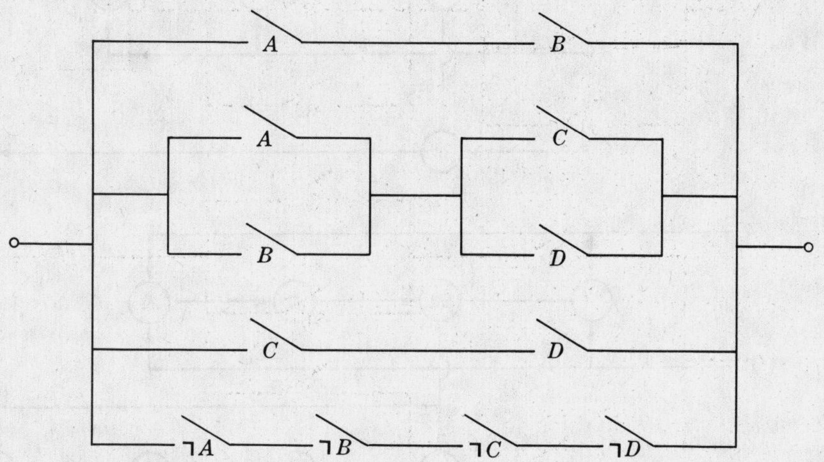

Use at most ten switches.

(*c*)

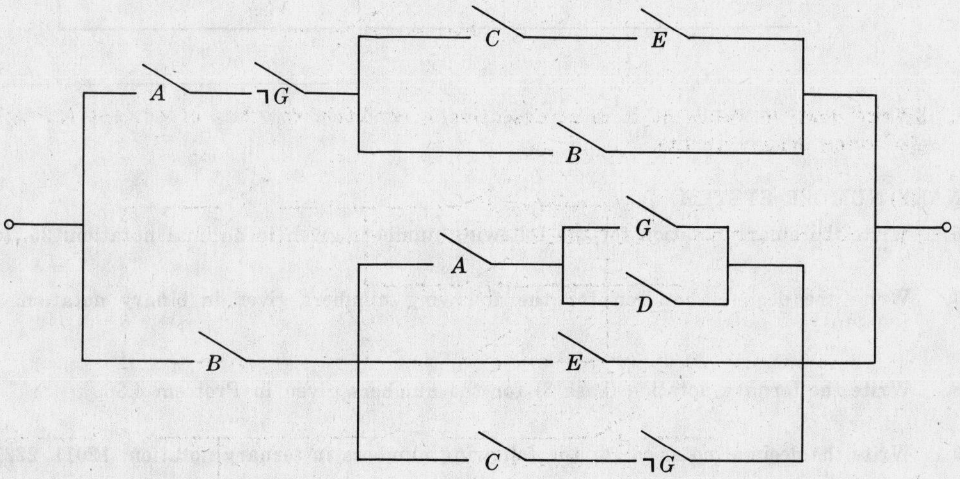

Use at most six switches.

4.31. Is there an equivalent bridge circuit simpler than the series-parallel circuit of Fig. 4-13 (Example 4.6)?

4.32. A light is to be controlled from two wall switches such that flicking one of the wall switches changes the state of the light ("on" to "off", or "off" to "on"). Construct a switching circuit that will allow current to flow to the light under the given condition. (*Hint*: Compare Example 4.6.)

4.33. A municipal board consists of the Mayor, President of the City Council, Comptroller, and three Borough Presidents. The Mayor has two votes and all the others one vote. A motion obtaining a majority passes except that any motion opposed by all three Borough Presidents fails. Write a switching circuit which will indicate passage of a motion.

LOGIC CIRCUITS

4.34. Construct logic circuits corresponding to the following statement forms.

(*a*) $(A \mathbin{\&} \neg B) \vee (B \mathbin{\&} (C \vee \neg A))$

(*b*) $(A \rightarrow B) \vee \neg C$

4.35. Write down statement forms corresponding to the following logic circuits.

(*a*)

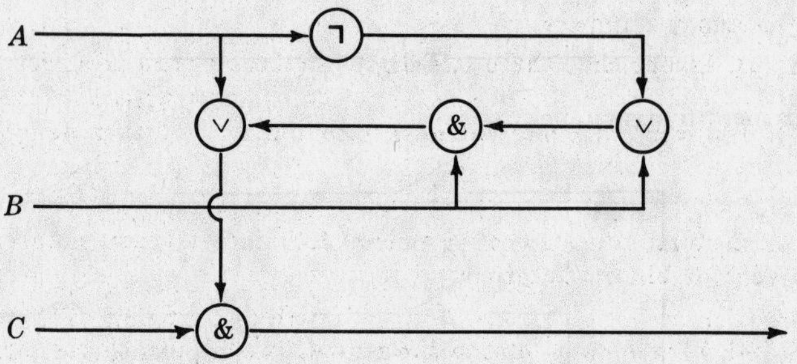

(*b*)

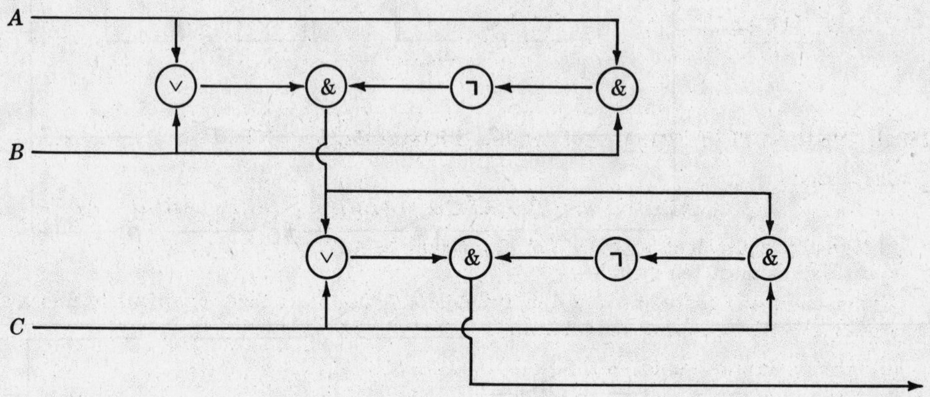

BINARY NUMBER SYSTEM

4.36. Write the binary notation for the following numbers given in decimal notation: 35, 74, 155, 320.

4.37. Write the decimal notation for the following numbers given in binary notation: 10110, 111011, 10001101.

4.38. Write the ternary notation (base 3) for the numbers given in Problem 4.36.

4.39. Write the decimal notation for the following numbers in ternary notation: 12011, 222110, 10110.

4.40. Solve Problem 4.36 for base 6 and base 8, instead of base 2.

4.41. Do the following additions in the binary system (and check by going over to the decimal system).

(*a*)
```
   11101
+   1011
```

(*b*)
```
   11000
+ 101110
```

4.42. Do the following multiplications in the binary system (and check by going over to the decimal system).

(*a*)
```
   11101
×   1011
```

(*b*)
```
   11000
×   1010
```

4.43. (*a*) Let a non-negative integer less than 10 be given in binary notation: $a_3a_2a_1a_0$. Letting A_i stand for "a_i is 1", construct a logic circuit producing the statement that the given integer is a perfect square.

(*b*) Same as (*a*), except that the resulting proposition states that the given integer is even.

(*c*) Same as (*a*), except that the resulting proposition states that the given integer is a perfect cube.

4.44. (*a*) Using half-adders and full adders, draw a logic circuit which carries out the addition of two four-digit binary numbers.

(*b*) Same as (*a*), except that three two-digit numbers are to be added.

4.45. Translate the following decimal integers into binary notation using the method of Problem 4.4: 27, 59, 124.

4.46. Translate the decimal numbers of Problem 4.45 into ternary notation, using a method analogous to the one given for binary notation.

4.47. Perform the following binary subtractions directly and also by the method indicated in Problem 4.6.

(*a*)		(*b*)		(*c*)	
	1100110		1110001		10101
	− 111011		− 1011100		− 11010

MINIMAL DISJUNCTIVE NORMAL FORMS. PRIME IMPLICANTS

4.48. Show that

$$\bar{A}BC\bar{D} \vee ABC\bar{D} \vee \bar{A}\bar{B}\bar{C}D \vee ABCD \vee \bar{A}BCD \vee \bar{A}\bar{B}CD \vee \bar{A}B\bar{C}D$$

is logically equivalent to $BC \vee \bar{A}D$ (cf. Example 4.17).

4.49. Under the same assumptions as in Problem 4.7, use don't care conditions to find a simple switching circuit for the following properties:

(*a*) the given number is odd;

(*b*) the given number is composite (i.e. has a divisor different from 1 and itself).

4.50. Which of the following dnf's are simpler than the dnf $AB\bar{C} \vee A\bar{B} \vee BC\bar{D}$?
(*a*) $A\bar{B} \vee \bar{A}B\bar{C}$; (*b*) $A \vee \bar{B} \vee CD \vee \bar{A}\bar{C}$; (*c*) $ABCD \vee A\bar{B}C$; (*d*) $AB\bar{C}D \vee \bar{A}C \vee \bar{A}\bar{B}$.

4.51. Which of the following are prime implicants of the dnf $AB\bar{C} \vee A\bar{B} \vee BC\bar{D}$?
(*a*) A, (*b*) $A\bar{C}$, (*c*) $AB\bar{D}$, (*d*) $BC\bar{D}$.
(*Note*: There are prime implicants which do not occur in this list.)

4.52. Show that Φ is logically equivalent to $\psi \vee \Phi$ if and only if ψ logically implies Φ.

4.53. Determine whether the fundamental conjunction $\bar{A}B$ is superfluous in
(*a*) $ABC \vee \bar{A}B \vee BC$; (*b*) $AC \vee B\bar{C} \vee \bar{A}B$.

4.54. Determine whether the fundamental conjunction $B\bar{C}$ is superfluous in
(*a*) $AB\bar{C}D \vee \bar{A}BCD \vee \bar{A}\bar{B}CD \vee B\bar{C}$; (*b*) $B\bar{C} \vee AB\bar{D} \vee BD \vee \bar{A}\bar{C}\bar{D}$.

4.55. Show that $\psi \vee \Phi$ is logically equivalent to $\alpha\psi \vee \Phi$ if and only if ψ logically implies $\alpha \vee \Phi$.

4.56. Determine whether the first occurrence of the literal $\bar{C}$ is superfluous in
(*a*) $\bar{A}B\bar{C} \vee A\bar{B} \vee C\bar{A}$; (*b*) $AC \vee B\bar{C} \vee A\bar{C}$.

4.57. Determine whether the first occurrence of the literal B is superfluous in
(*a*) $A\bar{B}C \vee BC \vee \bar{A}\bar{C}$; (*b*) $\bar{A}\bar{B}C \vee BC \vee AC$.

4.58. Find irredundant dnf's logically equivalent to

(*a*) $ABCD \vee \bar{A}BD \vee \bar{C}\bar{D} \vee A\bar{B}D \vee \bar{A}B\bar{C}$; (*b*) $AB\bar{D} \vee A\bar{C} \vee A\bar{B}D \vee \bar{B}\bar{C}\bar{D} \vee BD$.

4.59. Prove that a full dnf containing n letters is not a tautology if and only if it has fewer than 2^n disjuncts.

4.60. Find full dnf's logically equivalent to (*a*) $ABD \vee \bar{A}B\bar{C}$; (*b*) $\bar{A} \vee BC \vee AC$.

4.61. Carry out the proof of Theorem 4.4, using Lemmas 4.2-4.3.

4.62. Show that if $\psi\alpha$ logically implies Φ and $\psi\bar{\alpha}$ logically implies Φ, then ψ logically implies Φ.

4.63. Find all prime implicants of the following statement forms, using the Quine-McCluskey method.

(*a*) $(AB \leftrightarrow C) \,\&\, A\bar{C}$

(*b*) $AB\bar{C} \vee \bar{A}BC \vee \bar{A}\bar{B}\bar{C} \vee A\bar{B}C$

(*c*) $ABCD \vee \bar{A}\bar{B}\bar{C}\bar{D} \vee \bar{A}B\bar{C}D \vee \bar{A}\bar{B}C\bar{D} \vee A\bar{B}CD$

(*d*) $AB\bar{C}D\bar{E} \vee A\bar{B}CDE \vee AB\bar{C}DE \vee \bar{A}B\bar{C}DE \vee \bar{A}\bar{B}CD\bar{E} \vee \bar{A}\bar{B}C\bar{D}\bar{E}$

4.64. Do the irredundant dnf's of Problem 4.58 contain all their prime implicants?

4.65. Prove that a statement form is logically equivalent to the disjunction of all its prime implicants.

4.66. Find all the prime implicants of the dnf in (*a*) Problem 4.11, (*b*) Problem 4.51.

4.67. Construct the prime implicant tables for the dnf's in Problem 4.63*b, c, d*.

4.68. Find the minimal dnf's for the dnf's in Problem 4.63*b, c, d*, using the Quine-McCluskey method and prime implicant tables.

4.69. Find the minimal dnf's for the following dnf's, using the Quine-McCluskey method and prime implicant tables.

(*a*) $ABCDE \vee AB\bar{C}DE \vee A\bar{B}CDE \vee A\bar{B}C\bar{D}E \vee AB\bar{C}\bar{D}E \vee \bar{A}\bar{B}CDE$
$\vee \bar{A}B\bar{C}D\bar{E} \vee A\bar{B}\bar{C}\bar{D}E \vee \bar{A}\bar{B}\bar{C}\bar{D}E \vee A\bar{B}\bar{C}\bar{D}\bar{E}$

(*b*) $ABCD \vee AB\bar{C}D \vee A\bar{B}CD \vee \bar{A}\bar{B}CD \vee \bar{A}B\bar{C}D \vee \bar{A}BC\bar{D} \vee \bar{A}\bar{B}\bar{C}D \vee A\bar{B}\bar{C}\bar{D} \vee \bar{A}\bar{B}C\bar{D}$

(*c*) $ABCDEG \vee \bar{A}BCDEG \vee AB\bar{C}DEG \vee ABCD\bar{E}G \vee A\bar{B}\bar{C}DEG \vee \bar{A}BC\bar{D}EG \vee A\bar{B}CD\bar{E}G$
$\vee AB\bar{C}DE\bar{G} \vee \bar{A}\bar{B}\bar{C}DEG \vee \bar{A}B\bar{C}D\bar{E}G \vee A\bar{B}\bar{C}\bar{D}EG \vee A\bar{B}CD\bar{E}\bar{G} \vee \bar{A}\bar{B}C\bar{D}\bar{E}G$
$\vee A\bar{B}\bar{C}\bar{D}E\bar{G} \vee \bar{A}\bar{B}CD\bar{E}\bar{G} \vee A\bar{B}\bar{C}\bar{D}\bar{E}\bar{G} \vee \bar{A}B\bar{C}\bar{D}\bar{E}\bar{G}$

(*d*) $\bar{A}\bar{B}\bar{C}\bar{D}\bar{E} \vee \bar{A}\bar{B}C\bar{D}\bar{E} \vee \bar{A}\bar{B}CD\bar{E} \vee \bar{A}\bar{B}\bar{C}\bar{D}E \vee \bar{A}\bar{B}\bar{C}DE \vee \bar{A}\bar{B}CDE$
$\vee A\bar{B}CDE \vee ABCDE \vee AB\bar{C}\bar{D}E \vee AB\bar{C}DE$

4.70. Are the irredundant dnf's of Problem 4.58*a, b* minimal dnf's?

4.71. Give a full argument showing that, if Φ is a full dnf and, in the prime implicant table for Φ, every column contains an entry from a row corresponding to a fundamental conjunction in the core, then the disjunction of the members of the core is the unique minimal dnf for Φ (cf. Examples 4.25-4.26).

4.72. Verify the assertion in Example 4.29 that members of the secondary core must be a disjunct of every minimal dnf.

4.73. Find minimal dnf's for the following dnf's with don't care conditions. Use the Quine-McCluskey method and prime implicant tables.

(*a*) $ABCDE \vee A\bar{B}CDE \vee ABC\bar{D}\bar{E} \vee AB\bar{C}\bar{D}\bar{E} \vee \bar{A}\bar{B}CDE \vee \bar{A}B\bar{C}\bar{D}\bar{E}$, with don't care conditions $AB\bar{C}D\bar{E}$, $\bar{A}\bar{B}CD\bar{E}$, $A\bar{B}C\bar{D}\bar{E}$, $\bar{A}B\bar{C}D\bar{E}$, $\bar{A}BC\bar{D}\bar{E}$, $A\bar{B}\bar{C}\bar{D}\bar{E}$.

(*b*) $\bar{A}BCDE \vee \bar{A}B\bar{C}DE \vee \bar{A}B\bar{C}D\bar{E} \vee \bar{A}\bar{B}\bar{C}D\bar{E} \vee \bar{A}\bar{B}\bar{C}\bar{D}\bar{E} \vee A\bar{B}\bar{C}\bar{D}\bar{E}$, with don't care conditions $\bar{A}\bar{B}\bar{C}DE$, $A\bar{B}\bar{C}\bar{D}E$, $A\bar{B}\bar{C}D\bar{E}$, $AB\bar{C}\bar{D}\bar{E}$, $\bar{A}BC\bar{D}\bar{E}$, $\bar{A}\bar{B}\bar{C}\bar{D}\bar{E}$, $AB\bar{C}D\bar{E}$.

(*c*) $\bar{A}\bar{B}\bar{C}D\bar{E}\bar{G} \vee A\bar{B}C\bar{D}E\bar{G} \vee \bar{A}BCDEG \vee ABCDE\bar{G} \vee ABCD\bar{E}\bar{G} \vee ABC\bar{D}\bar{E}\bar{G} \vee A\bar{B}C\bar{D}\bar{E}\bar{G} \vee \bar{A}\bar{B}\bar{C}\bar{D}\bar{E}\bar{G} \vee ABCDEG$, with don't care conditions $\bar{A}\bar{B}CDEG$, $\bar{A}\bar{B}\bar{C}\bar{D}\bar{E}G$, $\bar{A}\bar{B}\bar{C}DE\bar{G}$, $\bar{A}B\bar{C}D\bar{E}\bar{G}$, $ABC\bar{D}\bar{E}G$, $AB\bar{C}\bar{D}\bar{E}G$, $\bar{A}\bar{B}CDE\bar{G}$, $AB\bar{C}\bar{D}E\bar{G}$, $AB\bar{C}\bar{D}\bar{E}\bar{G}$.

4.74. Show that if one fundamental conjunction ψ_1 includes another ψ_2, then ψ_2 is logically equivalent to $\psi_1 \vee \psi_2$.

4.75. By the consensus method, find all prime implicants of the following dnf's:

(*a*) $ABC \vee AB\bar{C}D \vee \bar{A}\bar{B}C \vee B\bar{C}\bar{D}$; (*b*) $ABCD \vee \bar{A}\bar{B} \vee B\bar{C}D \vee A\bar{B}\bar{C}\bar{D} \vee \bar{A}B\bar{C}\bar{D}$.

4.76. By the consensus method, find all prime implicants of the dnf's in:

(*a*) Problem 4.11, (*b*) Problem 4.13, (*c*) Problem 4.63*b*, (*d*) Problem 4.69*a*, *c*.

4.77. Check the solution to Problem 4.18 by expanding the original dnf into a full dnf and using the Quine-McCluskey method.

4.78. Apply the consensus method to find all minimal dnf's for:

(*a*) $AE \vee BCE \vee \bar{A}BC\bar{D} \vee \bar{A}\bar{B}\bar{C}\bar{D}E$; (*b*) $ABC \vee B\bar{C}\bar{D} \vee ACD \vee \bar{A}\bar{B}\bar{D}$; (*c*) $A\bar{C} \vee BC \vee \bar{B}\bar{D} \vee BD$.

Check your results by using the Quine-McCluskey method.

4.79. Apply the consensus method to find all minimal dnf's for:

(*a*) dnf in Problem 4.13; (*b*) dnf's in Problem 4.69*a*, *b*, *c*, *d*; (*c*) dnf's in Problem 4.75*a*, *b*.

KARNAUGH MAPS

4.80. Using Karnaugh maps, find the minimal dnf's for:

(*a*) $A\bar{B}C \vee AB\bar{C} \vee \bar{A}\bar{B}C \vee \bar{A}BC \vee \bar{A}B\bar{C}$

(*b*) $ABCD \vee A\bar{B}C\bar{D} \vee A\bar{B}CD \vee \bar{A}\bar{B}\bar{C}\bar{D} \vee \bar{A}BCD \vee AB\bar{C}D \vee \bar{A}B\bar{C}D$

(*c*) $A\bar{B}CD \vee A\bar{B}C\bar{D} \vee AB\bar{C}D \vee \bar{A}\bar{B}\bar{C}\bar{D} \vee \bar{A}BCD \vee \bar{A}B\bar{C}\bar{D} \vee \bar{A}B\bar{C}D$

(*d*) $ABCD \vee ABC\bar{D} \vee \bar{A}\bar{B}C\bar{D} \vee A\bar{B}\bar{C}\bar{D} \vee A\bar{B}\bar{C}D \vee \bar{A}\bar{B}CD \vee \bar{A}\bar{B}\bar{C}\bar{D}$
$\vee \bar{A}BCD \vee \bar{A}B\bar{C}\bar{D} \vee \bar{A}B\bar{C}D \vee \bar{A}\bar{B}\bar{C}D$

(*e*) $\bar{A}\bar{B}CDE \vee \bar{A}BCDE \vee ABC\bar{D}E \vee A\bar{B}C\bar{D}E \vee \bar{A}\bar{B}C\bar{D}E \vee A\bar{B}\bar{C}\bar{D}E \vee \bar{A}\bar{B}\bar{C}\bar{D}E \vee AB\bar{C}DE$
$\vee A\bar{B}\bar{C}DE \vee ABCD\bar{E} \vee \bar{A}BC\bar{D}\bar{E} \vee ABC\bar{D}\bar{E} \vee A\bar{B}C\bar{D}\bar{E}$
$\vee \bar{A}\bar{B}\bar{C}\bar{D}\bar{E} \vee \bar{A}B\bar{C}\bar{D}\bar{E} \vee AB\bar{C}D\bar{E}$

(*f*) $AB \vee \bar{A}CDE \vee \bar{A}B\bar{D} \vee \bar{C}D\bar{E} \vee A\bar{B}\bar{D}$

4.81. Solve Problem 4.48 by means of a Karnaugh map.

4.82. By use of Karnaugh maps, solve Problems 4.69*a*, *b*, *c*, *d* and 4.78*a*, *b*.

4.83. Using a Karnaugh map, find all minimal dnf's for

$$ABCDE \vee A\bar{B}CDE \vee \bar{A}BCDE \vee \bar{A}BC\bar{D}E \vee AB\bar{C}\bar{D}E \vee \bar{A}\bar{B}\bar{C}\bar{D}E \vee AB\bar{C}DE$$
$$\vee \bar{A}\bar{B}\bar{C}DE \vee \bar{A}B\bar{C}DE \vee ABCD\bar{E} \vee \bar{A}\bar{B}CD\bar{E} \vee \bar{A}BCD\bar{E} \vee ABC\bar{D}\bar{E}$$
$$\vee \bar{A}B\bar{C}\bar{D}\bar{E} \vee AB\bar{C}D\bar{E} \vee A\bar{B}\bar{C}D\bar{E} \vee \bar{A}B\bar{C}D\bar{E}$$

(Be sure that you are not using superfluous disjuncts.)

4.84. Draw a Karnaugh map for six statement letters, and try to use it to solve Problem 4.69*c* and Problem 4.16.

4.85. Using Karnaugh maps, find all minimal dnf's for the given statement forms, with the indicated don't care conditions.

(*a*) $ABC \vee \bar{A}BD \vee \bar{A}\bar{C}D \vee A\bar{B}C\bar{D}$, with the don't care conditions $AB\bar{C}\bar{D}$, $\bar{A}\bar{B}C\bar{D}$, $A\bar{B}CD$.

(*b*) $AB\bar{C}D\bar{E} \vee \bar{A}BCDE \vee A\bar{B}\bar{C}DE \vee \bar{A}BC\bar{D}E \vee \bar{A}\bar{B}\bar{C}D\bar{E} \vee ABCD\bar{E} \vee AB\bar{C}D\bar{E}$, with the don't care conditions $AB\bar{C}\bar{D}$, $\bar{A}\bar{B}CDE$.

(*c*) $ABC \vee \bar{A}BC\bar{D} \vee \bar{A}\bar{C}$, with the don't care conditions $A\bar{B}\bar{C}D$, $\bar{A}\bar{B}\bar{C}\bar{D}$, $\bar{A}\bar{B}CD$, $AB\bar{C}\bar{D}$.

4.86. Find all minimal dnf's or cnf's for the statement forms in

(*a*) Problems 4.13, 4.18, 4.19, 4.21, 4.22.

(*b*) Problems 4.63*b, c, d*, 4.69*a, b, c, d*, 4.78*a, b, c*, 4.80*a-f*.

4.87. Find all minimal dnf's or cnf's for the following statement forms, with the indicated don't care conditions.

(*a*) $A\bar{B}C\bar{D} \vee \bar{A}\bar{B}C\bar{D} \vee A\bar{B}\bar{C}\bar{D}$, with the don't care conditions $A\bar{B}CD$, $\bar{A}\bar{B}CD$, $A\bar{B}\bar{C}D$, $\bar{A}\bar{B}\bar{C}D$, $ABC\bar{D}$, $\bar{A}BC\bar{D}$, $AB\bar{C}D$, $\bar{A}B\bar{C}\bar{D}$, as in Problem 4.23.

(*b*) Same as Problem 4.24.

(*c*) Same as Problem 4.16.

(*d*) Same as Problem 4.14.

(*e*) Same as Problem 4.73*a, b, c*.

Chapter 5

Topics in the Theory of Boolean Algebras

5.1 LATTICES

A *lattice* is an ordered pair $\langle L, \leqq \rangle$ consisting of a non-empty set L together with a partial order $\leqq$ on L satisfying:

(L4) For any x and y in L, the set $\{x, y\}$ has both a least upper bound (lub) and a greatest lower bound (glb).

We have seen in Chapter 3 (Theorem 3.9) that a Boolean algebra $\mathcal{B}$ determines a lattice $\langle B, \leqq \rangle$, with $x \vee y$ and $x \wedge y$ as the lub and glb respectively. Therefore it should cause no confusion if, for any lattice $\langle L, \leqq \rangle$ and for any x and y in L, we use $x \vee y$ and $x \wedge y$ to denote the lub and glb of $\{x, y\}$, respectively.

Example 5.1.

The set $\{a, b, c, d, e, f\}$ is not a lattice with respect to the partial order pictured in Fig. 5-1. For, $\{a, b\}$ has no lub.

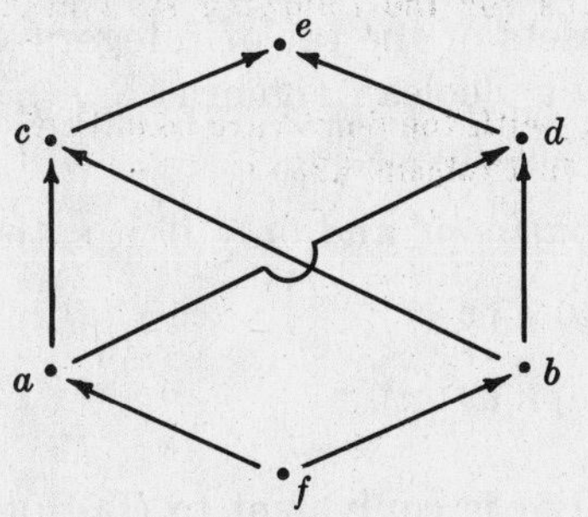

Fig. 5-1

Theorem 5.1. For any elements x, y, z of a lattice $\langle L, \leqq \rangle$:

(*a*) $x \wedge x = x$ and $x \vee x = x$ (Idempotence)

(*b*) $x \wedge y = y \wedge x$ and $x \vee y = y \vee x$ (Commutativity)

(*c*) $x \wedge (y \wedge z) = (x \wedge y) \wedge z$ and $x \vee (y \vee z) = (x \vee y) \vee z$ (Associativity)

(*d*) $x \wedge (x \vee y) = x$ and $x \vee (x \wedge y) = x$ (Absorption)

(*e*) $x \leqq y \leftrightarrow x \wedge y = x$ and $x \leqq y \leftrightarrow x \vee y = y$

(*f*) $x \leqq y \rightarrow (x \wedge z \leqq y \wedge z \;\&\; x \vee z \leqq y \vee z)$

Proof. (*a*) and (*b*) follow directly from the definition of lub and glb.

(*c*) First, notice that $x \wedge (y \wedge z) \leqq y \wedge z \leqq z$. Also, $x \wedge (y \wedge z) \leqq y \wedge z \leqq y$ and $x \wedge (y \wedge z) \leqq x$. Therefore by definition of glb, $x \wedge (y \wedge z) \leqq x \wedge y$. Thus since we have $x \wedge (y \wedge z) \leqq z$ and $x \wedge (y \wedge z) \leqq x \wedge y$, it follows by the definition of glb that $x \wedge (y \wedge z) \leqq (x \wedge y) \wedge z$.

Using this result twice, we have $(x \wedge y) \wedge z = z \wedge (x \wedge y) \leqq (z \wedge x) \wedge y = y \wedge (z \wedge x) \leqq (y \wedge z) \wedge x = x \wedge (y \wedge z)$. Thus $x \wedge (y \wedge z) \leqq (x \wedge y) \wedge z$ and $(x \wedge y) \wedge z \leqq x \wedge (y \wedge z)$, and therefore by (PO 3), $x \wedge (y \wedge z) = (x \wedge y) \wedge z$. A similar proof shows that

$$x \vee (y \vee z) = (x \vee y) \vee z$$

(*d*) Clearly, $x \leqq x$ and $x \leqq x \vee y$. Hence by definition of glb, $x \leqq x \wedge (x \vee y)$. But $x \wedge (x \vee y) \leqq x$. Therefore by (PO 3), $x \wedge (x \vee y) = x$. Similarly, $x \vee (x \wedge y) = x$.

(*e*) First, if $x \wedge y = x$, then $x = x \wedge y \leqq y$. Conversely, if $x \leqq y$, then, by definition of glb, $x \wedge y = x$. Similarly, $x \leqq y \leftrightarrow x \vee y = y$.

(*f*) Assume $x \leqq y$. Then

$$\begin{aligned}(x \wedge z) \wedge (y \wedge z) &= (x \wedge y) \wedge z \quad \text{(by (a), (b), (c))}\\ &= x \wedge z \quad \text{(by (e))}\end{aligned}$$

Therefore $x \wedge z \leqq y \wedge z$, by (*e*). Analogously, $(x \vee z) \vee (y \vee z) = (x \vee y) \vee z = y \vee z$, and therefore $x \vee z \leqq y \vee z$. ▶

By a *unit* 1 of a lattice $\langle L, \leqq \rangle$ we mean an upper bound of the whole set L. It is clear that, if a unit exists it is unique. By a *zero* 0 of $\langle L, \leqq \rangle$ we mean a lower bound of L, and clearly, if a zero exists it is unique. Obviously we have $0 \wedge x = 0$, $0 \vee x = x$, $x \vee 1 = 1$, $x \wedge 1 = x$ for all x in the lattice.

A lattice may lack a unit. For example, the set of all finite subsets of the set of integers, with respect to the partial order $\subseteq$, is a lattice without a unit. A lattice may lack a zero, e.g. the lattice of all cofinite subsets of the set of integers with respect to the partial order $\subseteq$. In the lattice determined by a Boolean algebra $\langle B, \wedge_{\mathcal{B}}, \vee_{\mathcal{B}}, '^{\mathcal{B}}, 0_{\mathcal{B}}, 1_{\mathcal{B}} \rangle$, $1_{\mathcal{B}}$ is the unit of the lattice and $0_{\mathcal{B}}$ is the zero of the lattice.

A lattice is said to be *distributive* if and only if it satisfies the following two laws:

(L5) $x \wedge (y \vee z) = (x \wedge y) \vee (x \wedge z)$;

(L6) $x \vee (y \wedge z) = (x \vee y) \wedge (x \vee z)$.

Theorem 5.2. In any lattice, (L5) is equivalent to (L6) (and therefore in the definition of distributive lattice it suffices to assume either (L5) or (L6)).

Proof. Assume (L5). Then

$$\begin{aligned}(x \vee y) \wedge (x \vee z) &= [(x \vee y) \wedge x] \vee [(x \vee y) \wedge z] = x \vee [(x \wedge z) \vee (y \wedge z)]\\ &= [x \vee (x \wedge z)] \vee (y \wedge z) = x \vee (y \wedge z)\end{aligned}$$

Therefore (L6) holds. The proof of (L5) from (L6) is similar and is left to the reader. ▶

The lattice determined by a Boolean algebra is distributive. (L5) and (L6) are simply Axioms (3) and (4) for Boolean algebras.

Example 5.2.

The lattice shown in Fig. 5-2 is not distributive. For,

$$d \wedge (b \vee c) = d \wedge 1 = d$$

while

$$(d \wedge b) \vee (d \wedge c) = 0 \vee 0 = 0$$

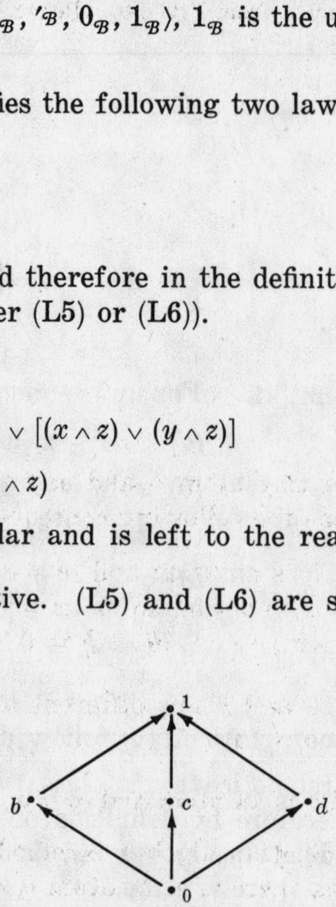

Fig. 5-2

A lattice with zero 0 and unit 1 is said to be *complemented* if, for any x in the lattice, there exists an *inverse* x' in the lattice such that $x \wedge x' = 0$ and $x \vee x' = 1$. Obviously the lattice determined by a Boolean algebra is complemented.

If a distributive lattice with zero and unit is complemented, then, for any x, the inverse x' is unique. To see this, note that the proof of Theorem 3.1 (uniqueness of complements in Boolean algebras) still is valid under the given assumption.

Theorem 5.3. A complemented distributive lattice $\langle L, \leqq \rangle$ with $0 \neq 1$ determines a Boolean algebra $\langle L, \wedge, \vee, ', 0, 1 \rangle$.

Proof. Axioms (1)-(2) were proved in Theorem 5.1(*b*). Axioms (3)-(4) are just the distributive laws. Axioms (5)-(6) have already been treated above. Axioms (7)-(8) follow from the fact that the lattice is complemented, and Axiom (9) is part of our hypothesis. ▶

5.2 ATOMS

A nonzero element b of a Boolean algebra is said to be an *atom* if and only if, for all elements x of the Boolean algebra, the condition $x \leqq b$ implies that $x = b$ or $x = 0$.

Example 5.3.

In the Boolean algebra $\mathcal{P}(A)$ of all subsets of a non-empty set A, the atoms are the singletons $\{x\}$, i.e. the sets consisting of a single element.

Example 5.4.

The Boolean algebra of all positive integral divisors of 70 (cf. Example 3.4) has as its atoms $\{2, 5, 7\}$, as is evident from Fig. 5-3. (Remember that the integer 1 is the zero element.)

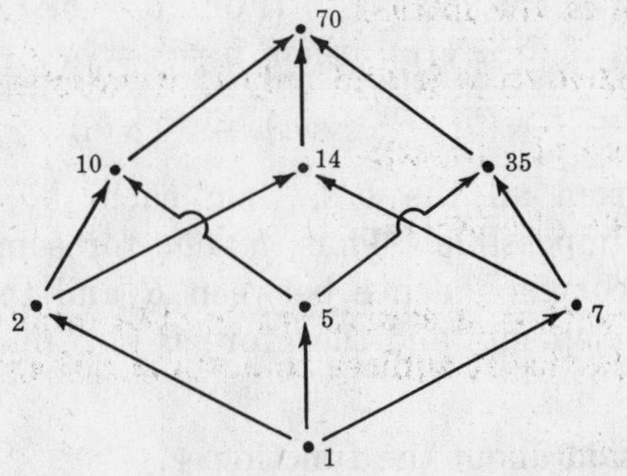

Fig. 5-3

For any atom b and any element x, either $b \wedge x = b$ or $b \wedge x = 0$ (since $b \wedge x \leqq b$). This has the following consequences:

(i) If b is an atom and $b \leqq x_1 \vee \cdots \vee x_n$, then $b \leqq x_i$ for some i. (For, if $b \not\leqq x_i$, then $b \wedge x_i \neq b$ and so $b \wedge x_i = 0$. Hence if $b \not\leqq x_i$ for all i, then $b = b \wedge (x_1 \vee \cdots \vee x_n) = (b \wedge x_1) \vee \cdots \vee (b \wedge x_n) = 0 \vee \cdots \vee 0 = 0$, which is a contradiction.)

(ii) If b and c are different atoms, then $b \wedge c = 0$. (For, if $b \wedge c \neq 0$, then $b = b \wedge c = c \wedge b = c$.)

(iii) If b is an atom and $b \not\leqq x$, then $b \leqq x'$. (For, $b \leqq 1 = x \vee x'$, and we then use (i).)

A Boolean algebra is called *atomic* if and only if, for every nonzero element x of the algebra, there is some atom b such that $b \leqq x$. The Boolean algebras of Examples 5.3-5.4 are both atomic.

Theorem 5.4. Every finite Boolean algebra is atomic.

Proof. Given a nonzero element x_0 of the algebra. Assume there is no atom b such that $b \leq x_0$. In particular, x_0 is not an atom and therefore there is some nonzero element x_1 such that $x_1 \leq x_0$ and $x_1 \neq x_0$, i.e. $0 < x_1 < x_0$. x_1 cannot be an atom; hence there is some nonzero element x_2 such that $x_2 < x_1$. Continuing in this way, we obtain a sequence $x_0, x_1, x_2, \ldots$ such that $x_0 > x_1 > x_2 > \cdots$.† All the terms of this sequence are distinct (by Theorem 3.8), contradicting the fact that there are only a finite number of elements in the algebra. ▶

Given an element x of a Boolean algebra $\mathcal{B}$, we define $\Psi(x)$ to be the set of all atoms b of $\mathcal{B}$ such that $b \leq x$. Clearly, $\Psi(0) = \emptyset$ and $\Psi(1)$ is the set A of all atoms of $\mathcal{B}$.

Lemma 5.5. In an atomic Boolean algebra $\mathcal{B}$, the function Ψ is one-one, i.e. if $x \neq y$, then $\Psi(x) \neq \Psi(y)$.

Proof. Assume $x \neq y$. Then $x \nleq y$ or $y \nleq x$; say, $x \nleq y$. Hence $x \wedge y' \neq 0$. Since $\mathcal{B}$ is atomic, there is an atom $b \leq x \wedge y'$. Then $b \leq x$ and so $b \in \Psi(x)$. However, $b \leq y'$ and therefore $b \notin \Psi(y)$. (For, if $b \leq y$, then $b \leq y \wedge y' = 0$, and b would have to be 0.) Hence $\Psi(x) \neq \Psi(y)$. ▶

Theorem 5.6. Every finite Boolean algebra $\mathcal{B}$ has 2^n elements, where the positive integer n is the number of elements in the set A of atoms of $\mathcal{B}$.

Proof. By Theorem 5.4, $\mathcal{B} = \langle B, \wedge, \vee, ', 0, 1\rangle$ is atomic, and, by Lemma 5.5, Ψ is a one-one function from B into the set $\mathcal{P}(a)$ of all subsets of A. Now let C be any subset of A. Since $\mathcal{B}$ is finite, so is A and therefore also C. Thus $C = \{b_1, \ldots, b_k\}$. Let $x = b_1 \vee \cdots \vee b_k$. Then $\Psi(x) = \{b_1, \ldots, b_k\} = C$. (For, $b_i \leq b_1 \vee \cdots \vee b_k = x$ for all i. Thus $C \subseteq \Psi(x)$. On the other hand, if $b \in \Psi(x)$, then $b \leq x = b_1 \vee \cdots \vee b_k$. Hence

$$b = b \wedge x = b \wedge (b_1 \vee \cdots \vee b_k) = (b \wedge b_1) \vee \cdots \vee (b \wedge b_k)$$

Now if b were different from all the b_i's, then each $b \wedge b_i = 0$ and we would have $b = 0 \vee \cdots \vee 0 = 0$ which is impossible. Thus $b = b_i$ for some i, i.e. $\Psi(x) \subseteq C$.) We have proved that Ψ is a one-one correspondence between B and the whole set $\mathcal{P}(A)$. Since A has n elements, $\mathcal{P}(A)$ has 2^n elements and therefore B also must have 2^n elements. ▶

Something more can be said about the function Ψ.

Theorem 5.7. If $\mathcal{B}$ is an atomic Boolean algebra, then the function Ψ is an isomorphism from $\mathcal{B}$ into the Boolean algebra $\mathcal{P}(A)$. If $\mathcal{B}$ is a finite Boolean algebra, then Ψ is an isomorphism from $\mathcal{B}$ onto $\mathcal{P}(A)$.

Proof. Remember that A is the set of atoms of $\mathcal{B}$, and $\Psi(x) = \{b : b \in A \;\&\; b \leq x\}$. We already know from Lemma 5.5 that Ψ is one-one. Next, we show that $\Psi(x') = \overline{\Psi(x)}$. For, on the one hand, if b is an atom and $b \leq x'$, then $b \nleq x$. Thus $\Psi(x') \subseteq \overline{\Psi(x)}$. On the other hand, if b is an atom and $b \nleq x$, then $b \leq x'$. Thus $\overline{\Psi(x)} \subseteq \Psi(x')$. Hence $\Psi(x') = \overline{\Psi(x)}$. Finally, we shall show that $\Psi(x \wedge y) = \Psi(x) \cap \Psi(y)$. On the one hand, if an atom $b \leq x \wedge y$, then $b \leq x$ and $b \leq y$. Thus $\Psi(x \wedge y) \subseteq \Psi(x) \cap \Psi(y)$. On the other hand, if b is an atom and $b \leq x \;\&\; b \leq y$, then $b \leq x \wedge y$. Thus $\Psi(x) \cap \Psi(y) \subseteq \Psi(x \wedge y)$. Hence $\Psi(x \wedge y) = \Psi(x) \cap \Psi(y)$. Hence Ψ is an isomorphism from B into $\mathcal{P}(A)$. When B is finite, the proof of

†We use the usual conventions: $x \geq y$ means $y \leq x$; $x > y$ means $y < x$; $x \nleq y$ means $\neg(x \leq y)$; $x \nless y$ means $\neg(x < y)$.

Theorem 5.6 shows that the range of Ψ is all of $\mathcal{P}(A)$. ▶

Corollary 5.8. Any two finite Boolean algebras with the same number of elements are isomorphic.

Proof. By the second part of Theorem 5.7 and Theorem 5.6, it suffices to show that, if A and C are finite sets with the same number of elements, then the Boolean algebras $\mathcal{P}(A)$ and $\mathcal{P}(C)$ are isomorphic. Let $A = \{a_1, \ldots, a_n\}$ and $C = \{c_1, \ldots, c_n\}$. Define the function Θ from $\mathcal{P}(A)$ into $\mathcal{P}(C)$ as follows: for any subset $\{a_{j1}, \ldots, a_{jk}\}$ of A, let $\Theta(\{a_{j1}, \ldots, a_{jk}\}) = \{c_{j1}, \ldots, c_{jk}\}$. It is obvious that Θ is the required isomorphism. ▶

The second part of Theorem 5.7 shows that any finite Boolean algebra is isomorphic with a Boolean algebra of *all* subsets of a set A. This turns out not to be true for arbitrary infinite Boolean algebras, although, as we shall see later, any Boolean algebra is isomorphic with a field of sets (i.e. to a subalgebra of the Boolean algebra of all subsets of a set).

5.3 SYMMETRIC DIFFERENCE. BOOLEAN RINGS

In a Boolean algebra $\mathcal{B}$ we define the operation $+$ of *symmetric difference* as follows:

$$x + y = (x \wedge y') \vee (x' \wedge y)$$

In the Boolean algebra $\mathcal{P}(A)$ of all subsets of a non-empty set A, $x + y = x \Delta y$ (cf. Section 2.6). In the Boolean algebra of statement bundles (cf. Example 3.5), $[\mathbf{A}] + [\mathbf{B}] = [\mathbf{A} + \mathbf{B}]$, where the second $+$ is the exclusive-or connective.

Theorem 5.9. The operation $+$ has the following properties.

(*a*) $x + y = y + x$

(*b*) $x + 0 = x$

(*c*) $x + x' = 1$

(*d*) $x + (y + z) = (x + y) + z$

(*e*) $x \wedge (y + z) = (x \wedge y) + (x \wedge z)$

(*f*) $x + x = 0$

(*g*) $x + y = x + z \rightarrow y = z$

(*h*) $1 + x = x'$

(*i*) $x + y = z \rightarrow y = x + z$

(*j*) $x = z \leftrightarrow x + z = 0$

Proof.

(*a*) $x + y = (x \wedge y') \vee (x' \wedge y)$

$$y + x = (y \wedge x') \vee (y' \wedge x) = (y' \wedge x) \vee (y \wedge x') = (x \wedge y') \vee (x' \wedge y)$$

(*b*) $x + 0 = (x \wedge 0') \vee (x' \wedge 0) = (x \wedge 1) \vee 0 = x \wedge 1 = x$

(*c*) $x + x' = (x \wedge (x')') \vee (x' \wedge x') = (x \wedge x) \vee x' = x \vee x' = 1$

(*d*)
$$\begin{aligned} x + (y + z) &= x + ((y \wedge z') \vee (y' \wedge z)) \\ &= (x \wedge [(y \wedge z') \vee (y' \wedge z)]') \vee (x' \wedge [(y \wedge z') \vee (y' \wedge z)]) \\ &= [x \wedge (y' \vee z) \wedge (y \vee z')] \vee [(x' \wedge y \wedge z') \vee (x' \wedge y' \wedge z)] \\ &= [x \wedge ((y' \wedge z') \vee (y \wedge z))] \vee [(x' \wedge y \wedge z') \vee (x' \wedge y' \wedge z)] \\ &= (x \wedge y' \wedge z') \vee (x \wedge y \wedge z) \vee (x' \wedge y \wedge z') \vee (x' \wedge y' \wedge z) \end{aligned}$$

On the other hand, $(x+y)+z = z+(x+y)$. But, to calculate $z+(x+y)$, we use the equation just found for $x+(y+z)$ after substituting z for x, x for y, and y for z. We obtain

$$\begin{aligned} z+(x+y) &= (z \wedge x' \wedge y') \vee (z \wedge x \wedge y) \vee (z' \wedge x \wedge y') \vee (z' \wedge x' \wedge y) \\ &= (x \wedge y' \wedge z') \vee (x \wedge y \wedge z) \vee (x' \wedge y \wedge z') \vee (x' \wedge y' \wedge z) \\ &= x+(y+z) \end{aligned}$$

(*e*) $x \wedge (y+z) = x \wedge ((y \wedge z') \vee (y' \wedge z)) = (x \wedge y \wedge z') \vee (x \wedge y' \wedge z)$.

$$\begin{aligned} (x \wedge y)+(x \wedge z) &= [(x \wedge y) \vee (x \wedge z)'] \vee [(x \wedge y)' \wedge (x \wedge z)] \\ &= [(x \wedge y) \wedge (x' \vee z')] \vee [(x' \vee y') \wedge (x \wedge z)] \\ &= [y \wedge (x \wedge (x' \vee z'))] \vee [((x' \vee y') \wedge x) \wedge z)] \\ &= [y \wedge x \wedge z'] \vee [x \wedge y' \wedge z] = (x \wedge y \wedge z') \vee (x \wedge y' \wedge z) \end{aligned}$$

(*f*) $x+x = (x \wedge x') \vee (x' \wedge x) = 0 \vee 0 = 0$.

(*g*) Assume $x+y = x+z$. Then $x+(x+y) = x+(x+z)$. Hence

$$(x+x)+y = (x+x)+z, \qquad 0+y = 0+z, \qquad y = z$$

(*h*) Add x to both sides of (*c*) and use (*f*).

(*i*) Add x to both sides of $x+y = z$ and use (*f*).

(*j*) Taking $y=0$ in (*i*) yields $x=z \rightarrow x+z=0$. If we exchange y and z in (*i*) and set $y=0$, we obtain $x+z=0 \rightarrow x=z$. ▶

By a *ring*, we mean a structure $\mathcal{R} = \langle R, +, \times, 0\rangle$, where R is a set, $+$ and $\times$ are binary operations on R, and 0 is an element of R, such that

(1) $(x+y)+z = x+(y+z)$;

(2) $x+y = y+x$;

(3) $x+0 = x$;

(4) for any x, there is a unique element $(-x)$ such that $x+(-x) = 0$;

(5) $(x \times y) \times z = x \times (y \times z)$;

(6) $x \times (y+z) = (x \times y)+(x \times z)$;

(7) $(y+z) \times x = (y \times x)+(z \times x)$.

A ring $\mathcal{R}$ is said to be *commutative* if and only if, in addition,

(8) $x \times y = y \times x$.

A ring $\mathcal{R}$ is said to have a *unit element* if and only if there is an element 1 in R such that

(9) $x \times 1 = 1 \times x = x$.

(Clearly, there cannot be another unit element u, for we would then have $u = 1 \times u = 1$.)

In Theorem 5.9 we have already verified that a Boolean algebra determines the commutative ring $\langle B, +, \wedge, 0\rangle$ with unit element 1. This enables us to apply the highly developed algebraic theory of rings to the study of Boolean algebras. But we also can give a more precise characterization of Boolean algebras in terms of rings, in the following way.

A ring $\mathcal{R} = \langle R, +, \times, 0\rangle$ is said to be a *Boolean ring* if and only if it satisfies the identity

$$x^2 = x$$

for all x. (Here we employ the usual abbreviation: $x^2 = x \times x$.)

Theorem 5.10. Let $\mathcal{R} = \langle R, +, \times, 0\rangle$ be a Boolean ring. Then

(*a*) $x + x = 0$

(*b*) $x = -x$

(*c*) $x + y = 0 \leftrightarrow x = y$

(*d*) $x \times y = y \times x$ (Thus the ring must be commutative.)

Proof. First, we observe that for an arbitrary ring the cancellation law $x + y = x + z \rightarrow y = z$ holds. For, if $x + y = x + z$, then

$$\begin{aligned}(-x) + (x + y) &= (-x) + (x + z)\\ ((-x) + x) + y &= ((-x) + x) + z\\ 0 + y &= 0 + z\\ y &= z\end{aligned}$$

From the cancellation law it follows that

$$x = x + z \rightarrow z = 0 \tag{1}$$

For, if $x = x + z$, then $x + 0 = x + z$, and the cancellation law yields $z = 0$.

(*a*) $x + x = (x + x) \times (x + x) = x^2 + x^2 + x^2 + x^2 = x + x + x + x$. By (*1*), $x + x = 0$.

(*b*) Since $x + x = 0$, $x = (-x)$ by the uniqueness assumption for $(-x)$ (cf. Axiom (4) for rings).

(*c*) If $x + y = 0$, then, again using $x + x = 0$, we conclude that $x = y$ by the uniqueness assumption of Axiom (4).

(*d*) $(x + y) = (x + y) \times (x + y) = x^2 + (x \times y) + (y \times x) + y^2 = (x + y) + (x \times y) + (y \times x)$. By (*1*), $0 = (x \times y) + (y \times x)$. Hence by (*c*), $x \times y = y \times x$. ▶

Theorem 5.11. Let $\mathcal{R} = \langle R, +, \times, 0\rangle$ be a Boolean ring with unit element $1 \neq 0$. If we define

$$x' = 1 + x, \qquad x \wedge y = x \times y, \qquad x \vee y = x + y + (x \times y)$$

then $\mathcal{B} = \langle R, \wedge, \vee, ', 0, 1\rangle$ is a Boolean algebra.

Proof. We must verify Axioms (1)-(9) for Boolean algebras.

(1) $x \vee y = y \vee x$.

$$x \vee y = x + y + (x \times y) = y + x + (y \times x) = y \vee x$$

(2) $x \wedge y = y \wedge x$. This is just (*d*) of Theorem 5.10.

(3) $x \wedge (y \vee z) = (x \wedge y) \vee (x \wedge z)$.

$$x \wedge (y \vee z) = x \times (y + z + (y \times z)) = (x \times y) + (x \times z) + (x \times (y \times z))$$

$$\begin{aligned}(x \wedge y) \vee (x \wedge z) &= (x \times y) + (x \times z) + ((x \times y) \times (x \times z))\\ &= (x \times y) + (x \times z) + (x^2 \times y \times z)\\ &= (x \times y) + (x \times z) + (x \times y \times z)\end{aligned}$$

(4) $x \vee (y \wedge z) = (x \vee y) \wedge (x \vee z)$.

$$x \vee (y \wedge z) = x + (y \times z) + (x \times y \times z)$$

$$\begin{aligned}
(x \vee y) \wedge (x \vee z) &= [(x \vee y) \wedge x] \vee [(x \vee y) \wedge z] \quad \text{by (3)} \\
&= [(x \wedge x) \vee (y \wedge x)] \vee [(x \wedge z) \vee (y \wedge z)] \quad \text{by (3)} \\
&= [x \vee (y \times x)] \vee [(x \times z) \vee (y \times z)] \\
&= [x + (x \times y) + (x \times x \times y)] \vee [(x \times z) + (y \times z) + (x \times y \times z^2)] \\
&= [x + (x \times y) + (x \times y)] \vee [(x \times z) + (y \times z) + (x \times y \times z)] \\
&= x \vee [(x \times z) + (y \times z) + (x \times y \times z)] \\
&= x + (x \times z) + (y \times z) + (x \times y \times z) + (x \times [(x \times z) + (y \times z) + (x \times y \times z)]) \\
&= x + (x \times z) + (y \times z) + (x \times y \times z) \\
&\qquad + [(x \times (x \times z)) + (x \times (y \times z)) + (x \times (x \times y \times z))] \quad \text{by (3)} \\
&= x + (x \times z) + (y \times z) + (x \times y \times z) + (x \times z) + (x \times y \times z) + (x \times y \times z) \\
&= x + (y \times z) + (x \times y \times z) + [(x \times z) + (x \times z)] + [(x \times y \times z) + (x \times y \times z)] \\
&= x + (y \times z) + (x \times y \times z)
\end{aligned}$$

(5) $x \vee 0 = 0$.

$$x \vee 0 = x + 0 + (x \times 0) = x + 0 + 0 = x$$

(Note that we have used the fact that, in any ring, $x \times 0 = 0$. To see this fact, observe that $x \times 0 = x \times (0 + 0) = (x \times 0) + (x \times 0)$; and then by *(1)* in the proof of Theorem 5.10, $x \times 0 = 0$.)

(6) $x \wedge 1 = x$.

This is just $x \times 1 = x$, which follows from the definition of a unit element.

(7) $x \vee x' = 1$.

$$\begin{aligned}
x \vee x' &= x + x' + (x \times x') = x + (1 + x) + (x \times (1 + x)) \\
&= 1 + (x + x) + ((x \times 1) + (x \times x)) = 1 + 0 + (x + x) = 1 + 0 + 0 = 1
\end{aligned}$$

(8) $x \wedge x' = 0$.

$$x \wedge x' = x \times (1 + x) = (x \times 1) + (x \times x) = x + x = 0$$

(9) $0 \neq 1$. This is an assumption of the theorem. ▶

Thus we see that a Boolean ring with nonzero unit element determines a Boolean algebra, and vice versa any Boolean algebra determines a Boolean ring with nonzero unit element (essentially Theorem 5.9).

5.4 ALTERNATIVE AXIOMATIZATIONS

There are numerous axiom systems for Boolean algebras.† The following system is a variation of one due to Byrne [101].

A *Byrne algebra* is a structure $\langle B, \wedge, ', 0 \rangle$ where B is a set, $\wedge$ is a binary operation on B, $'$ is a singulary operation on B, and 0 is an element of B, satisfying the postulates:

†The one we have used (Axioms (1)-(9)) is due to Huntington [121]. For systems proposed up to 1933, cf. Huntington [122]. For later work, cf. Sikorski [148], p. 1, footnote 1.

(B1) $x \wedge y = y \wedge x$

(B2) $x \wedge (y \wedge z) = (x \wedge y) \wedge z$

(B3) $x \wedge x = x$

(B4) $x \wedge y' = 0 \leftrightarrow x \wedge y = x$

(B5) $0 \neq 0'$

Let us introduce a few definitions.

Definitions.

$$1 \text{ for } 0'$$
$$x \vee y \text{ for } (x' \wedge y')'$$
$$x \leq y \text{ for } x \wedge y = x$$

From what we have already proved, it follows that, for any Boolean algebra $\langle B, \wedge, \vee, ', 0, 1\rangle$, the structure $\langle B, \wedge, ', 0\rangle$ is a Byrne algebra. The converse is established in the following theorem.

Theorem 5.12. For any Byrne algebra $\langle B, \wedge, ', 0\rangle$, the structure $\langle B, \wedge, \vee, ', 0, 1\rangle$ is a Boolean algebra, where $\vee$ and 1 are defined as above. In particular,

(*a*) $x \wedge x' = 0$

(*b*) $x \wedge y' = 0 \leftrightarrow x \leq y$

(*c*) $x \leq x$

(*d*) $x \leq y \;\&\; y \leq x \rightarrow x = y$

(*e*) $x \leq y \;\&\; y \leq z \rightarrow x \leq z$

(*f*) $x \wedge y \leq x$

(*g*) $x \wedge 0 = 0$

(*h*) $x'' = x$

(*i*) $x \wedge y = (x' \vee y')'$

(*j*) $x \vee y = y \vee x$

(*k*) $x \vee (y \vee z) = (x \vee y) \vee z$

(*l*) $x \vee x = x$

(*m*) $x \leq y \leftrightarrow y' \leq x'$

(*n*) $x \vee y' = 1 \leftrightarrow x \vee y = x$

(*o*) Duality: Any theorem in the language of Byrne algebras (i.e. involving the symbols $\wedge, ', 0$) is transformed into another theorem when we replace $\wedge$ by $\vee$, and 0 by 1. Under this replacement, the defined term $x \vee y$ (i.e. $(x' \wedge y')'$) becomes $(x' \vee y')'$, which is equal to $x \wedge y$, and the defined term 1 (i.e. $0'$) becomes $1'$, which is equal to 0 (by (*h*)). Thus the dual of a theorem is a theorem.

(*p*) $x \leq y \leftrightarrow x \vee y = y$. (Hence the dual of $x \leq y$ is equivalent to $y \leq x$.)

(*q*) $x \wedge 1 = x$

(*r*) $x \vee 0 = x$

(*s*) $x \vee x' = 1$

(*t*) $x \leq x \vee y$

(*u*) $x \vee (x \wedge y) = x \wedge (x \vee y) = x$

(*v*) $x \leq y \rightarrow (x \wedge z \leq y \wedge z \;\&\; x \vee z \leq y \vee z)$

(w) $(x \leq z \;\&\; y \leq z) \rightarrow x \vee y \leq z$

(x) $(z \leq x \;\&\; z \leq y) \rightarrow z \leq x \wedge y$

(y) $x \wedge (x' \vee y) = x \wedge y$

(z_1) $x \wedge (y \vee z) = (x \wedge y) \vee (x \wedge z)$

(z_2) $x \vee (y \wedge z) = (x \vee y) \wedge (x \vee z)$

(z_3) Axioms (1)-(9) for Boolean algebras hold.

Proof.

(*a*) $x \wedge x = x$. Hence by (B4), $x \wedge x' = 0$.

(*b*) This follows immediately from (B4) and the definition of $\leq$.

(*c*) This follows immediately from (B3) and the definition of $\leq$.

(*d*) Assume $x \leq y \;\&\; y \leq x$. Then $x \wedge y = x \;\&\; y \wedge x = y$. By (B1), $x = y$.

(*e*) Assume $x \leq y \;\&\; y \leq z$. Then $x \wedge y = x \;\&\; y \wedge z = y$. Hence $x \wedge z = (x \wedge y) \wedge z = x \wedge (y \wedge z) = x \wedge y = x$. Thus $x \leq z$.

(*f*) $(x \wedge y) \wedge x = (x \wedge x) \wedge y = x \wedge y$. Thus $x \wedge y \leq x$.

(*g*) $x \wedge 0 = x \wedge (x \wedge x') = (x \wedge x) \wedge x' = x \wedge x' = 0$.

(*h*) $x'' \wedge x' = x' \wedge x'' = 0$ (by (*a*)). Therefore $x'' \leq x$ (by (*b*)). Likewise, $x''' \leq x'$ and $x'''' \leq x''$. Hence $x'''' \leq x$ (by (*e*)). Therefore $x'''' \wedge x' = 0$ (by (*b*)). Hence $x' \leq x'''$ (by (*b*)), and therefore $x' = x'''$ (by (*d*)). Thus $x \wedge x''' = 0$ (by (*a*)), and then $x \leq x''$ (by (*b*)). Therefore $x = x''$ (by (*d*)).

(*i*) $x' \vee y' = (x'' \wedge y'')' = (x \wedge y)'$. Hence $(x' \vee y')' = (x \wedge y)'' = x \wedge y$.

(*j*) $x \vee y = (x' \wedge y')' = (y' \wedge x')' = y \vee x$.

(*k*) $x \vee (y \vee z) = (x' \wedge (y' \wedge z')'')' = (x' \wedge (y' \wedge z'))'$.
$(x \vee y) \vee z = z \vee (x \vee y) = (z' \wedge (x' \wedge y'))' = (x' \wedge (y' \wedge z'))'$.

(*l*) $x \vee x = (x' \wedge x')' = x'' = x$.

(*m*) Assume $x \leq y$. Then $x \wedge y' = 0$ (by (*b*)). Therefore $y' \wedge x'' = 0$, and, by (*b*), $y' \leq x'$. Conversely, if $y' \leq x'$, then $x'' \leq y''$. Hence $x \leq y$.

(*n*) $y \leq x \leftrightarrow x' \leq y'$ (by (*m*)). Therefore $x' \wedge y = 0 \leftrightarrow x' \wedge y' = x'$. Hence $(x' \wedge y)' = 0' \leftrightarrow (x' \wedge y')' = x''$ and so $x \vee y' = 1 \leftrightarrow x \vee y = x$.

(*o*) Changing $\wedge$ to $\vee$ and 0 to 1 transforms the axioms for Byrne algebras into theorems. ((B1) becomes (*j*); (B2) becomes (*k*); (B3) becomes (*l*); (B4) becomes (*n*); and (B5) becomes $1 \neq 1'$ which by (*h*) is equivalent to $1 \neq 0$.) Hence if we make these changes in all propositions of a proof, we obtain a proof of the transformed theorem.

(*p*)
$$\begin{aligned} x \leq y &\leftrightarrow y' \leq x' \\ &\leftrightarrow y' \wedge x' = y' \\ &\leftrightarrow (y' \wedge x')' = y'' \\ &\leftrightarrow x \vee y = y \end{aligned}$$

(*q*) $0 \leq x'$, by (*g*). $x \leq 1$, by (*m*) and (*h*). Therefore $x \wedge 1 = x$.

(*r*) This is the dual of (*q*).

(*s*) This is the dual of (*a*).

(*t*) This is the dual of (*f*).

(u) $x \wedge y \leq x$, by (f). Hence by (p), $(x \wedge y) \vee x = x$. The dual of this is $(x \vee y) \wedge x = x$.

(v) Assume $x \leq y$. Then $x \wedge y = x$. Hence $(x \wedge z) \wedge (y \wedge z) = (x \wedge y) \wedge z = x \wedge z$. Thus $x \wedge z \leq y \wedge z$. Also, since $x \leq y$, we have, by (p), $x \vee y = y$. Hence $(x \vee z) \vee (y \vee z) = (x \vee y) \vee z = y \vee z$, and so by (p), $x \vee z \leq y \vee z$.

(w) Assume $x \leq z$ & $y \leq z$. Then $z = z \vee x = z \vee y$. Therefore $z \vee (x \vee y) = (z \vee x) \vee y = z \vee y = z$. Hence $x \vee y \leq z$.

(x) This is the dual of (w).

(y) $x \wedge (x' \vee y) = x \wedge (x'' \wedge y')' = x \wedge (x \wedge y')'$. Hence

$$(x \wedge (x' \vee y)) \wedge y' = (x \wedge y') \wedge (x \wedge y')' = 0$$

Then by (B4), $x \wedge (x' \vee y) = (x \wedge (x' \vee y)) \wedge y = x \wedge ((x' \vee y) \wedge y) = x \wedge y$ (by (u)).

(z_1) First, $y \leq y \vee z$ & $z \leq y \vee z$ (by (t)). Therefore by (v), $x \wedge y \leq x \wedge (y \vee z)$ & $x \wedge z \leq x \wedge (y \vee z)$. Hence by (w), $(x \wedge y) \vee (x \wedge z) \leq x \wedge (y \vee z)$. On the other hand,

$$\begin{aligned} x \wedge (y \vee z) \wedge ((x \wedge y) \vee (x \wedge z))' &= x \wedge (y \vee z) \wedge (x \wedge y)' \wedge (x \wedge z)' \\ &= x \wedge (y \vee z) \wedge (x' \vee y') \wedge (x' \vee z') \\ &= (y \vee z) \wedge (x \wedge (x' \vee y')) \wedge (x \wedge (x' \vee z')) \\ &= (y \vee z) \wedge (x \wedge y') \wedge (x \wedge z') \quad \text{(by (y))} \\ &= x \wedge (y \vee z) \wedge (y' \wedge z') \\ &= x \wedge (y' \wedge z')' \wedge (y' \wedge z') \\ &= x \wedge 0 = 0 \end{aligned}$$

Hence $x \wedge (y \vee z) \leq (x \wedge y) \vee (x \wedge z)$, by (b)). Now apply (d).

(z_2) is the dual of (z_1).

(z_3) All the Axioms (1)-(9) for a Boolean algebra already have been proved. ▸

5.5 IDEALS

An *ideal* of a Boolean algebra $\mathcal{B} = \langle B, \wedge, \vee, ', 0, 1\rangle$ is a non-empty subset J of B such that:

(i) $(x \in J \text{ \& } y \in J) \rightarrow x \vee y \in J$

(ii) $(x \in J \text{ \& } y \in B) \rightarrow x \wedge y \in J$

It is clear that (ii) is equivalent to

(ii′) $(x \in J \text{ \& } y \leq x) \rightarrow y \in J$

For, assume (ii) and let $x \in J$ & $y \leq x$. Then since $y \leq x$, we have $y = x \wedge y$. Hence by (ii), $y \in J$. Conversely, assume (ii′) and let $x \in J$ & $y \in B$. Since $x \wedge y \leq x$, it follows by (ii′) that $x \wedge y \in J$.

Notice that 0 belongs to every ideal, since $0 \leq x$ for all x.

Example 5.5. $\{0\}$ is an ideal.

Example 5.6. B is itself an ideal.

Every ideal different from B is called a *proper ideal*. In particular, $\{0\}$ is a proper ideal.

Note: An ideal J is proper if and only if $1 \notin J$. For, if $1 \notin J$, then $J \subset B$. Conversely, if $1 \in J$ and if $y \in B$, then $y \leq 1$. Hence by (ii′), $y \in B$. Thus $J = B$.

Example 5.7.

If A is a non-empty set and if $\mathcal{P}(A)$ is the Boolean algebra of all subsets of A, then the set J of all finite subsets of A is an ideal. J is proper if and only if A is infinite. (More generally, if $\mathfrak{m}$ is any infinite cardinal number, the set of all subsets of cardinality less than $\mathfrak{m}$ is an ideal.)

Example 5.8.

Given $u \in B$, the set J_u of all $v \leq u$ is an ideal. For, if $v_1 \leq u$ and $v_2 \leq u$, then $v_1 \vee v_2 \leq u$; and if $v \leq u$ and $y \leq v$, then $y \leq u$. The ideal J_u is called the *principal ideal generated by u*.

Theorem 5.13. Given any subset C of a Boolean algebra $\mathcal{B}$, the intersection W of all ideals J containing C (i.e. such that $C \subseteq J$) is itself an ideal containing C. W is said to be *generated by* C, and is denoted Gen (C).

Proof. There is at least one ideal containing C, namely B itself. Assume x and y are in W, and z is in B. If J is any ideal containing C, then $x \in J$ & $y \in J$. Hence $x \vee y \in J$. Likewise, $x \wedge z \in J$. Thus $x \vee y$ and $x \wedge z$ are in W, and therefore W is an ideal. ▶

Theorem 5.14. Given any subset C of a Boolean algebra, the ideal Gen (C) consists of all elements of the form

$$(y_1 \wedge x_1) \vee \cdots \vee (y_k \wedge x_k), \qquad k \geq 1$$

where $x_1, \ldots, x_k \in C$ and $y_1, \ldots, y_k$ are arbitrary elements of B.

Proof. Let D be the set of elements of the given form. The join of any two elements of D is clearly again of the same form and therefore also in D. In addition, for any $y \in B$, the meet

$$\begin{aligned} y \wedge ((y_1 \wedge x_1) \vee \cdots \vee (y_k \wedge x_k)) &= (y \wedge (y_1 \wedge x_1)) \vee \cdots \vee (y \wedge (y_k \wedge x_k)) \\ &= ((y \wedge y_1) \wedge x_1) \vee \cdots \vee ((y \wedge y_k) \wedge x_k) \end{aligned}$$

is again in D. Thus D is an ideal. Since $x = 1 \wedge x$, every member x of C is in D. Every member $(y_1 \wedge x_1) \vee \cdots \vee (y_k \wedge x_k)$ of D belongs to any ideal J containing C, for, since $x_i \in C$, it follows that $y_i \wedge x_i \in J$ and therefore that $(x_1 \wedge y_1) \vee \cdots \vee (x_k \wedge y_k) \in J$. Hence D is the intersection of all ideals containing C. ▶

Theorem 5.15. Given any subset C of a Boolean algebra $\mathcal{B}$, the ideal Gen (C) consists of the set E of all y such that

$$y \leq x_1 \vee \cdots \vee x_k$$

where $x_1, \ldots, x_k$ are arbitrary members of C.

Proof. If $y \leq x_{i_1} \vee \cdots \vee x_{i_k}$ and $z \leq x_{j_1} \vee \cdots \vee x_{j_m}$, then

$$y \vee z \leq x_{i_1} \vee \cdots \vee x_{i_k} \vee x_{j_1} \vee \cdots \vee x_{j_m}$$

and if $y \leq x_1 \vee \cdots \vee x_k$ and $v \leq y$, then $v \leq x_1 \vee \cdots \vee x_k$. Thus E is an ideal. In addition, if x is in C, then $x \leq x$ and therefore x is in E. Clearly, every member of E belongs to every ideal containing C. Hence $E = \text{Gen}(C)$. ▶

Corollary 5.16. If C is any subset of a Boolean algebra $\mathcal{B}$, then the ideal Gen (C) generated by C is a proper ideal if and only if

$$x_1 \vee \cdots \vee x_k \neq 1$$

for all $x_1, \ldots, x_k$ in C.

Proof. Note first that $1 \leq u$ is equivalent to $1 = u$, for any u. Hence by Theorem 5.15, $1 \in \text{Gen}(C)$ if and only if $1 = x_1 \vee \cdots \vee x_k$ for some $x_1, \ldots, x_k$ in C. But, $1 \in \text{Gen}(C)$ if and only if Gen (C) is not a proper ideal. ▶

Theorem 5.17. If J is an ideal of a Boolean algebra $\mathcal{B}$, and if $y \in B$, then the ideal $\text{Gen}(J \cup \{y\})$, generated by J plus y, consists of all elements of the form

$$(z \wedge y) \vee x$$

where $z \in B$ and $x \in J$.

Proof. Let H be the set of all elements $(z \wedge y) \vee x$, where $z \in B$ and $x \in J$. First, $y \in H$, since $y = (1 \wedge y) \vee 0$. Also, if $v \in J$, then $v = (0 \wedge y) \vee v \in H$. Thus $J \cup \{y\} \subseteq H$. In addition,

$$((z_1 \wedge y) \vee x_1) \vee ((z_2 \wedge y) \vee x_2) \;=\; ((z_1 \vee z_2) \wedge y) \vee (x_1 \vee x_2) \in H$$

Also, if $x \in J$, then

$$w \wedge ((z \wedge y) \vee x) \;=\; ((w \wedge z) \wedge y) \vee (w \wedge x) \in H$$

Hence H is an ideal. Finally, if I is an ideal containing $J \cup \{y\}$, and if $x \in J$, then $(z \wedge y) \vee x \in I$; thus $I \supseteq H$. ▶

Corollary 5.18. If J is an ideal of a Boolean algebra $\mathcal{B}$, and if $y \in B \sim J$, then the ideal $\text{Gen}(J \cup \{y\})$ generated by J plus y is a proper ideal if and only if $x \vee y \neq 1$ for all x in J, i.e. if and only if, for all x in J, $y' \not\leq x$.

Proof. By Corollary 5.16, $\text{Gen}(J \cup \{y\})$ is proper if and only if $x_1 \vee \cdots \vee x_k \vee y \neq 1$ for all $x_1, \ldots, x_k$ in J. (Note that if $x_1 \vee \cdots \vee x_k \vee y \neq 1$, then $x_1 \vee \cdots \vee x_k \neq 1$.) But since J is an ideal, $x = x_1 \vee \cdots \vee x_k$ is also in J. Hence the indicated condition is equivalent to saying that $x \vee y \neq 1$ for all x in J. The additional remark follows from the fact that $x \vee y = 1$ is equivalent to $y' \leq x$. ▶

Definition. An ideal M of a Boolean algebra $\mathcal{B}$ is said to be *maximal* if and only if M is a proper ideal and there is no proper ideal J of $\mathcal{B}$ such that $M \subset J$.

Theorem 5.19. Given a proper ideal M of a Boolean algebra $\mathcal{B}$. Then M is maximal if and only if, for any y in B, either $y \in M$ or $y' \in M$.

Proof.

(*a*) Assume M is maximal. Assume $y \notin M$. We must show that $y' \in M$. Let $I = \text{Gen}(M \cup \{y\})$. Then I is an ideal of $\mathcal{B}$ such that $M \subset I$. Hence by the maximality of M, $I = B$. Therefore by Corollary 5.18, $y' \leq x$ for some x in M. Since M is an ideal, $y' \in M$.

(*b*) Assume that $y \in M$ or $y' \in M$ for all y in B. Assume $M \subset J$, where J is an ideal. We must prove that $J = B$. Let $y \in J \sim M$. Since $y \notin M$, $y' \in M$. Hence $y' \in J$. Then $1 = y \vee y' \in J$. Therefore $J = B$. ▶

Definition. An ideal J in a Boolean algebra $\mathcal{B}$ is *prime* if and only if, for any x and y in B, $(x \notin J \;\&\; y \notin J) \rightarrow x \wedge y \notin J$.

Theorem 5.20. A proper ideal J in a Boolean algebra $\mathcal{B}$ is maximal if and only if it is prime.

Proof.

(*a*) Assume J is maximal. Assume $x \notin J \;\&\; y \notin J$. Then by Theorem 5.19, $x' \in J$ and $y' \in J$. Hence $x' \vee y' \in J$. Since J is proper, $x \wedge y = (x' \vee y')' \notin J$. Thus J is prime.

(*b*) Assume J prime. Given $y \in B$. By Theorem 5.19 it suffices to show that $y \in J$ or $y' \in J$. Assume $y \notin J$ and $y' \notin J$. Since J is prime, $0 = y \wedge y' \notin J$ which is a contradiction. ▶

Theorem 5.21. The maximal principal ideals are the principal ideals $J_{u'}$, where u is an atom.

Proof.

(*a*) Assume u is an atom. To prove $J_{u'}$ maximal, we shall use Theorem 5.19. Assume $y \in B$ and $y \notin J_{u'}$. Then $y \nleq u'$. Hence $u \nleq y'$. Since u is an atom, $u \leq y$ (by Remark (iii) on page 135). Hence $y' \leq u'$, i.e. $y' \in J_{u'}$.

(*b*) Assume $J_{u'}$ maximal. To prove that u is an atom, we assume $v \leq u$ and we must show that $v = 0$ or $v = u$. Assume $v \neq u$. Since $J_{u'}$ is maximal, $v \in J_{u'}$ or $v' \in J_{u'}$. Hence $v \leq u'$ or $v' \leq u'$, i.e. $v \leq u'$ or $u \leq v$. But $u \nleq v$, since $v \leq u$ & $v \neq u$. Hence $v \leq u'$. Since $v \leq u$ and $v \leq u'$, it follows that $v \leq u \wedge u' = 0$. Therefore $v = 0$. ▶

Example 5.9.

In the Boolean algebra $\mathcal{P}(A)$ of all subsets of a non-empty set A, the atoms are the singletons $\{a\}$, where $a \in A$. Hence a maximal principal ideal consists of all subsets X of A such that $a \notin X$.

Example 5.10.

In a finite Boolean algebra $\mathcal{B}$, every maximal ideal M is principal. For, there are a finite number of atoms $a_1, \ldots, a_k$ in B, and $a_1 \vee a_2 \vee \cdots \vee a_k = 1$. Hence there is some atom a which is not in M. Then $M = J_{a'}$. To see this, observe first that since $a \notin M$, then $a' \in M$. Therefore $J_{a'} \subseteq M$. On the other hand, if $y \in M$, then $a \nleq y$ since $a \notin M$. Since a is an atom and $a \nleq y$, it follows that $a \leq y'$, and so $y \leq a'$. Therefore $M \subseteq J_{a'}$.

5.6 QUOTIENT ALGEBRAS

Let J be an ideal of a Boolean algebra $\mathcal{B}$.

Definition. $x \equiv_J y$ if and only if $x + y \in J$.

Recall that $x + y = (x \wedge y') \vee (x' \wedge y)$ (cf. Section 5.3). Hence, since J is an ideal, $x \in J$ & $y \in J \rightarrow x + y \in J$.

Theorem 5.22. If J is an ideal of a Boolean algebra $\mathcal{B}$,

(*a*) $x \equiv_J x$.

(*b*) $x \equiv_J y \rightarrow y \equiv_J x$.

(*c*) $x \equiv_J y$ & $y \equiv_J z \rightarrow x \equiv_J z$.

(*d*) $(x \equiv_J y$ & $a \equiv_J b) \rightarrow (x' \equiv_J y'$ & $x \wedge a \equiv_J y \wedge b$ & $x \vee a \equiv_J y \vee b)$.

Proof.

(*a*) $x + x = 0 \in J$.

(*b*) $x + y = y + x$.

(*c*) $x + z = x + 0 + z = x + (y + y) + z = (x + y) + (y + z)$.

(*d*) Assume $x \equiv_J y$ and $a \equiv_J b$. Since $x' + y' = x + y$, it follows that $x' \equiv_J y'$. Next, notice that for any z, $(x \wedge z) \equiv_J (y \wedge z)$, since $(x \wedge z) + (y \wedge z) = (x + y) \wedge z$. Hence

$$x \wedge a \equiv_J y \wedge a = a \wedge y \equiv_J b \wedge y = y \wedge b$$

Lastly,

$$x \vee a = (x' \wedge a')' \equiv_J (y' \wedge b')' = y \vee b$$ ▶

Definition. $$[x] = \{y : x \equiv_J y\}$$

$[x]$ is called the *equivalence class* of x modulo J. Clearly, $x \in [x]$.

Theorem 5.23. (*a*) $[x] = [z] \leftrightarrow x \equiv_J z$.

(*b*) $[x] \neq [z] \rightarrow [x] \cap [z] = \emptyset$.

Proof.

(*a*) First, assume $[x] = [z]$. Since $x \in [x]$, we obtain $x \in [z]$, i.e. $x \equiv_J z$. Second, assume $x \equiv_J z$. Then for any y,

$$y \in [x] \rightarrow x \equiv_J y \rightarrow z \equiv_J y \rightarrow y \in [z]$$

Hence $[x] \subseteq [z]$. Similarly, $[z] \subseteq [x]$. Therefore $[x] = [z]$.

(*b*) $y \in [x] \cap [z] \rightarrow (x \equiv_J y \;\&\; z \equiv_J y) \rightarrow x \equiv_J z \rightarrow [x] = [z]$. ▶

Given an ideal J of the Boolean algebra $\mathcal{B}$. Let B/J be the set of equivalence classes modulo J. We define operations $\wedge_J$, $\vee_J$, $'_J$ on B/J as follows:

$$[x] \wedge_J [y] = [x \wedge y]$$

$$[x] \vee_J [y] = [x \vee y]$$

$$[x]'_J = [x']$$

These definitions are meaningful by virtue of Theorem 5.22(*d*). For example, if X and Y are equivalence classes modulo J, we select any $x \in X$ and any $y \in Y$. If $x_1 \in X$ and $y_1 \in Y$, then $[x \wedge y] = [x_1 \wedge y_1]$ by virtue of Theorem 5.22(*d*) and Theorem 5.23(*a*). Thus $[x \wedge y]$ does not depend upon the particular x chosen in X or the particular y chosen in Y. This shows that our definition of $\wedge_J$ makes sense.

Let 0_J stand for $[0]$, and let 1_J stand for $[1]$. Clearly, $0_J = J$ and $1_J = \{y : y' \in J\}$.

Let $\mathcal{B}/J$ stand for $\langle B/J, \wedge_J, \vee_J, '_J, 0_J, 1_J \rangle$.

Theorem 5.24. If J is a proper ideal of a Boolean algebra $\mathcal{B}$, then $\mathcal{B}/J$ is a Boolean algebra (called the *quotient algebra* of $\mathcal{B}$ by J).

Proof. We must check the axioms for Boolean algebras.

(1) $[x] \vee [y] = [x \vee y] = [y \vee x] = [y] \vee [x]$.

Axioms (2)-(8) are proved in a similar manner. (This is left as an exercise for the reader.)

(9) $0_J \neq 1_J$, since $0 \not\equiv_J 1$. (This follows from the equation $0 + 1 = 1$ and the assumption that J is a *proper* ideal.) ▶

Example 5.11.

Let J be the ideal $\{0\}$. Then $x \equiv_J y$ if and only if $x = y$. Thus the elements of B/J are the singletons $\{x\}$, where $x \in B$. The function $f(x) = \{x\}$ is an isomorphism between $\mathcal{B}$ and $\mathcal{B}/\{0\}$.

Theorem 5.25. $B/J = \{0_J, 1_J\}$ if and only if J is a maximal ideal.

Proof. If J is maximal, then, for any x, either $x \in J$ or $x' \in J$. Hence $[x] = 0_J$ or $[x] = 1_J$. Conversely, assume that for every x in B, $[x] = 0_J$ or $[x] = 1_J$. Hence $x \in J$ or $x' \in J$. Therefore by Theorem 5.19, J is maximal. ▶

Remark. If $x \in B$, then $[x] = x + J$, where $x + J$ stands for $\{x + u : u \in J\}$. For, on the one hand, $y \in [x] \rightarrow x + y \in J \rightarrow x + y = u$ for some u in $J \rightarrow y = x + u$ for some $u \in J$. On the other hand, if $y = x + u$ for some u in J, then $y + x = u \in J$, and therefore $y \in [x]$.

5.7 THE BOOLEAN REPRESENTATION THEOREM

The theory of Boolean algebras is intended to be a generalization of the algebra of sets. We already have proved (Theorem 5.7) that every finite Boolean algebra is isomorphic to the Boolean algebra of all subsets of some set A (namely, A is the set of atoms). This is not true for all Boolean algebras. For, in Problem 5.16 we cite an example of an atomless Boolean algebra $\mathcal{B}$; and since the Boolean algebra $\mathcal{P}(A)$ of all subsets of any set A is atomic, $\mathcal{B}$ cannot be isomorphic to $\mathcal{P}(A)$. However, we shall show in what follows that every Boolean algebra is isomorphic to some field of sets.

To this end, we shall need the following general mathematical principle.

Zorn's Lemma: Given a set Z of sets such that, for every $\subset$-chain C in Z,† the union $\bigcup_{A \in C} A$ is also in Z. Then there is an $\subset$-maximal set M in Z, i.e. $M \in Z$, and, if A is any set in Z, then $M \not\subset A$.

A proof of Zorn's Lemma (based upon the use of a more transparent assumption, the so-called axiom of choice) is given in Appendix C.

Theorem 5.26 (*Maximal Ideal Theorem*). If J is a proper ideal in a Boolean algebra $\mathcal{B}$, then there is a maximal ideal M in $\mathcal{B}$ such that $J \subseteq M$ (i.e. every proper ideal can be extended to a maximal ideal).

Proof. Assume J is a proper ideal. Let Z be the class of all proper ideals K in $\mathcal{B}$ such that $J \subseteq K$. Now assume that C is an $\subset$-chain in Z. Then $\bigcup_{I \in C} I$ is a proper ideal (cf. Problem 5.34) containing J. Hence by Zorn's Lemma there is a maximal set M in Z. But $M \supseteq J$ and M is a maximal ideal in $\mathcal{B}$ (for, if $M^{\#}$ is any proper ideal such that $M \subseteq M^{\#}$, then $M^{\#} \in Z$ and hence $M = M^{\#}$). ▶

Example 5.12.

Consider the Boolean algebra $\mathcal{P}(A)$ of all subsets of an infinite set A. Let J be the ideal of all finite subsets of A. By Theorem 5.26, J can be extended to a maximal ideal M. M cannot be a principal ideal. (For, a principal maximal ideal consists of all subsets of A not containing some fixed element b of A (cf. Example 5.9), while every singleton $\{b\}$, being finite, belongs to J and therefore to M. Hence if M were principal, we would have $A = \{b\} \cup (A \sim \{b\}) \in M$, and then M would not be a proper ideal.) No way is known for describing such an ideal M in a constructive way, i.e. no property is known for which the sets satisfying this property form a maximal ideal M containing J as a subset.

Corollary 5.27. Every Boolean algebra has at least one maximal ideal.

Proof. The set $\{0\}$ is an ideal. Hence by Theorem 5.26, $\{0\}$ can be extended to a maximal ideal. ▶

Corollary 5.28. If C is any subset of a Boolean algebra $\mathcal{B}$ such that for any $u_1, \ldots, u_n$ in C, $u_1 \vee \cdots \vee u_n \neq 1$, then there is a maximal ideal M containing C.

Proof. By Corollary 5.16, the ideal Gen(C) generated by C is a proper ideal. Then by Theorem 5.26 there is a maximal ideal M containing Gen(C), and therefore also containing C. ▶

---

†By an $\subset$-chain C in Z we mean a subset of Z such that, if $A \in C$ and $B \in C$ and $A \neq B$, then either $A \subset B$ or $B \subset A$. More generally, if R is a binary relation on a set W, then an R-chain in W is a subset of W on which R is transitive, connected, and antisymmetric (i.e. $xRy \,\&\, yRx \rightarrow x = y$).

Corollary 5.29. If x is a nonzero element of a Boolean algebra $\mathcal{B}$, there is a maximal ideal M not containing the element x.

Proof. $x' \neq 1$. Hence by Corollary 5.28 there is a maximal ideal M containing x'. Hence $x \notin M$. (Otherwise, $1 = x \vee x' \in M$.) ▶

Theorem 5.30. (*Stone's Representation Theorem*). Every Boolean algebra $\mathcal{B}$ is isomorphic to a field of sets.

Proof. For each x in B, let $\Xi(x)$ be the set of all maximal ideals of $\mathcal{B}$ such that $x \notin M$. Clearly, $\Xi(1)$ is the set $\mathcal{M}$ of all maximal ideals of $\mathcal{B}$, while $\Xi(0)$ is the empty set. For any $x \neq 0$, $\Xi(x)$ is non-empty, by Corollary 5.29. Also, $\Xi(x') = \overline{\Xi(x)}$, since every maximal ideal M contains precisely one of the elements x and x'. Finally, $\Xi(x \vee y) = \Xi(x) \cup \Xi(y)$. (First, if $x \notin M$, then $x \vee y \notin M$, since $x \leq x \vee y$. Likewise, if $y \notin M$, then $x \vee y \notin M$. Hence $\Xi(x) \cup \Xi(y) \subseteq \Xi(x \vee y)$. Conversely, if $x \in M$ & $y \in M$, then $x \vee y \in M$. Hence if $x \vee y \notin M$, then $x \notin M$ or $y \notin M$. Therefore $\Xi(x \vee y) \subseteq \Xi(x) \cup \Xi(y)$.) Thus Ξ is an isomorphism of $\mathcal{B}$ into the Boolean algebra $\mathcal{P}(\mathcal{M})$ of all subsets of the set $\mathcal{M}$ of all maximal ideals. The range of Ξ is a field of sets isomorphic to $\mathcal{B}$. ▶

Corollary 5.31. For any sentence **A** of the theory of Boolean algebras, **A** holds for all Boolean algebras if and only if **A** holds for all fields of sets.

(By a sentence of the theory of Boolean algebras we mean either an equality $\tau = \sigma$, where τ and σ are Boolean expressions, or an expression obtained from such equalities by applying the logical connectives and the quantifiers "for all x" and "there exists an x".)

5.8 INFINITE MEETS AND JOINS

Given a Boolean algebra $\mathcal{B} = \langle B, \wedge, \vee, ', 0, 1\rangle$ and a subset A of B. If the least upper bound (lub) of A exists, it is denoted $\bigvee_{x \in A} x$ and called the *join* of A. Thus in order for y to be the join of A, it is necessary and sufficient that:

(*a*) $x \leq y$ for all $x \in A$.

(*b*) For any v, if $x \leq v$ for all $x \in A$, then $y \leq v$.

If the greatest lower bound (glb) of A exists, it is denoted $\bigwedge_{x \in A} x$ and called the *meet* of A.

Notice that

$$\bigvee_{x \in \emptyset} x = 0, \qquad \bigwedge_{x \in \emptyset} x = 1$$

$$\bigvee_{x \in B} x = 1, \qquad \bigwedge_{x \in B} x = 0$$

Definition. The Boolean algebra $\mathcal{B}$ is said to be *complete* if and only if $\bigvee_{x \in A} x$ and $\bigwedge_{x \in A} x$ exist for all subsets A of B, i.e. every subset of B has both a lub and a glb.

Example 5.13.

In the field $\mathcal{P}(K)$ of all subsets of a non-empty set K,

$$\bigvee_{x \in A} x = \bigcup_{x \in A} x \quad \text{and} \quad \bigwedge_{x \in A} x = \bigcap_{x \in A} x$$

Thus $\mathcal{P}(K)$ is a complete Boolean algebra.

Theorem 5.32. (*De Morgan's Laws*)

$$(a)\quad \bigvee_{x \in A} x = \Big(\bigwedge_{x \in A} (x')\Big)'$$

$$(b)\quad \bigwedge_{x \in A} x = \Big(\bigvee_{x \in A} (x')\Big)'$$

(Each equation is taken to mean that, if one side has meaning, then so does the other, and they are equal.)

Proof. (*a*) Assume $\bigvee_{x \in A} x$ exists, and let $y = (\bigvee_{x \in A} x)'$. Now if $u \in A$, then $u \le \bigvee_{x \in A} x$. Hence $y \le u'$. Thus y is a lower bound of the set W of all x', where $x \in A$. Now assume v is a lower bound of W. Thus $v \le x'$ for all $x \in A$. Hence $x \le v'$ for all $x \in A$. Thus $\bigvee_{x \in A} x \le v'$. Hence $v \le (\bigvee_{x \in A} x)' = y$. We have shown that $y = \bigwedge_{x \in A} (x')$. Hence $\bigvee_{x \in A} x = y' = (\bigwedge_{x \in A} (x'))'$. On the other hand, assume $\bigwedge_{x \in A} (x')$ exists. Let $b = (\bigwedge_{x \in A} (x'))'$. If $u \in A$, then $\bigwedge_{x \in A} (x') \le u'$. Hence $u \le b$. Thus b is an upper bound of A. If c is an upper bound of A, then $x \le c$ for all $x \in A$; this implies that $c' \le x'$ for all $x \in A$. Hence $c' \le \bigwedge_{x \in A} (x')$, and $b = (\bigwedge_{x \in A} (x'))' \le c$. Therefore $b = \bigvee_{x \in A} x$. The proof of (*b*) is similar. ▶

Corollary 5.33. If all subsets of B have a meet (respectively, a join), then B is complete.

Example 5.14.

Let $\mathcal{B}$ be the Boolean algebra of all finite and cofinite sets of positive integers. Let A be the set of all sets of the form $\{2n\}$, where n is a positive integer, i.e. A is the set of singletons of the positive even integers. Then A has no join. For, if u were equal to $\bigvee_{x \in A} x$, then u would have to contain all positive even integers. But, since u would be cofinite, u would also have to contain all but finitely many odd positive integers. Then any proper subset of u obtained by removing an odd integer also would be an upper bound of A, contradicting the assumption that u is the least upper bound of A. Thus $\mathcal{B}$ is not complete.

Example 5.15.

Let A be the field of sets consisting of all finite sets of positive integers and all sets $N \sim X$, where N is the set of all *non-negative* integers and X is any finite set of positive integers. A is a subfield of the field $\mathcal{P}(N)$ of all subsets of N. Let C be the set of all singletons of the form $\{n\}$, where n is a positive integer. Then the join of C in the Boolean algebra $\mathcal{P}(N)$ is $N \sim \{0\}$. However, the join of C in the Boolean algebra A is N. (Notice that $N \sim \{0\}$ does not belong to A.) This illustrates two facts:

(1) The join of a set in a field of sets is not necessarily the union. (In Example 5.15, the union of C is $N \sim \{0\}$ but the join in A is N.)

(2) If A_1 is a Boolean subalgebra of a Boolean algebra A_2, and if Y is a set of elements of A_1, then the join of Y in A_1 (if it exists) is not necessarily the join of Y in A_2 (if it exists).

Of course the same facts hold for meets as well as joins.

Because of the fact illustrated in (2), we shall, if necessary, designate the join in a Boolean algebra $\mathcal{B}$ of a set Y of elements by $\bigvee^{\mathcal{B}}_{x \in Y} x$, and the meet by $\bigwedge^{\mathcal{B}}_{x \in Y} x$.

Definition. By a *complete field of sets* we mean a field of sets such that, for any subset A of the field, the union and intersection of the sets in A are also in the field. The field $\mathcal{P}(K)$ of all subsets of a non-empty set K is an example of a complete field of sets. Clearly, a complete field of sets is a complete Boolean algebra, with union and intersection serving as join and meet.

5.9 DUALITY

If $\mathcal{B} = \langle B, \wedge, \vee, ', 0, 1\rangle$ is a Boolean algebra, then its *dual* $\mathcal{B}^* = \langle B, \vee, \wedge, ', 1, 0\rangle$ is also a Boolean algebra. In fact, the function $f(x) = x'$ is an isomorphism between $\mathcal{B}$ and $\mathcal{B}^*$ (cf. Problem 5.42).

Theorem 5.34. (*Duality Principle*). For any Boolean sentence **A**, the *dual* formula **A***, obtained from **A** by exchanging 0 and 1 and exchanging $\wedge$ and $\vee$, will be true for $\mathcal{B}$ if and only if **A** is true for $\mathcal{B}$.

Proof. **A** is true for $\mathcal{B}$ if and only if it is true for the isomorphic algebra $\mathcal{B}^*$. But the interpretation of **A** with respect to the model $\mathcal{B}^*$ is the same as the interpretation of **A*** with respect to $\mathcal{B}$. ▶

There are various extensions of the Duality Principle. The transformation from **A** to the dual **A*** exchanges $\leq$ and $\geq$. (For, $x \leq y \leftrightarrow x \wedge y = x$, and the dual of $x \wedge y = x$ is $x \vee y = x$, which is equivalent to $x \geq y$.) In addition, the taking of duals interchanges the general notions of meet and join.

Example 5.16.

The second part of De Morgan's Laws (Theorem 5.32)

$$(b) \qquad \bigwedge_{x \in A} x = \left(\bigvee_{x \in A} (x')\right)'$$

is the dual of the first part:

$$(a) \qquad \bigvee_{x \in A} x = \left(\bigwedge_{x \in A} (x')\right)'$$

Hence our proof of (*a*) automatically is also a proof of (*b*).

Example 5.17.

From $\bigvee_{x \in \emptyset} x = 0$, it follows by duality that $\bigwedge_{x \in \emptyset} x = 1$. Similarly, $\bigwedge_{x \in B} x = 0$ follows from $\bigvee_{x \in B} x = 1$.

5.10. INFINITE DISTRIBUTIVITY

Theorem 5.35. (*a*) $x \wedge \bigvee_{u \in A} u = \bigvee_{u \in A} (x \wedge u)$.

(*b*) $x \vee \bigwedge_{u \in A} u = \bigwedge_{u \in A} (x \vee u)$.

(These equations are intended to mean that, if the left side is meaningful, then the right side is also meaningful and the two sides are equal.)

Proof. Since (*b*) is the dual of (*a*), it suffices to prove (*a*). As a preliminary, note that

$$a \wedge b \leq c \rightarrow b \leq a' \vee c \qquad (5.1)$$

(For, if $a \wedge b \leq c$, then $b = (a' \wedge b) \vee (a \wedge b) \leq a' \vee c$.) By taking the dual of (*5.1*) and changing a to a', we obtain

$$c \leq a' \vee b \rightarrow a \wedge c \leq b \qquad (5.2)$$

Assume now that $\bigvee_{u \in A} u$ exists. If $v \in A$, then $v \leq \bigvee_{u \in A} u$ and therefore

$$x \wedge v \leq x \wedge \bigvee_{u \in A} u$$

Thus $x \wedge \bigvee_{u \in A} u$ is an upper bound of the set Z of all $x \wedge u$, where $u \in A$. Now assume w is an upper bound of the set Z. Then for all $u \in A$, $x \wedge u \leq w$; and therefore by (*5.1*), $u \leq x' \vee w$. Hence $\bigvee_{u \in A} u \leq x' \vee w$. By (*5.2*), $x \wedge \bigvee_{u \in A} u \leq w$. Thus $x \wedge \bigvee_{u \in A} u = \bigvee_{u \in A} (x \wedge u)$. ▶

Remark. If the right-hand side of (*a*) or (*b*) of Theorem 5.35 is meaningful, the left-hand side need not be. For example, if $x = 0$, then $\bigvee_{u \in A} (x \wedge u) = 0$, but $\bigvee_{u \in A} u$ need not exist if $\mathcal{B}$ is not complete.

By means of the ordinary distributive laws, one obtains identities of the form

$$(x_{11} \vee x_{12} \vee \cdots \vee x_{1k_1}) \wedge (x_{21} \vee x_{22} \vee \cdots \vee x_{2k_2}) \wedge \cdots \wedge (x_{m1} \vee x_{m2} \vee \cdots \vee x_{mk_m})$$

$$= \bigvee_{1 \leq j_i \leq k_i} (x_{1j_1} \wedge x_{2j_2} \wedge \cdots \wedge x_{mj_m})$$

For example,

$$(x_{11} \vee x_{12}) \wedge (x_{21} \vee x_{22} \vee x_{23}) = (x_{11} \wedge x_{21}) \vee (x_{11} \wedge x_{22}) \vee (x_{11} \wedge x_{23}) \vee (x_{12} \wedge x_{21})$$

$$\vee (x_{12} \wedge x_{22}) \vee (x_{12} \wedge x_{23})$$

One even can extend these identities to the following infinite case.

$$\left(\bigvee_{i \in A_1} x_{1i}\right) \wedge \left(\bigvee_{i \in A_2} x_{2i}\right) \wedge \cdots \wedge \left(\bigvee_{i \in A_k} x_{ki}\right)$$

$$= \bigvee_{j_i \in A_i} (x_{1j_1} \wedge x_{2j_2} \wedge \cdots \wedge x_{kj_k}) \qquad (5.3)$$

where the join on the right is taken over all possible terms $x_{1j_1} \wedge x_{2j_2} \wedge \cdots \wedge x_{kj_k}$, with $j_1 \in A_1, \ldots, j_k \in A_k$, and where we assume that all the joins on the left-hand side exist. This identity is proved by induction on k, using Theorem 5.35(*a*) (plus generalized associativity; cf. Problem 5.37).

For any sets S and W, let S^W stand for the set of all functions from W into S. Assume given a function assigning to each $w \in W$ and $s \in S$ an element $x_{w,s}$ of a given Boolean algebra $\mathcal{B}$. Consider

$$\bigwedge_{w \in W} \left(\bigvee_{s \in S} x_{w,s}\right) = \bigvee_{f \in S^W} \left(\bigwedge_{w \in W} x_{w,f(w)}\right) \qquad (5.4)$$

where the join on the right extends over all functions $f \in S^W$.

Definition. If $\mathfrak{m}$ and $\mathfrak{n}$ are cardinal numbers, the Boolean algebra $\mathcal{B}$ is said to be $(\mathfrak{m}, \mathfrak{n})$-*distributive* if and only if, whenever W has cardinal number $\mathfrak{m}$ and S has cardinal number $\mathfrak{n}$ and $x_{w,s}$ is any assignment of elements of B such that the left-hand side of (*5.4*) makes sense and each term $\bigwedge_{w \in W} x_{w,f(w)}$ on the right-hand side makes sense, then the right-hand side makes sense and the equation (*5.4*) holds.

$\mathcal{B}$ is said to be *completely distributive* if and only if $\mathcal{B}$ is $(\mathfrak{m}, \mathfrak{n})$-distributive for all cardinals $\mathfrak{m}$ and $\mathfrak{n}$. $\mathcal{B}$ is said to be $\mathfrak{m}$-*distributive* if and only if $\mathcal{B}$ is $(\mathfrak{m}, \mathfrak{m})$-distributive.

We have seen above (equation (*5.3*)) that if $\mathfrak{m}$ is finite every Boolean algebra is $(\mathfrak{m}, \mathfrak{n})$-distributive, no matter what $\mathfrak{n}$ is. Obviously $(\mathfrak{m}, \mathfrak{n})$-distributivity also holds when $\mathfrak{n} = 1$. However, if $\mathfrak{m}$ is infinite and $\mathfrak{n} \geq 2$, then a Boolean algebra need not be $(\mathfrak{m}, \mathfrak{n})$-distributive, even when $\mathfrak{n}$ is finite.

It is easy to see that, if $\mathcal{B}$ is $(\mathfrak{m}, \mathfrak{n})$-distributive, then

(a) $\mathfrak{p} \leq \mathfrak{m} \rightarrow \mathcal{B}$ is $(\mathfrak{p}, \mathfrak{n})$-distributive;

(b) $\mathfrak{p} \leq \mathfrak{n} \rightarrow \mathcal{B}$ is $(\mathfrak{m}, \mathfrak{p})$-distributive.

To verify (a), it suffices to choose a set W^* such that $W \cap W^* = \emptyset$ and $W \cup W^*$ has cardinality $\mathfrak{m}$, and then to extend the given assignment by letting $x_{w,s} = 1$ for all $w \in W^*$. To prove (b), we need only choose a set S^* so that $S \cap S^* = \emptyset$ and $S \cup S^*$ has cardinality $\mathfrak{n}$, and then extend the given assignment by letting $x_{w,s} = 0$ for all $s \in S^*$.

From (a) and (b) it follows that $\mathcal{B}$ is completely distributive if and only if $\mathcal{B}$ is $\mathfrak{m}$-distributive for all $\mathfrak{m}$.

It can be shown that $\mathcal{B}$ is $\mathfrak{m}$-distributive if and only if $\mathcal{B}$ is $(\mathfrak{m}, 2)$-distributive (cf. [140] and [149]).

Example 5.18.

The field $\mathcal{P}(K)$ of all subsets of a non-empty set K is completely distributive. This follows from the fact that

$$\bigcap_{w \in W} \left(\bigcup_{s \in S} x_{w,s} \right) = \bigcup_{f \in S^W} \left(\bigcap_{w \in W} x_{w,f(w)} \right) \tag{5.5}$$

always holds for arbitrary sets $x_{w,s}$.

Remark. By the Duality Principle, it is not necessary to give separate consideration to the dual of (5.4):

$$\bigvee_{w \in W} \left(\bigwedge_{s \in S} x_{w,s} \right) = \bigwedge_{f \in S^W} \left(\bigvee_{w \in W} x_{w,f(w)} \right)$$

5.11 $\mathfrak{m}$-COMPLETENESS

Let $\mathfrak{m}$ be an infinite cardinal number. A Boolean algebra $\mathcal{B}$ is said to be $\mathfrak{m}$-*complete* if and only if every subset of B having cardinal number $\leq \mathfrak{m}$ possesses a least upper bound (lub) and a greatest lower bound (glb).

By De Morgan's Laws (Theorem 5.32), for $\mathfrak{m}$-completeness it suffices to know only that every subset of B having cardinal number $\leq \mathfrak{m}$ possesses a lub (or that every subset of B having cardinal number $\leq \mathfrak{m}$ possesses a glb).

If $\aleph_0$ is the cardinal number of the set of integers, it is customary to use the term *σ-algebra* instead of *$\aleph_0$-complete Boolean algebra.* Thus $\mathcal{B}$ is a σ-algebra if and only if every denumerable subset of B has a lub.

We shall use the term *$\mathfrak{m}$-complete field of sets* for a field of sets $\mathcal{F}$ such that any subset of $\mathcal{F}$ of cardinality $\leq \mathfrak{m}$ has its union in $\mathcal{F}$. In addition, by a σ-field of sets we mean a field of sets closed under denumerable unions.

Clearly, an $\mathfrak{m}$-complete field of sets is an $\mathfrak{m}$-complete Boolean algebra, and a σ-field of sets is a σ-algebra.

Example 5.19.

Let K be a non-denumerable set. The field $\mathcal{F}$ of all subsets of K which are either countable (i.e. either finite or denumerable) or co-countable (i.e. their complement is countable) is a σ-field, but not a complete field of sets (nor a complete Boolean algebra). For, let A be a subset of K such that both A and its complement are non-denumerable. Then for each $x \in A$, the set $\{x\} \in \mathcal{F}$. However, the union of all the sets $\{x\}$, where $x \in A$, is equal to A, which does not belong to $\mathcal{F}$.

Example 5.20.

The field $\mathcal{G}$ of all finite and cofinite subsets of an infinite set K is not a σ-algebra. For, let A be an infinite subset of K whose complement is also infinite. Then the union of all sets $\{x\}$, where $x \in A$, is infinite but not cofinite. This shows that we do not have a σ-field. To see that $\mathcal{G}$ is not even a σ-algebra, observe that the same set E of all sets $\{x\}$, where $x \in A$, cannot have a lub in $\mathcal{G}$. For, if C were such a set, then $x \in A \rightarrow \{x\} \subseteq C$. Hence $A \subseteq C$, and so C would be infinite. Hence C must be cofinite, and therefore C must intersect the complement $\bar{A}$ of A. If we choose $y \in C \cap \bar{A}$, then $C \sim \{y\}$ would be an upper bound of E in $\mathcal{G}$, contradicting the assumption that C is the lub of E.

Solved Problems

LATTICES

5.1. In each of the following diagrams, a partial order $\preccurlyeq$ of a set A is represented. For which of them is $\langle A, \preccurlyeq \rangle$ a lattice? A distributive lattice? A complemented lattice?

(a)

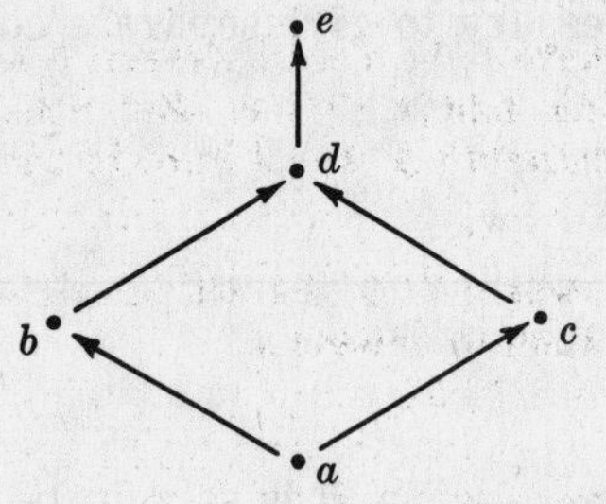

(c)

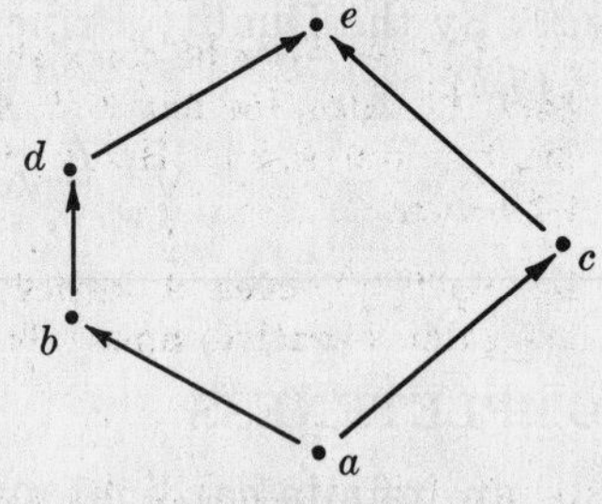

(b)

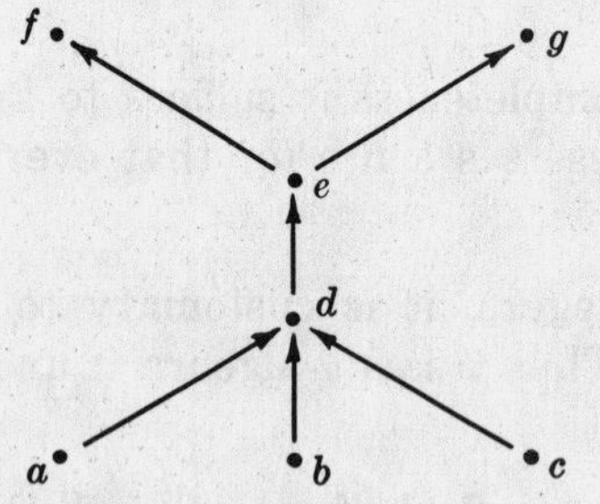

(d)

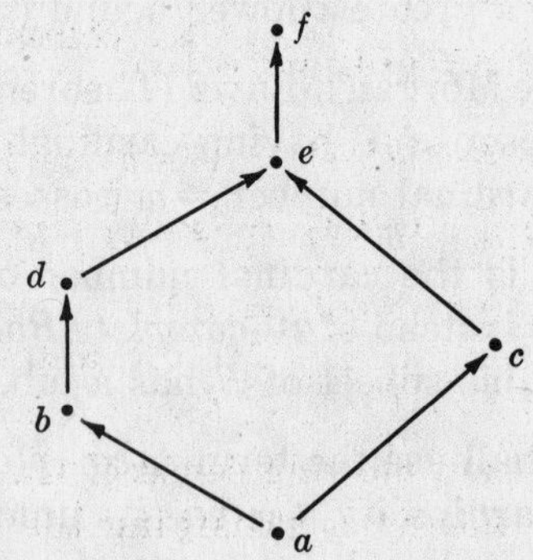

Solution:

(a) This is a distributive lattice with zero a and unit e, but it is not a complemented lattice. (For example, there is no y corresponding to c such that $c \vee y = e$ and $c \wedge y = a$.)

(b) This is not a lattice. (For example, f and g have no lub.)

(c) This is a complemented lattice with zero a and unit e, but it is not distributive. (For example, $d \wedge (b \vee c) = d$, while $(d \wedge b) \vee (d \wedge c) = b$.)

(d) This is a lattice with zero a and unit f, but it is not complemented (c has no complement) and not distributive ($d \wedge (b \vee c) = d \wedge e = d$, while $(d \wedge b) \vee (d \wedge c) = b \vee a = b$.)

5.2. Prove that any finite lattice has a zero element and a unit element.

Solution:

Let $a_1, \ldots, a_n$ be the elements of the lattice, and let $b = a_1 \vee \cdots \vee a_n$. Then b is a unit. For, $a_i \leqq b$ for each i. Similarly, $a_1 \wedge \cdots \wedge a_n$ is a zero.

5.3. (a) Given a set A totally ordered by a binary relation $\leqq$. Prove that $\langle A, \leqq \rangle$ is a distributive lattice. When is $\langle A, \leqq \rangle$ a complemented lattice?

(b) Give an example of a distributive lattice lacking both zero and unit elements.

Solution:

(a) For any x and y in A, either $x \leqq y$ or $y \leqq x$. Then $\max(x, y)$, the larger of x and y, is obviously the lub $x \vee y$, while $\min(x, y)$, the smaller of x and y, is the glb $x \wedge y$. Thus $\langle A, \leqq \rangle$ is a lattice. We must now prove the distributive law (L5) which becomes

$$\min(x, \max(y, z)) = \max(\min(x, y), \min(x, z))$$

First, $\max(y, z) \geqq z$ and $\max(y, z) \geqq y$. Hence $\min(x, \max(y, z)) \geqq \min(x, z)$ and $\min(x, \max(y, z)) \geqq \min(x, y)$. Therefore $\min(x, \max(y, z)) \geqq \max(\min(x, y), \min(x, z))$. Now assume $\min(x, \max(y, z)) > \max(\min(x, y), \min(x, z))$, and we shall derive a contradiction. It follows from our assumption that $x > \max(\min(x, y), \min(x, z))$. Therefore $x > \min(x, y)$ and $x > \min(x, z)$. Hence $x > y$ and $x > z$. Consequently $\max(\min(x, y), \min(x, z)) = \max(y, z)$ and $\min(x, \max(y, z)) = \max(y, z)$, contradicting our assumption. The other distributive law (L6) must hold by virtue of Theorem 5.2.

For $\langle A, \leqq \rangle$ to be complemented, there would have to be a least element 0 and a greatest element 1. Also, for any x in A, $\max(x, x') = 1$ and $\min(x, x') = 0$. But either $x = \max(x, x')$ or $x = \min(x, x')$. Hence $x = 1$ or $x = 0$. Therefore A would have to contain at most two elements.

(b) By Part (a), such a lattice is given by $\langle I, \leqq \rangle$, where I is the set of all integers (positive, zero, and negative) and $\leqq$ is the usual order relation on integers.

5.4. Show that lattices can be characterized by the six laws:

(a) $x \vee y = y \vee x$
(b) $x \wedge y = y \wedge x$
} commutative laws

(c) $x \vee (y \vee z) = (x \vee y) \vee z$
(d) $x \wedge (y \wedge z) = (x \wedge y) \wedge z$
} associative laws

(e) $x \vee (x \wedge y) = x$
(f) $x \wedge (x \vee y) = x$
} absorption laws

in the following sense: if $\langle L, \wedge, \vee \rangle$ is a structure such that $\wedge$ and $\vee$ are binary operation on the set L satisfying the laws (a)-(f), and if we define $x \leqq y$ by $x \wedge y = x$, then $\langle L, \leqq \rangle$ is a lattice with $\text{lub}(\{x, y\}) = x \vee y$ and $\text{glb}(\{x, y\}) = x \wedge y$.

Solution:

We already know, by Theorem 5.1, that any lattice satisfies (a)-(f). Conversely, assume that $\langle L, \wedge, \vee \rangle$ is a structure satisfying (a)-(f) and define $x \leqq y$ by $x \wedge y = x$. Notice that $x \leqq y \leftrightarrow x \vee y = y$. (For, assume $x \leqq y$, i.e. $x \wedge y = x$. Then by (e) and (b), $y = y \vee (y \wedge x) = y \vee x = x \vee y$. Conversely, if $y = x \vee y$, then, by (f), $x = x \wedge (x \vee y) = x \wedge y$, i.e. $x \leqq y$.) Now,

(i) $x \leqq x$, i.e. $x \wedge x = x$. (For, by (f) and (e), $x = x \wedge (x \vee (x \wedge x)) = x \wedge x$.)

(ii) $(x \leqq y \;\&\; y \leqq z) \rightarrow x \leqq z$. (For, we are given $x \wedge y = x$ and $y \wedge z = y$. Then $x \wedge z = (x \wedge y) \wedge z = x \wedge (y \wedge z) = x \wedge y = x$, i.e. $x \leqq z$.)

(iii) $(x \leq y \;\&\; y \leq x) \rightarrow x = y$. (For, $x = x \wedge y = y \wedge x = y$.) Thus $\leq$ is a partial order on L.

(iv) $x \wedge y = \text{glb}(\{x, y\})$. (For, $(x \wedge y) \wedge x = x \wedge y$ implies $x \wedge y \leq x$. Likewise, $x \wedge y \leq y$. Thus $x \wedge y$ is a lower bound of $\{x, y\}$. Assume z is any lower bound of $\{x, y\}$, i.e.

$$z \wedge x = z \quad \& \quad z \wedge y = z$$

Then $z \wedge (x \wedge y) = (z \wedge x) \wedge y = z \wedge y = z$, i.e. $z \leq x \wedge y$.)

(v) $x \vee y = \text{lub}(\{x, y\})$. This is proved in a manner similar to that of (iv).

Remark. Since (*a*) and (*b*), (*c*) and (*d*), and (*e*) and (*f*) are duals of each other, the result we have just demonstrated yields a Duality Principle for lattices.

5.5. (*a*) Show that the following inequalities hold in any lattice.

(i) $(x \wedge y) \vee (x \wedge z) \leq x \wedge (y \vee z)$

(ii) $x \vee (y \wedge z) \leq (x \vee y) \wedge (x \vee z)$

(*b*) Show that each of the following inequalities is a necessary and sufficient condition for a lattice to be distributive.

(iii) $x \wedge (y \vee z) \leq (x \wedge y) \vee (x \wedge z)$

(iv) $(x \vee y) \wedge (x \vee z) \leq x \vee (y \wedge z)$

(v) $x \wedge (y \vee z) \leq (x \wedge y) \vee z$

Solution:

(*a*) By duality, it suffices to prove (i). We have

$$x \wedge y \leq x \wedge (y \vee z) \quad \text{and} \quad x \wedge z \leq x \wedge (y \vee z)$$

Hence $(x \wedge y) \vee (x \wedge z) \leq x \wedge (y \vee z)$.

(*b*) For (iii) we just use (i), and for (iv) we use (ii). For (v) we first assume distributivity. Then

$$x \wedge (y \vee z) \leq (x \wedge y) \vee (x \wedge z) \leq (x \wedge y) \vee z$$

Conversely, let us assume that (v) holds. Since $a \wedge (b \vee c) \leq a$, it follows by (v) that

$$\begin{aligned} a \wedge (b \vee c) \leq a \wedge ((a \wedge b) \vee c) &= a \wedge (c \vee (a \wedge b)) \\ &\leq (a \wedge c) \vee (a \wedge b) \quad \text{by (v)} \\ &= (a \wedge b) \vee (a \wedge c) \end{aligned}$$

Thus we have shown that (iii) holds and therefore that the lattice is distributive.

5.6. Prove that, in any lattice,

$$z \leq x \rightarrow (x \wedge y) \vee z \leq x \wedge (y \vee z)$$

Solution:

Assume $z \leq x$. Hence $z \leq x \wedge (y \vee z)$. Also, $x \wedge y \leq x \wedge (y \vee z)$. Therefore

$$(x \wedge y) \vee z \leq x \wedge (y \vee z)$$

5.7. Let us call a lattice *modular* if and only if it obeys the following law:

$$z \leq x \rightarrow x \wedge (y \vee z) \leq (x \wedge y) \vee z$$

(*a*) Show that a lattice L is modular if and only if it obeys the law

$$z \leq x \rightarrow (x \wedge y) \vee z = x \wedge (y \vee z)$$

(*b*) Prove that distributivity implies modularity.

Solution:

(*a*) This follows immediately from Problem 5.6.

(*b*) This is a direct consequence of Problem 5.5(*b*(v)).

5.8. Determine which of the lattices appearing in Problems 5.1(*c, d*) and Example 5.2 are modular.

Solution:

The lattice of Problem 5.1(*c*) is

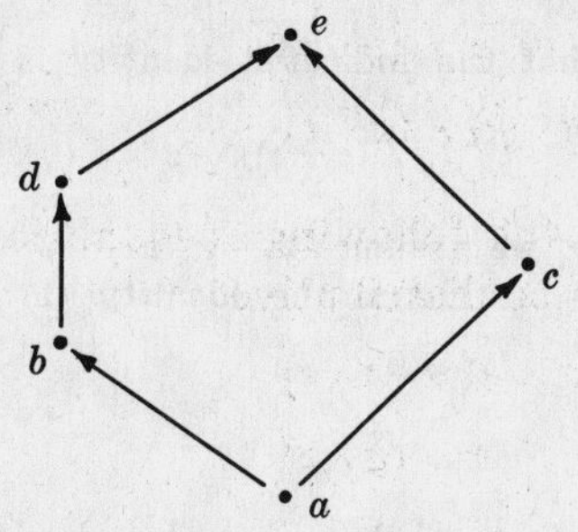

This is not modular, since $b \leq d$ and $b \vee (c \wedge d) \neq (b \vee c) \wedge d$. (Namely, $b \vee (c \wedge d) = b \vee a = b$ and $(b \vee c) \wedge d = e \wedge d = d$.)

The lattice of Problem 5.1(*d*) is

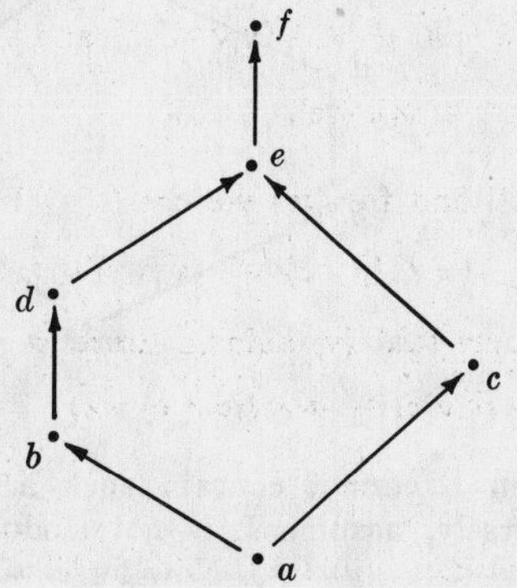

This is not modular since it contains the lattice of Problem 5.5(*c*) as a sublattice†, and we have just seen that the latter is not modular.

The lattice of Example 5.2 is

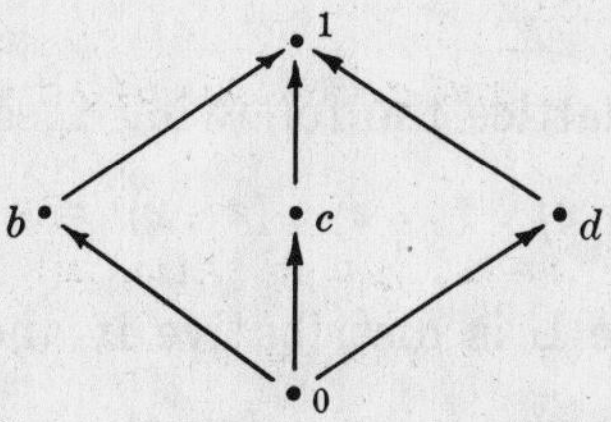

This lattice is modular. To see this, we must verify

$$z \leq x \rightarrow x \wedge (y \vee z) = (x \wedge y) \vee z$$

†By a *sublattice* of a lattice $\langle L, \leq \rangle$ we mean a lattice determined by a subset of L closed under $\wedge$ and $\vee$.

It is obvious that if $z = x$, the result $x \wedge (y \vee x) \leqq (x \wedge y) \vee x$ is an immediate consequence of the absorption laws. Therefore we may assume that $z < x$. It is obvious that this implies that $z = 0$ or $x = 1$. But if $z = 0$ the inequality reduces to $x \wedge y \leqq x \wedge y$, and if $x = 1$ the inequality reduces to $y \vee z \leqq y \vee z$.

5.9. Show that a lattice is modular if and only if it satisfies the identity

$$(z \wedge (x \vee y)) \vee y \;=\; (z \vee y) \wedge (x \vee y)$$

Solution:

Assume L is modular. In the condition for modularity of Problem 5.7(*a*), substitute z for y, y for z and $x \vee y$ for x. The antecedent then becomes the true statement $y \leqq x \vee y$, and we obtain

$$(z \wedge (x \vee y)) \vee y \;=\; (x \vee y) \wedge (z \vee y)$$

Conversely, assume that the indicated identity is true. Exchanging y and z, we obtain the identity

$$(y \wedge (x \vee z)) \vee z \;=\; (y \vee z) \wedge (x \vee z) \qquad (1)$$

Now, to prove modularity, we assume $z \leqq x$ and we have to prove that $(y \wedge x) \vee z = x \wedge (y \vee z)$. But since $z \leqq x$, $z \vee x = x$. Hence the identity (*1*) becomes $(y \wedge x) \vee z = (y \vee z) \wedge x$, which is precisely what is required.

5.10. Prove that a lattice L is modular if and only if it does not contain a sublattice isomorphic to the lattice of Problem 5.1(*c*),

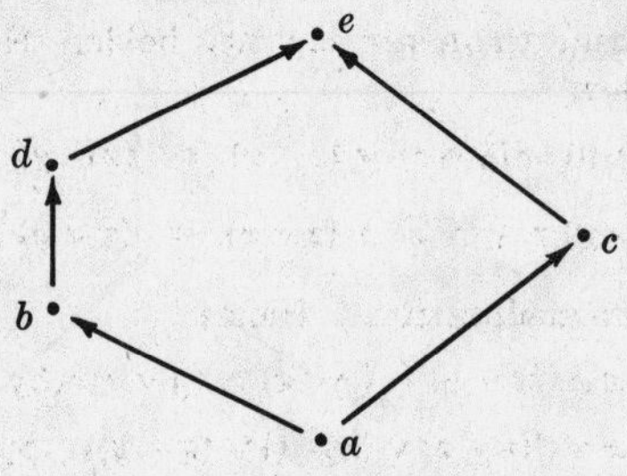

Solution:

If L is modular, then L cannot contain such a sublattice, since the latter is not modular (cf. Problem 5.8). Conversely, assume L is not modular. Then there exist elements x, y, z such that $z \leqq x$ and $z \vee (y \wedge x) < (z \vee y) \wedge x$. Now let $a = y \wedge x$, $b = z \vee (y \wedge x)$, $c = y$, $d = (z \vee y) \wedge x$, $e = z \vee y$. We leave it as an exercise for the reader to show that a, b, c, d and e are pairwise distinct and that $d \wedge c = b \wedge c = a$, $b \vee d = d$, and $c \vee d = c \vee b = e$. This shows that L has a sublattice isomorphic to the one in the diagram above.

5.11. (*a*) Show that in any lattice the following inequality holds.

$$(x \wedge y) \vee (y \wedge z) \vee (z \wedge x) \;\leqq\; (x \vee y) \wedge (y \vee z) \wedge (z \vee x)$$

(*b*) Show that a lattice L is distributive if and only if the following identity holds.

$$(x \wedge y) \vee (y \wedge z) \vee (z \wedge x) \;=\; (x \vee y) \wedge (y \vee z) \wedge (z \vee x)$$

(*c*) Prove that a lattice is distributive if and only if it has no sublattice isomorphic with either the lattice of Problem 5.1(*c*), Fig. 5-4 below, or the lattice of Example 5.2, Fig. 5-5 below.

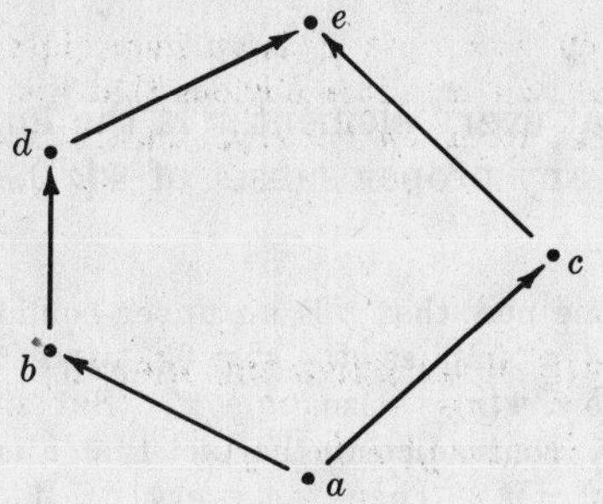

Fig. 5-4

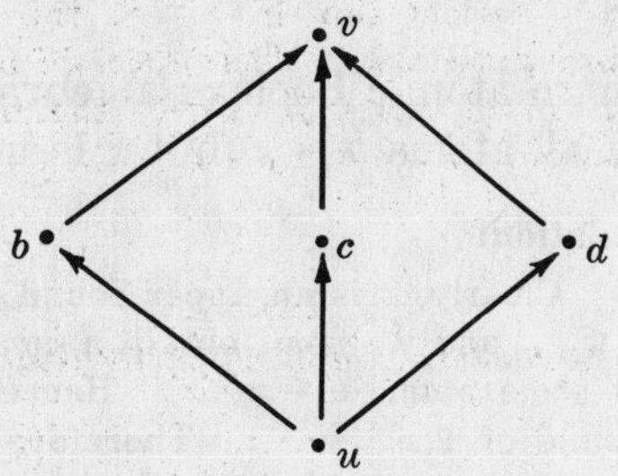

Fig. 5-5

Solution:

(*a*) From $x \wedge y \leqq y \vee z$ and $x \wedge y \leqq z \vee x$, we infer

$$x \wedge y \leqq (x \vee y) \wedge (y \vee z) \wedge (z \vee x)$$

Similarly, we obtain $$y \wedge z \leqq (x \vee y) \wedge (y \vee z) \wedge (z \vee x)$$

and $$z \wedge x \leqq (x \vee y) \wedge (y \vee z) \wedge (z \vee x)$$

Hence $$(x \wedge y) \vee (y \wedge z) \vee (z \wedge x) \leqq (x \vee y) \wedge (y \vee z) \wedge (z \vee x)$$

(*b*) First, assume the lattice is distributive. Then

$$\begin{aligned}(x \vee y) \wedge (y \vee z) \wedge (z \vee x) &= ((x \vee y) \wedge (y \vee z) \wedge z) \vee ((x \vee y) \wedge (y \vee z) \wedge x) \\ &= ((x \vee y) \wedge z) \vee ((y \vee z) \wedge x) \\ &= ((x \wedge z) \vee (y \wedge z)) \vee ((y \wedge x) \vee (z \wedge x)) \\ &= (x \wedge y) \vee (y \wedge z) \vee (z \wedge x)\end{aligned}$$

Conversely, assume the given identity holds. Now let us prove that the lattice is modular. Assume $x \leqq z$. Then

$$(x \wedge y) \vee (y \wedge z) \vee (z \wedge x) = (x \wedge y) \vee (y \wedge z) \vee x = x \vee (y \wedge z)$$

and $$(x \vee y) \wedge (y \vee z) \wedge (z \vee x) = (x \vee y) \wedge (y \vee z) \wedge z = (x \vee y) \wedge z$$

Thus we have shown modularity. Hence

$$\begin{aligned}x \wedge (y \vee z) &= x \wedge (x \vee y) \wedge (x \vee z) \wedge (y \vee z) \quad \text{(by absorption)} \\ &= x \wedge ((y \wedge z) \vee (z \wedge x) \vee (x \wedge y)) \quad \text{(by the assumed identity)} \\ &= [x \wedge (y \wedge z)] \vee ((z \wedge x) \vee (x \wedge y)) \quad \text{(by modularity and } x \geqq (z \wedge x) \vee (x \wedge y)) \\ &= [x \wedge (y \wedge z)] \vee [(x \wedge y) \vee (x \wedge z)] \\ &= (x \wedge y) \vee (x \wedge z) \quad \text{(since } x \wedge (y \wedge z) \leqq (x \wedge y) \vee (x \wedge z))\end{aligned}$$

Thus the lattice is distributive.

(*c*) Clearly, if a lattice is distributive it has no sublattices of either of the two indicated forms, since the latter are both non-distributive (cf. Example 5.2 and Problem 5.8). Conversely, assume the lattice non-distributive. If it is not modular, then the result follows by Problem 5.10. So we may assume modularity. By parts (*a*) and (*b*) of this problem, there exist elements x, y, z such that

$$(x \wedge y) \vee (y \wedge z) \vee (z \wedge x) < (x \vee y) \wedge (y \vee z) \wedge (z \vee x)$$

Let
$$\begin{aligned}u &= (x \wedge y) \vee (y \wedge z) \vee (z \wedge x), \\ v &= (x \vee y) \wedge (y \vee z) \wedge (z \vee x), \\ d &= u \vee (x \wedge v) = (u \vee x) \wedge v, \\ b &= u \vee (y \wedge v) = (u \vee y) \wedge v, \\ c &= u \vee (z \wedge v) = (u \vee z) \wedge v.\end{aligned}$$

We leave it as an exercise for the reader to show that v, d, b, c, u are all distinct and that $d \wedge b = b \wedge c = c \wedge d = u$ and $d \vee b = b \vee c = c \vee d = v$.

ATOMS

5.12. In an atomic Boolean algebra $\mathcal{B}$, prove that every element x is the lub of the set $\Psi(x)$ of all atoms $b \leq x$ (but x is not the lub of any proper subset of $\Psi(x)$).

Solution:

Clearly, x is an upper bound of $\Psi(x)$. Assume now that z is an upper bound of $\Psi(x)$ such that $x \not\leq z$, and we shall obtain a contradiction. $x \not\leq z$ implies $x \wedge z' \neq 0$. Since $\mathcal{B}$ is atomic, there is some atom $b \leq x \wedge z'$. Hence $b \leq x$, i.e. $b \in \Psi(x)$. Also, $b \leq z'$. But since z is an upper bound of $\Psi(x)$, $b \leq z$. Therefore $b \leq z \wedge z' = 0$, contradicting the fact that b is an atom. Lastly, assume x is the lub of $W \subset \Psi(x)$. Let $b \in \Psi(x) \sim W$. Then for every $c \in W$, $c \leq x \wedge b'$ (since $c \wedge (x \wedge b') = (c \wedge x) \wedge b' = c \wedge b' = c$, by properties (ii)-(iii), page 135, of atoms). Hence $x \wedge b'$ is an upper bound of W and therefore $x \leq x \wedge b'$. But $b \leq x$. This implies $b \leq b'$ and therefore $b = 0$, which is a contradiction.

5.13. (*a*) In an atomic Boolean algebra, show that 1 is the lub of the set of all atoms. In particular, when the algebra has a finite number of atoms $a_1, \ldots, a_k$, $1 = a_1 \vee \cdots \vee a_k$.

(*b*) Prove that an atomic Boolean algebra is finite if and only if its set of atoms is finite.

Solution:

(*a*) This is an immediate corollary of Problem 5.12. The additional remark follows from the fact that $a_1 \vee \cdots \vee a_k$ is the lub of $\{a_1, \ldots, a_k\}$.

(*b*) When an algebra is finite, then its set of atoms must be finite. Conversely, assume that there are only finitely many atoms $a_1, \ldots, a_k$. By Problem 5.12, every element x of the algebra is the lub of all the atoms $b \leq x$, and therefore x is of the form $a_{j_1} \vee \cdots \vee a_{j_m}$ where $j_1 < \cdots < j_m \leq k$. But since there are only a finite number of joins of that form, there can be only a finite number of elements in the algebra.

5.14. Show that any infinite Boolean algebra $\mathcal{B}$ contains an infinite set of pairwise-disjoint elements.

Solution:

Case 1. Assume $\mathcal{B}$ is atomic. Then $\mathcal{B}$ has infinitely many atoms, by Problem 5.13(*b*). If x and y are distinct atoms, then $x \wedge y = 0$ (by property (ii) of atoms).

Case 2. $\mathcal{B}$ is not atomic. Then there is an element $x_0 \neq 0$ such that x_0 contains no atom. Hence there is some x_1 such that $0 < x_1 < x_0$. Similarly, there is some x_2 such that $0 < x_2 < x_1$. Proceeding in this manner, we obtain (using the axiom of choice) an infinite sequence $x_0, x_1, x_2, \ldots$ such that $x_0 > x_1 > x_2 > \cdots$. Let $y_0 = x_0 \sim x_1$, $y_1 = x_1 \sim x_2$, $y_2 = x_2 \sim x_3$, ... Then $y_i \wedge y_j = 0$ whenever $i \neq j$.

5.15. Exhibit a Boolean algebra which is not isomorphic to any Boolean algebra of the form $\mathcal{P}(A)$.

Solution:

Consider the Boolean algebra $\mathcal{B}$ of statement bundles, based upon the propositional calculus (cf. Example 3.5). $\mathcal{B}$ is denumerable. For, since there are denumerably many statement forms, there can only be a countable number of statement bundles. However, distinct statement letters determine distinct statement bundles, since distinct statement letters are not logically equivalent. Hence there are denumerably many statement bundles. Assume $\mathcal{B}$ is isomorphic to some $\mathcal{P}(A)$. Then A must be infinite, for otherwise $\mathcal{P}(A)$ would be finite and so would $\mathcal{B}$. Thus $\mathcal{P}(A)$ is denumerable (since it is isomorphic with $\mathcal{B}$) and A is infinite. By Cantor's Theorem (Problem 2.22), A must have smaller cardinality than $\mathcal{P}(A)$. A is equinumerous with a subset of $\mathcal{P}(A)$, and therefore A must be denumerable. But then A would be equinumerous with $\mathcal{P}(A)$ (since both are denumerable), contradicting Cantor's Theorem. (What we have shown, by means of Cantor's Theorem, is that no denumerable Boolean algebra can be isomorphic with any $\mathcal{P}(A)$, and we also exhibited a particular denumerable Boolean algebra.)

5.16. Show that the Boolean algebra $\mathcal{B}$ of statement bundles (cf. Example 3.5) is atomless.

Solution:

Assume [**A**] is an atom. Then $[\mathbf{A}] \neq 0_{\mathcal{B}}$, and **A** cannot be a contradiction. Let A_i be any statement letter not occurring in **A**. Then **A** does not logically imply A_i, for we can assign A_i the value F and assign the statement letters in **A** suitable values so that **A** is T. In addition, A_i & **A** is not a contradiction, since we can make A_i T and, at the same time, make **A** T. Also, A_i & **A** logically implies **A**. Thus $0_{\mathcal{B}} < [A_i \,\&\, \mathbf{A}] < [\mathbf{A}]$, contradicting the assumption that [**A**] is an atom.

This problem provides another way of solving Problem 5.15, since any Boolean algebra $\mathcal{P}(A)$ is atomic.

SYMMETRIC DIFFERENCE. BOOLEAN RINGS.

5.17. In any Boolean algebra, prove

(a) $x + y = (x \vee y) \wedge (x \wedge y)'$

(b) $x + (x \vee y) = x' \wedge y$

(c) $y + (x \wedge y) = x' \wedge y$

(d) $x \vee y = x + y + (x \wedge y)$.

Solution:

(a) $$(x \vee y) \wedge (x \wedge y)' = (x \vee y) \wedge (x' \vee y')$$
$$= (x \wedge x') \vee (x \wedge y') \vee (y \wedge x') \vee (y \wedge y')$$
$$= (x \wedge y') \vee (x' \wedge y) = x + y$$

(b) $$x + (x \vee y) = (x \vee (x \vee y)) \wedge (x \wedge (x \vee y))' \quad \text{(by (a))}$$
$$= (x \vee y) \wedge x' = (x \wedge x') \vee (y \wedge x') = y \wedge x' = x' \wedge y$$

(c) $$y + (x \wedge y) = (y \vee (x \wedge y)) \wedge (y \wedge (x \wedge y))' \quad \text{(by (a))}$$
$$= y \wedge (x \wedge y)' = y \wedge (x' \vee y') = (y \wedge x') \vee (y \wedge y') = x' \wedge y$$

(d) $$(x + y) + (x \vee y) = y + (x + (x \vee y)) = y + (x' \wedge y) \quad \text{(by (b))}$$
$$= x'' \wedge y \quad \text{(by (c))}$$
$$= x \wedge y$$

Thus $(x + y) + (x \vee y) = x \wedge y$. Hence $x \vee y = x + y + (x \wedge y)$, by Theorem 5.9(i).

5.18. Given a Boolean algebra $\mathcal{B} = \langle B, \wedge, \vee, ', 0, 1\rangle$, we have seen that $\mathcal{B}$ determines a Boolean ring with unit $r(\mathcal{B}) = \langle B, +, \wedge, 0\rangle$. Theorem 5.11 tell us that, starting with a Boolean ring $\mathcal{R} = \langle R, +, \times, 0\rangle$ with unit element $1 \neq 0$, and defining $x' = 1 + x$, $x \wedge y = x \times y$, and $x \vee y = x + y + (x \times y)$, we obtain a Boolean algebra

$$b(\mathcal{R}) = \langle R, \wedge, \vee, ', 0, 1\rangle$$

Show that these transformations are inverses of each other in the sense that $b(r(\mathcal{B})) = \mathcal{B}$ and $r(b(\mathcal{R})) = \mathcal{R}$.

Solution:

Start with a Boolean algebra $\mathcal{B} = \langle B, \wedge_{\mathcal{B}}, \vee_{\mathcal{B}}, '^{\mathcal{B}}, 0_{\mathcal{B}}, 1_{\mathcal{B}}\rangle$. Then $r(\mathcal{B}) = \langle B, +_{\mathcal{B}}, \wedge_{\mathcal{B}}, 0_{\mathcal{B}}\rangle$ with unit element $1_{\mathcal{B}}$. Let $\mathcal{C} = b(r(\mathcal{B}))$. By definition of $\mathcal{C}$,

$$x'^{\mathcal{C}} = 1_{\mathcal{B}} +_{\mathcal{B}} x = x'^{\mathcal{B}} \quad \text{(by Theorem 5.9(}h\text{))}$$
$$x \wedge_{\mathcal{C}} y = x \wedge_{\mathcal{B}} y$$
$$x \vee_{\mathcal{C}} y = x +_{\mathcal{B}} y +_{\mathcal{B}} (x \wedge_{\mathcal{B}} y) = x \vee_{\mathcal{B}} y \quad \text{(by Problem 5.17(}d\text{))}$$

Thus $b(r(\mathcal{B})) = \mathcal{B}$.

Now let us start with a Boolean ring $\mathcal{R} = \langle R, +_{\mathcal{R}}, \times_{\mathcal{R}}, 0_{\mathcal{R}} \rangle$ with unit element $1_{\mathcal{R}} \neq 0_{\mathcal{R}}$. Then in $\mathcal{D} = b(\mathcal{R})$,

$$x'^{\mathcal{D}} = 1 +_{\mathcal{R}} x$$

$$x \wedge_{\mathcal{D}} y = x \times_{\mathcal{R}} y$$

$$x \vee_{\mathcal{D}} y = x +_{\mathcal{R}} y +_{\mathcal{R}} (x \times_{\mathcal{R}} y)$$

Let $\mathcal{S} = r(\mathcal{D}) = r(b(\mathcal{R}))$.

$$\begin{aligned}
x +_{\mathcal{S}} y &= (x \wedge_{\mathcal{D}} y'^{\mathcal{D}}) \vee_{\mathcal{D}} (x'^{\mathcal{D}} \wedge_{\mathcal{D}} y) \\
&= (x \times_{\mathcal{R}} (1 +_{\mathcal{R}} y)) \vee_{\mathcal{D}} ((1 +_{\mathcal{R}} x) \times_{\mathcal{R}} y) \\
&= [x \times_{\mathcal{R}} (1 +_{\mathcal{R}} y)] +_{\mathcal{R}} [(1 +_{\mathcal{R}} x) \times_{\mathcal{R}} y] +_{\mathcal{R}} [(x \times_{\mathcal{R}} (1 +_{\mathcal{R}} y)) \times_{\mathcal{R}} ((1 +_{\mathcal{R}} x) \times_{\mathcal{R}} y)] \\
&= [x +_{\mathcal{R}} (x \times_{\mathcal{R}} y)] +_{\mathcal{R}} [y +_{\mathcal{R}} (x \times_{\mathcal{R}} y)] +_{\mathcal{R}} (x \times_{\mathcal{R}} y \times_{\mathcal{R}} (1 +_{\mathcal{R}} x) \times_{\mathcal{R}} (1 +_{\mathcal{R}} y)) \\
&= x +_{\mathcal{R}} y +_{\mathcal{R}} [(x \times_{\mathcal{R}} y) \times_{\mathcal{R}} (1 +_{\mathcal{R}} x +_{\mathcal{R}} y +_{\mathcal{R}} (x \times_{\mathcal{R}} y))] \\
&= x +_{\mathcal{R}} y +_{\mathcal{R}} (x \times_{\mathcal{R}} y) +_{\mathcal{R}} x \times_{\mathcal{R}} (x \times_{\mathcal{R}} y) +_{\mathcal{R}} y \times_{\mathcal{R}} (x \times_{\mathcal{R}} y) +_{\mathcal{R}} (x \times_{\mathcal{R}} y)^2 \\
&= x +_{\mathcal{R}} y +_{\mathcal{R}} (x \times_{\mathcal{R}} y) +_{\mathcal{R}} (x \times_{\mathcal{R}} y) +_{\mathcal{R}} (x \times_{\mathcal{R}} y) +_{\mathcal{R}} (x \times_{\mathcal{R}} y) \\
&= x +_{\mathcal{R}} y
\end{aligned}$$

Also, $x \times_{\mathcal{S}} y = x \wedge_{\mathcal{D}} y = x \times_{\mathcal{R}} y$. Thus $r(b(\mathcal{R})) = \mathcal{R}$.

5.19. Solutions of Equations.

(*a*) Show that any equation $\tau = \sigma$ is equivalent to an equation of the form $\rho = 0$.

(*b*) Show that a finite system of equations $\tau_1 = 0, \ldots, \tau_n = 0$ is equivalent to a single equation $\sigma = 0$.

(*c*) Find necessary and sufficient conditions for the existence of a solution u of an equation $\tau(u, u_1, \ldots, u_k) = 0$, and, when there is a solution, find them all.

(*d*) Show that $\tau(u, u_1, \ldots, u_k) = 0$ has a unique solution in u if and only if $\tau = u + \rho$, where ρ does not contain u.

Solution:

(*a*) $\tau = \sigma$ if and only if $(\tau \wedge \sigma') \vee (\tau' \wedge \sigma) = 0$, by Theorem 5.9(*j*).

(*b*) $\tau_1 = 0 \;\&\; \ldots \;\&\; \tau_n = 0 \;\leftrightarrow\; \tau_1 \vee \cdots \vee \tau_n = 0$.

(*c*) Write τ in disjunctive normal form. Grouping those terms involving u and those involving u', we obtain

$$\tau = (u \wedge \sigma_1) \vee (u' \wedge \sigma_2)$$

Hence

$$\begin{aligned}
\tau = 0 &\leftrightarrow (u \wedge \sigma_1) \vee (u' \wedge \sigma_2) = 0 \\
&\leftrightarrow u \wedge \sigma_1 = 0 \;\&\; u' \wedge \sigma_2 = 0 \\
&\leftrightarrow u \leq \sigma_1' \;\&\; \sigma_2 \leq u \\
&\leftrightarrow \sigma_2 \leq u \leq \sigma_1'
\end{aligned}$$

Hence there is a solution if and only if $\sigma_2 \leq \sigma_1'$, and the solutions are all u such that $\sigma_2 \leq u \leq \sigma_1'$. Hence the set of solutions consists of all $(\sigma_2 \vee w) \wedge \sigma_1'$, for all values of w. (For, if $\sigma_2 \leq u \leq \sigma_1'$, then $(\sigma_2 \vee u) \wedge \sigma_1' = u \wedge \sigma_1' = u$. Conversely, if $\sigma_2 \leq \sigma_1'$, then, for any w, $\sigma_2 \leq (\sigma_2 \vee w) \wedge \sigma_1' \leq \sigma_1'$.) If we wish to solve for the remaining variables, we then solve the inequality $\sigma_2 \leq \sigma_1'$, which is equivalent to the equality $\sigma_2 \wedge \sigma_1 = 0$.

(*d*) By (*c*), if there is a unique solution, $\sigma_2 = \sigma_1'$. (Otherwise, σ_2 and σ_1' would be distinct solutions.) Hence

$$\tau = (u \wedge \sigma_1) \vee (u' \wedge \sigma_1') = u + \sigma_1' = u + (1 + \sigma_1)$$

Conversely, if $\tau = u + \rho$, then $\tau = 0 \rightarrow u = \rho$.

5.20. Find necessary and sufficient conditions for existence of a solution and find all solutions of the following equations.

(a) $u \vee z = w$, (b) $u \wedge z = u \wedge w$.

Solution:

(a)
$$\begin{aligned} u \vee z = w \quad &\leftrightarrow \quad [(u \vee z) \wedge w'] \vee [(u \vee z)' \wedge w] = 0 \\ &\leftrightarrow \quad (u \wedge w') \vee (z \wedge w') \vee (u' \wedge z' \wedge w) = 0 \\ &\leftrightarrow \quad (u \wedge w') \vee (u \wedge z \wedge w') \vee (u' \wedge z \wedge w') \vee (u' \wedge z' \wedge w) = 0 \\ &\leftrightarrow \quad (u \wedge [w' \vee (z \wedge w')]) \vee (u' \wedge [(z \wedge w') \vee (z' \wedge w)]) = 0 \\ &\leftrightarrow \quad (u \wedge \underbrace{w'}_{\sigma_1}) \vee (u' \wedge \underbrace{(z + w)}_{\sigma_2}) = 0 \end{aligned}$$

$$\begin{aligned} \text{A solution exists} \quad &\leftrightarrow \quad \sigma_2 \leqq \sigma_1' \\ &\leftrightarrow \quad z + w \leqq w \\ &\leftrightarrow \quad (z + w) \wedge w' = 0 \\ &\leftrightarrow \quad (z \wedge w') + (w \wedge w') = 0 \\ &\leftrightarrow \quad z \wedge w' = 0 \end{aligned}$$

When $z \wedge w' = 0$, all solutions are of the form $((z + w) \vee x) \wedge w$ for arbitrary x. But

$$[(z + w) \wedge w] \vee [x \wedge w] = (w \wedge z') \vee (x \wedge w) = w \wedge (x \vee z')$$

The equation $z \wedge w' = 0$ always has solutions z and the solutions are all elements $y \wedge w$ for arbitrary y. Thus

$$z = y \wedge w$$

$$\begin{aligned} u = w \wedge (x \vee z') &= w \wedge (x \vee (y \wedge w)') = w \wedge (x \vee y' \vee w') \\ &= w \wedge (x \vee y') \quad \text{for arbitrary } x, y \text{ and } w \end{aligned}$$

(b)
$$\begin{aligned} u \wedge z = u \wedge w \quad &\leftrightarrow \quad [(u \wedge z) \wedge (u \wedge w)'] \vee [(u \wedge w) \wedge (u \wedge z)'] = 0 \\ &\leftrightarrow \quad [(u \wedge z) \wedge (u' \vee w')] \vee [(u \wedge w) \wedge (u' \vee z')] = 0 \\ &\leftrightarrow \quad (u \wedge z \wedge w') \vee (u \wedge w \wedge z') = 0 \\ &\leftrightarrow \quad u \wedge [(z \wedge w') \vee (w \wedge z')] = 0 \\ &\leftrightarrow \quad u \wedge (w + z) = 0 \end{aligned}$$

A solution exists if and only if $0 \leqq (w + z)'$. Hence a solution always exists. The solutions are $u = x \wedge (w + z)'$ for all x, w, z.

5.21. Find all solutions of the system

$$\begin{cases} u = (z \wedge w') \vee v & (a) \\ z \leqq (v \wedge w) \vee u & (b) \\ w \vee v = z' \vee u & (c) \end{cases}$$

Solution:

(a) is equivalent to

$$(u \wedge [(z' \vee w) \wedge v']) \vee (u' \wedge [(z \wedge w') \vee v]) = 0 \qquad (a')$$

(b) is equivalent to

$$z \wedge [(v' \vee w') \wedge u'] = 0 \qquad (b')$$

(c) is equivalent to

$$[(w \vee v) \wedge z \wedge u'] \vee [w' \wedge v' \wedge (z' \vee u)] = 0 \qquad (c')$$

Hence we must solve the system (a'), (b'), (c'). By Problem 5.19(*b*), this system is equivalent to the single equation

$$(u \wedge [(z' \vee w) \wedge v']) \vee (u' \wedge [(z \wedge w') \vee v]) \vee (z \wedge [(v' \vee w') \wedge u']) \vee [(w \vee v) \wedge z \wedge u'] \vee [w' \wedge v' \wedge (z' \vee u)] = 0$$

which is equivalent to

$$(u \wedge \{[(z' \vee w) \wedge v'] \vee [w' \wedge v']\}) \vee (u' \wedge \{[(z \wedge w') \vee v] \vee [z \wedge (v' \vee w')] \vee [(w \vee v) \wedge z] \vee [w' \wedge v' \wedge z']\}) = 0$$

Hence σ_1 is $[(z' \vee w) \wedge v'] \vee [w' \wedge v']$, and σ_2 is

$$[(z \wedge w') \vee v] \vee [z \wedge (v' \vee w')] \vee [(w \vee v) \wedge z] \vee [w' \wedge v' \wedge z']$$

By easy calculation, $\sigma_2 = v \vee z \vee w'$ and $\sigma_1 = v'$. Hence a solution exists if and only if $v \vee z \vee w' \leq v$. But the latter equation is equivalent to $v' \wedge (z \vee w') = 0$. By Problem 5.20(*b*), substituting v' for u, $z \vee w'$ for z, and 0 for w, we obtain the solution $v' = x \wedge (z \vee w')' = x \wedge (z' \wedge w)$, i.e. $v = x' \vee z \vee w'$. The solution for u is $u = (\sigma_2 \vee y) \wedge \sigma_1' = (v \vee z \vee w' \vee y) \wedge v = v$. Thus the solutions are $u = v = x' \vee z \vee w'$ for arbitrary x, z, w.

AXIOMATIZATIONS

5.22. In our axiom system for Boolean algebra (cf. Section 3.2), prove the independence of each of Axioms (3)-(4) and (7)-(9). Show that each of Axioms (1)-(2) and (5)-(6) is not independent. (A member **A** of a system $\mathcal{W}$ of axioms is said to be *independent* if and only if **A** is not provable from the set $\mathcal{W}$ − **A** of the other axioms.)

Solution:

Axiom (1) is derived as follows:

(*a*) $x \wedge x = x$ and $x \vee x = x$. (This is Theorem 3.3 on p. 54. The proof given there does not use Axiom (1).)

(*b*) $[y \wedge x = z \wedge x \;\&\; y \wedge x' = z \wedge x'] \rightarrow y = z$. (This is Theorem 3.5(v) on p. 55. The proof given there does not use Axiom (1).)

(*c*) $x \vee 1 = 1$.

Proof. $x \vee 1 = (x \vee 1) \wedge 1 = (x \vee 1) \wedge (x \vee x') = x \vee (1 \wedge x') = x \vee (x \wedge 1) = x \vee x' = 1$.

(*d*) $x = 0 \vee x$.

Proof. $x = x \wedge 1 = x \wedge (x' \vee 1) = (x \wedge x') \vee (x \wedge 1) = 0 \vee x$.

(*e*) $x \wedge 0 = 0$.

Proof. $0 = x \wedge x' = x \wedge (0 \vee x') = (x \wedge 0) \vee (x \wedge x') = (x \wedge 0) \vee 0 = x \wedge 0$.

(*f*) $(x \vee y) \wedge x = x$.

Proof. $(x \vee y) \wedge x = x \wedge (x \vee y) = (x \vee 0) \wedge (x \vee y) = x \vee (0 \wedge y) = x \vee (y \wedge 0) = x \vee 0 = x$.

(*g*) $(x \wedge y) \vee x = x$.

Proof. $(x \wedge y) \vee x = (x \wedge y) \vee (x \wedge 1) = x \wedge (y \vee 1) = x \wedge 1 = x$.

(*h*) $x \vee y = y \vee x$.

Proof. By (*b*), it suffices to prove: (i) $(x \vee y) \wedge x = (y \vee x) \wedge x$ and (ii) $(x \vee y) \wedge x' = (y \vee x) \wedge x'$.

(i) $(x \vee y) \wedge x = x$ (by (*f*)).
$(y \vee x) \wedge x = x \wedge (y \vee x) = (x \wedge y) \vee (x \wedge x) = (x \wedge y) \vee x = x$.

(ii) $(x \vee y) \wedge x' = x' \wedge (x \vee y) = (x' \wedge x) \vee (x' \wedge y) = (x \wedge x') \vee (x' \wedge y) = 0 \vee (x' \wedge y) = x' \wedge y$.
$(y \vee x) \wedge x' = x' \wedge (y \vee x) = (x' \wedge y) \vee (x' \wedge x) = (x' \wedge y) \vee (x \wedge x') = (x' \wedge y) \vee 0 = x' \wedge y$.

Axiom (2) is derived in a similar manner.

For the independence of Axiom (3), $x \wedge (y \vee z) = (x \wedge y) \vee (x \wedge z)$, use $0' = 1$, $1' = 0$, and

$\vee$	0	1
0	0	1
1	1	1

$\wedge$	0	1
0	1	0
1	0	1

Note that $0 \wedge (0 \vee 1) = 0 \wedge 1 = 0$, but $(0 \wedge 0) \vee (0 \wedge 1) = 1 \vee 0 = 1$.

For the independence of Axiom (4), $x \vee (y \wedge z) = (x \vee y) \wedge (x \vee z)$, use $0' = 1$, $1' = 0$, and

$\vee$	0	1
0	0	1
1	1	0

$\wedge$	0	1
0	0	0
1	0	1

For the independence of Axiom (7), $x \vee x' = 1$, let the domain of the model be $\mathcal{P}(A)$, where A is any non-empty set, take $\wedge$ to be $\cap$, and $\vee$ to be $\cup$. Let 0 be $\emptyset$ and let 1 be A. However, let $x' = 0$ for all x.

For the independence of Axiom (8), use the same model as for Axiom (7), except that $x' = A = 1$ for all x.

For the independence of Axiom (9), use the model $\{\emptyset\}$, with $\wedge$, $\vee$, $'$ as $\cap$, $\cup$, and $^-$ $(\emptyset \cap \emptyset = \emptyset \cup \emptyset = \bar{\emptyset} = \emptyset)$; $0 = 1 = \emptyset$.

To show that Axiom (5), $x \vee 0 = x$, is provable from the rest, note first that $x \vee 1 = 1$ for all x. For,

$$1 \;=\; x \vee x' \;=\; x \vee (x' \wedge 1) \;=\; (x \vee x') \wedge (x \vee 1) \;=\; 1 \wedge (x \vee 1) \;=\; x \vee 1$$

Hence $$x \vee 0 \;=\; x \vee (x \wedge x') \;=\; (x \wedge 1) \vee (x \wedge x') \;=\; x \wedge (1 \vee x') \;=\; x \wedge 1 \;=\; x$$

To show that Axiom (6), $x \wedge 1 = x$, follows from the rest, "dualize" the proof just given for the axiom $x \vee 0 = x$. Thus, first, $x \wedge 0 = 0$ for all x. For,

$$0 \;=\; x \wedge x' \;=\; x \wedge (x' \vee 0) \;=\; (x \wedge x') \vee (x \wedge 0) \;=\; 0 \vee (x \wedge 0) \;=\; x \wedge 0$$

Hence $$x \wedge 1 \;=\; x \wedge (x \vee x') \;=\; (x \vee 0) \wedge (x \vee x') \;=\; x \vee (0 \wedge x) \;=\; x \vee 0 \;=\; x$$

Detailed verification that the examples in the independence proofs satisfy the remaining axioms is left to the reader.

IDEALS

5.23. If C is a subset of a Boolean algebra $\mathcal{B}$ and if an ideal J contains C and is contained in every ideal containing C, show that J is the ideal Gen (C) generated by C.

Solution:

We have to show that J is equal to the intersection H of all ideals containing C. Since J is contained in every ideal containing C, it follows that $J \subseteq H$. On the other hand, since J is itself an ideal containing C, $H \subseteq J$. Therefore, $J = H$.

5.24. If J is an ideal of a Boolean algebra $\mathcal{B}$ and $y \in B$, prove that Gen $(J \cup \{y\})$ is a proper ideal if and only if $y' \notin J$.

Solution:

Assume $y' \in J$. Hence $1 = y \vee y' \in$ Gen $(J \cup \{y\})$, and therefore Gen $(J \cup \{y\})$ is not a proper ideal. Conversely, assume that Gen $(J \cup \{y\})$ is not a proper ideal. Then $1 \in$ Gen $(J \cup \{y\})$, and by Theorem 5.17 there exist $z \in B$ and $x \in J$ such that $1 = (z \wedge y) \vee x$. Hence

$$y' \;=\; y' \wedge 1 \;=\; y' \wedge ((z \wedge y) \vee x) \;=\; (y' \wedge (z \wedge y)) \vee (y' \wedge x) \;=\; y' \wedge x$$

But since $x \in J$, $y' \wedge x \in J$ and therefore $y' \in J$.

5.25. By a congruence relation R on a Boolean algebra $\mathcal{B}$ we mean a binary relation R on B satisfying the following properties.

(a) xRx (reflexivity)

(b) $xRy \rightarrow yRx$ (symmetry)

(c) $(xRy \;\&\; yRz) \rightarrow xRz$ (transitivity)

(d) $xRy \rightarrow (x'Ry' \;\&\; (x \wedge z)R(y \wedge z))$.

Define $J_R = \{x : xR0\}$. Prove: (i) J_R is an ideal. (ii) J_R is a proper ideal if and only if $\neg(0R1)$. (iii) $x \equiv_{J_R} y$ if and only if xRy.

Solution:

(i) By (a), $0 \in J_R$. Assume $x \in J_R$ and $y \in J_R$. Hence $xR0$ and $yR0$. By (d) $x'R1$, and, again by (d), $(x' \wedge y')Ry'$. But from (d), $yR0$ implies $y'R1$. Hence by (c), $(x' \wedge y')R1$, and, again by (d), $(x' \wedge y')'R0$, i.e. $x \vee y \in J_R$. Now assume that $x \in J_R$ and $z \in B$. Then $xR0$ and, by (d), $(x \wedge z)R0$, i.e. $x \wedge z \in J_R$.

(ii) This follows immediately from the fact that an ideal is proper if and only if it does not contain 1.

(iii) Since $+$ is definable in terms of the meet and complement, it follows by (d) that $xRy \rightarrow (x+y)R(y+y)$, i.e. $xRy \rightarrow (x+y)R0$. But $(x+y)R0$ is equivalent to $x+y \in J_R$, which in turn is equivalent by definition to $x \equiv_{J_R} y$.

5.26. A subset F of a Boolean algebra $\mathcal{B}$ is said to be a *filter* if and only if: (i) F is non-empty; (ii) $(x \in F \;\&\; y \in F) \rightarrow x \wedge y \in F$; (iii) $x \in F \;\&\; y \in B \rightarrow x \vee y \in F$. By an *ultrafilter* we mean a proper filter which is contained in no other proper filter. Prove:

(a) F is a filter if and only if $F' = \{x' : x \in F\}$ is an ideal.

(b) F is an ultrafilter if and only if F' is a maximal ideal.

(c) Assumption (iii) in the definition of filter may be replaced by

$$\text{(iii}'\text{)} \quad x \in F \;\&\; x \leq y \rightarrow y \in F$$

Solution:

(a) Assume F is a filter. Given $x \in F'$, $y \in F'$, $z \in B$. Then $x' \in F$ and $y' \in F$. Hence $x' \wedge y' \in F$ and $x' \vee z' \in F$. Therefore $x \vee y = (x' \wedge y')' \in F'$ and $x \wedge z = (x' \vee z')' \in F'$. Thus F' is an ideal. The converse is left as an exercise for the reader.

(b) This is an immediate consequence of (a).

(c) The equivalence between (iii) and (iii′) follows from the equivalence between $x \leq y$ and $x \vee y = y$.

5.27. Call a Boolean algebra $\mathcal{B}$ *simple* if and only if $\{0\}$ is the only proper ideal. Prove that $\mathcal{B}$ is simple if and only if $B = \{0, 1\}$.

Solution:

Clearly, if $B = \{0,1\}$, then $\{0\}$ is the only proper ideal. Conversely, assume $\mathcal{B}$ is simple. Let x be any element of B different from 1. Then the principal ideal J_x is a proper ideal, since $1 \not\leq x$. Since $\mathcal{B}$ is simple, $J_x = \{0\}$. But $x \in J_x$ and therefore $x = 0$, i.e. $B = \{0, 1\}$.

5.28. In a Boolean algebra $\mathcal{B}$, a set C of ideals of $\mathcal{B}$ is said to be an $\subset$-*chain* of ideals if and only if, for any J_1 and J_2 in C, either $J_1 \subseteq J_2$ or $J_2 \subseteq J_1$. (This amounts to saying that the relation $\subseteq$ totally orders C.) Prove that the union of an $\subset$-chain of ideals is again an ideal, and, if each ideal in C is proper, so is the union.

Solution:

Let $H = \bigcup_{J \in C} J$, where C is an $\subset$-chain of ideals. Given x and y in H, and z in B. Then $x \in J_1$ and $y \in J_2$ for some J_1 and J_2 in C. Since C is an $\subset$-chain, either $J_1 \subseteq J_2$ or $J_2 \subseteq J_1$, say, $J_1 \subseteq J_2$. Hence $x \in J_2$ and $y \in J_2$. Since J_2 is an ideal, $x \vee y \in J_2$ and $x \wedge z \in J_2$. But since $J_2 \subseteq H$, we obtain: $x \vee y \in H$ and $x \wedge z \in H$. Hence H is an ideal. If each ideal in C is proper, then $1 \notin J$ for each J in C. Hence $1 \notin H$ and therefore H is also proper.

5.29. If $\mathcal{R} = \langle R, +, \times, 0\rangle$ is a commutative ring, then a non-empty subset J of R is called a *ring-theoretic ideal* if and only if

(i) $(x \in J \ \&\ y \in J) \rightarrow x - y \in J$;

(ii) $(x \in J \ \&\ z \in R) \rightarrow x \times z \in J$.

Prove that if J is a subset of a Boolean algebra $\mathcal{B} = \langle B, \wedge, \vee, ', 0, 1\rangle$, and $\mathcal{R} = \langle B, +, \wedge, 0\rangle$ is the corresponding Boolean ring, then J is an ideal of $\mathcal{B}$ if and only if J is a ring-theoretic ideal of $\mathcal{R}$.

Solution:

Notice that, for Boolean rings, $(-y) = y$, and therefore we may replace $x - y$ in condition (i) by $x + y$. Now assume that J is an ideal in $\mathcal{B}$. We already know that $(x \in J \ \&\ y \in J) \rightarrow x + y \in J$, which is condition (i), while condition (ii) reads $(x \in J \ \&\ z \in B) \rightarrow x \wedge z \in J$, which is part of the definition of an ideal. Conversely, assume that J is a ring-theoretic ideal. By (ii), $(x \in J \ \&\ z \in B) \rightarrow x \wedge z \in J$. Now it remains to show that $(x \in J \ \&\ y \in J) \rightarrow x \vee y \in J$. So assume $x \in J \ \&\ y \in J$. By (ii), $x \wedge y \in J$, and, since $x \vee y = x + y + (x \wedge y)$, we may conclude by (i) that $x \vee y \in J$.

5.30. Show that, if J is an ideal of a Boolean algebra $\mathcal{B}$, then $x \equiv_J y$ if and only if there exists some element z in J such that $x \vee z = y \vee z$.

Solution:

Assume $x \equiv_J y$, i.e. $x + y \in J$. Let $z = x + y$. Then

$$x \vee z = x \vee (x+y) = x \vee y = y \vee (x+y) = y \vee z$$

Conversely, assume $x \vee z = y \vee z$ for some z in J. Then

$$x \wedge y' \leq (x \vee z) \wedge y' = (y \vee z) \wedge y' = z \wedge y' \leq z$$

Hence $x \wedge y' \in J$. Also,

$$x' \wedge y \leq x' \wedge (y \vee z) = x' \wedge (x \vee z) = x' \wedge z \leq z$$

Hence $x' \wedge y \in J$. Therefore $x + y = (x \wedge y') \vee (x' \wedge y) \in J$.

QUOTIENT ALGEBRAS

5.31. Let $\mathcal{B}$ be the Boolean algebra $\mathcal{P}(A)$, where A is some infinite non-empty set, and let J be the ideal of finite subsets of A. Prove that the quotient algebra $\mathcal{B}/J$ is atomless.

Solution:

Given an element $[X]$ of $\mathcal{B}/J$ such that $[X] \neq [0]$. Hence X is infinite. Then there is an infinite set Y such that $Y \subseteq X$ and $X \sim Y$ is infinite. (To see this, enumerate a subset of X, $\{a_1, a_2, \ldots\}$, and let $Y = \{a_1, a_3, a_5, \ldots\}$.) Since Y is infinite, $[Y] \neq 0_{\mathcal{B}/J}$. Also, since $Y \subseteq X$, $[Y] \leq [X]$. However, since $X \sim Y$ is infinite, $X + Y \notin J$, i.e. $X \not\equiv_J Y$. Hence $[X] \neq [Y]$. Thus $0_{\mathcal{B}/J} < [Y] < [X]$, and therefore $[X]$ cannot be an atom.

5.32. Given Boolean algebras $\mathcal{A} = \langle A, \wedge_{\mathcal{A}}, \vee_{\mathcal{A}}, '^{\mathcal{A}}, 0_{\mathcal{A}}, 1_{\mathcal{A}} \rangle$ and $\mathcal{B} = \langle B, \wedge_{\mathcal{B}}, \vee_{\mathcal{B}}, '^{\mathcal{B}}, 0_{\mathcal{B}}, 1_{\mathcal{B}} \rangle$. A function f from A into B is called a *homomorphism* from $\mathcal{A}$ *into* $\mathcal{B}$ if and only if

(i) $f(x'^{\mathcal{A}}) = (f(x))'^{\mathcal{B}}$;

(ii) $f(x \wedge_{\mathcal{A}} y) = f(x) \wedge_{\mathcal{B}} f(y)$.

If such a function has its range equal to all of B, then it is called a homomorphism from $\mathcal{A}$ *onto* $\mathcal{B}$, and $\mathcal{B}$ is called a *homomorphic image* of $\mathcal{A}$. A homomorphism f is an *isomorphism* of $\mathcal{A}$ *into* $\mathcal{B}$ if and only if f is one-one, and f is called an isomorphism of $\mathcal{A}$ *onto* $\mathcal{B}$ if and only if f is an isomorphism of $\mathcal{A}$ into $\mathcal{B}$ and the range of f is B. We say that $\mathcal{A}$ and $\mathcal{B}$ are *isomorphic* if and only if there is an isomorphism of $\mathcal{A}$ onto $\mathcal{B}$. If f is any function from A into B, by the *kernel* K_f we mean the set $\{x : x \in A \ \&\ f(x) = 0_{\mathcal{B}}\}$. Prove:

(*a*) If f is a homomorphism from $\mathcal{A}$ into $\mathcal{B}$, then

(1) $f(x \vee_{\mathcal{A}} y) = f(x) \vee_{\mathcal{B}} f(y)$;

(2) $f(x +_{\mathcal{A}} y) = f(x) +_{\mathcal{B}} f(y)$;

(3) $f(0_{\mathcal{A}}) = 0_{\mathcal{B}}$ and $f(1_{\mathcal{A}}) = 1_{\mathcal{B}}$.

(*b*) If f is a homomorphism from $\mathcal{A}$ into $\mathcal{B}$, the range $f(A)$ determines a subalgebra of $\mathcal{B}$.

(*c*) If f is a homomorphism from $\mathcal{A}$ into $\mathcal{B}$, and C determines a subalgebra of $\mathcal{B}$, then $f^{-1}(C) = \{x : x \in A \ \&\ f(x) \in C\}$ determines a subalgebra of $\mathcal{A}$.

(*d*) The identity mapping I_A is an isomorphism from $\mathcal{A}$ onto $\mathcal{A}$. (Hence the relation "isomorphic" is reflexive.)

(*e*) If f is an isomorphism from $\mathcal{A}$ onto $\mathcal{B}$, then the inverse function f^{-1} from B onto A is an isomorphism from $\mathcal{B}$ onto $\mathcal{A}$. (Hence the relation "isomorphic" is symmetric.)

(*f*) If f is a homomorphism from $\mathcal{A}$ into $\mathcal{B}$, and g is a homomorphism from $\mathcal{B}$ into an algebra $\mathcal{C} = \langle C, \wedge_{\mathcal{C}}, \vee_{\mathcal{C}}, '^{\mathcal{C}}, 0_{\mathcal{C}}, 1_{\mathcal{C}} \rangle$, then the composition $g \circ f$ is a homomorphism from $\mathcal{A}$ into $\mathcal{C}$. In particular, if f and g are isomorphisms onto, then so is $g \circ f$. (Hence the relation "isomorphic" is transitive.)

(*g*) A homomorphism f from $\mathcal{A}$ into $\mathcal{B}$ is an isomorphism of $\mathcal{A}$ into $\mathcal{B}$ if and only if the kernel $K_f = \{0_{\mathcal{A}}\}$.

(*h*) If J is an ideal of $\mathcal{A}$, then the function $f(x) = x + J$ for all x in A is a homomorphism from $\mathcal{A}$ onto $\mathcal{A}/J$ (called the *natural* homomorphism from $\mathcal{A}$ onto $\mathcal{A}/J$).

(*i*) If $\mathcal{B}$ is a homomorphic image of $\mathcal{A}$, then there is an ideal J of $\mathcal{A}$ such that $\mathcal{B}$ and $\mathcal{A}/J$ are isomorphic.

Solution:

(*a*)
$$\begin{aligned} f(x \vee_{\mathcal{A}} y) &= f((x'^{\mathcal{A}} \wedge_{\mathcal{A}} y'^{\mathcal{A}})'^{\mathcal{A}}) \\ &= (f(x'^{\mathcal{A}} \wedge_{\mathcal{A}} y'^{\mathcal{A}}))'^{\mathcal{B}} = (f(x'^{\mathcal{A}}) \wedge_{\mathcal{B}} f(y'^{\mathcal{A}}))'^{\mathcal{B}} \\ &= (f(x)'^{\mathcal{B}} \wedge_{\mathcal{B}} f(y)'^{\mathcal{B}})'^{\mathcal{B}} = f(x) \vee_{\mathcal{B}} f(y) \end{aligned}$$

The proof for $+$ is similar, since $+$ is definable in terms of $\wedge, \vee, '$. Now

$$f(0_{\mathcal{A}}) = f(x \wedge_{\mathcal{A}} x'^{\mathcal{A}}) = f(x) \wedge_{\mathcal{B}} f(x'^{\mathcal{A}}) = f(x) \wedge_{\mathcal{B}} f(x)'^{\mathcal{B}} = 0_{\mathcal{B}}$$

Hence $f(1_{\mathcal{A}}) = f(0_{\mathcal{A}}'^{\mathcal{A}}) = f(0_{\mathcal{A}})'^{\mathcal{B}} = 0_{\mathcal{B}}'^{\mathcal{B}} = 1_{\mathcal{B}}$.

(*b*) From now on we shall omit subscripts $\mathcal{A}$ and $\mathcal{B}$, wherever this is not likely to cause confusion. Assume now that f is a homomorphism from $\mathcal{A}$ into $\mathcal{B}$. Assume $u, v \in f(A)$. Then $u = f(x)$ and $v = f(y)$ for some x, y in A. Hence $u \wedge v = f(x) \wedge f(y) = f(x \wedge y) \in f(A)$. Similarly, $u' = f(x)' = f(x') \in f(A)$. Hence $f(A)$ determines a subalgebra of $\mathcal{B}$.

(*c*) Assume $x, y \in f^{-1}(C)$. Then $f(x \wedge y) = f(x) \wedge f(y) \in C$, since C is closed under $\wedge$. Similarly, $f(x') = f(x)' \in C$, since C is closed under complementation. Thus $x \wedge y \in f^{-1}(C)$ and $x' \in f^{-1}(C)$.

(*d*) This is obvious.

(*e*) Given $u, v \in B$. Then

$$f(f^{-1}(u \wedge v)) = u \wedge v = f(f^{-1}(u)) \wedge f(f^{-1}(v)) \;=\; f(f^{-1}(u) \wedge f^{-1}(v))$$

Since f is one-one, $f^{-1}(u \wedge v) = f^{-1}(u) \wedge f^{-1}(v)$. Similarly, $f(f^{-1}(u')) = u' = (ff^{-1}(u))' = f((f^{-1}(u))')$. Since f is one-one, $f^{-1}(u') = (f^{-1}(u))'$.

(*f*)

$$\begin{aligned}(g \circ f)(x \wedge y) &= g(f(x \wedge y)) = g(f(x) \wedge f(y)) \\ &= g(f(x)) \wedge g(f(y)) = (g \circ f)(x) \wedge (g \circ f)(y)\end{aligned}$$

Similarly, $$(g \circ f)(x') = g(f(x')) = g(f(x)') = (g(f(x)))' = ((g \circ f)(x))'$$

(*g*) Assume f is one-one, and let $x \in K_f$. Then $f(x) = 0 = f(0)$. Since f is one-one, $x = 0$. Conversely, assume $K_f = \{0\}$, and assume $f(x) = f(y)$. Then $f(x+y) = f(x) + f(y) = f(x) + f(x) = 0$. Thus $x + y \in K_f$, but, since $K_f = \{0\}$, $x + y = 0$, which is equivalent to $x = y$.

(*h*)

$$f(x \wedge y) = (x \wedge y) + J = [x \wedge y] = [x] \wedge [y] = f(x) \wedge f(y)$$

Similarly, $$f(x') = x' + J = [x'] = [x]' = f(x)'$$

(*i*) Assume f is a homomorphism from $\mathcal{A}$ onto $\mathcal{B}$. Let $J = K_f$. J is an ideal. (For, if $x, y \in J$ and $z \in A$, then $f(x \vee y) = f(x) \vee f(y) = 0 \vee 0 = 0$, and $f(x \wedge z) = f(x) \wedge f(z) = 0 \wedge f(z) = 0$.) For any x in A, we define $F([x]) = f(x)$. This definition is independent of the choice of the particular representative x in $[x]$, since

$$x + y \in J \;\rightarrow\; f(x+y) = 0 \;\rightarrow\; f(x) + f(y) = 0 \;\rightarrow\; f(x) = f(y)$$

Now $$F([x] \wedge [w]) = F([x \wedge w]) = f(x \wedge w) = f(x) \wedge f(w) = F([x]) \wedge F([w])$$

Similarly, $$F([x]') = F([x']) = f(x') = f(x)' = F([x])'$$

To see that F is one-one, we check that the kernel of F is $\{0\}$. Assume $F([x]) = 0$. Then $f(x) = 0$. Hence $x \in K_f$, and therefore $[x] = 0$. That the range of F is B follows from the fact that the range of f is B.

5.33. (*a*) Let b be a nonzero element of a Boolean algebra $\mathcal{B}$. Let B_b denote $\{x : x \leq b\}$. Define $u^{\#} = b \sim u$ for every u in B_b. Then show that $\langle B_b, \wedge, \vee, \#, 0, b\rangle$ is a Boolean algebra (denoted $\mathcal{B} \mid b$).

(*b*) Let b be a nonzero element of a Boolean algebra $\mathcal{B}$. Let J be the principal ideal $J_{b'}$ generated by b', i.e. $J = \{x : x \leq b'\}$. Define $\phi(u) = [u] = u + J$ for every $u \leq b$. Prove that ϕ is an isomorphism of $\mathcal{B} \mid b$ onto $\mathcal{B}/J$.

Solution:

(*a*) Since the operations $\wedge$, $\vee$ and $\#$ are closed in B_b, Axioms (1)-(4) are automatically satisfied. Axioms (5) and (9) are obvious. Axiom (6) becomes $x \wedge b = x$ which holds for all x in B_b. Axiom (7) reads $x \vee (b \sim x) = b$ for all x in B_b, which is obvious. Finally, Axiom (8) becomes $x \wedge (b \sim x) = 0$ which holds for all x.

(b) $$\phi(x \wedge y) \;=\; [x \wedge y] \;=\; [x] \wedge [y] \;=\; \phi(x) \wedge \phi(y)$$

Since $b + 1 = b' \in J$, we have $b + J = [1]$. Hence

$$\phi(x^{\#}) \;=\; \phi(b \sim x) \;=\; \phi(b \wedge x') \;=\; [b] \wedge [x'] \;=\; [1] \wedge [x]' \;=\; [x]' \;=\; (\phi(x))'$$

Thus ϕ is a homomorphism. To see that ϕ is one-one, assume that u is in the kernel K_ϕ. Then $[u] = 0_{\mathcal{B}/J} = J$, i.e. $u \in J$. Hence $u \leqq b'$. But $u \leqq b$. Therefore $u = 0$. Hence $K = \{0\}$, and ϕ is one-one. Assume now that $[v] \in \mathcal{B}/J$. Let $u = v \wedge b$. Then $[u] = [v] \wedge [b] = [v] \wedge [1] = [v]$, and $u \in \mathcal{B} \mid b$. Hence $[v]$ is in the range of ϕ. Therefore ϕ is an isomorphism of $\mathcal{B} \mid b$ onto $\mathcal{B}/J$.

BOOLEAN REPRESENTATION THEOREM

5.34. Show that every proper ideal J of a Boolean algebra $\mathcal{B}$ is equal to the intersection H of all maximal ideals containing it.

Solution:

By Theorem 5.26, there is a maximal ideal containing J. Now $J \subseteq H$. We must show that $H \subseteq J$. Assume $x \notin J$. Then by Problem 5.24, Gen $(J \cup \{x'\})$ is a proper ideal. Hence by Theorem 5.26 there is a maximal ideal M containing Gen $(J \cup \{x'\})$. Therefore $J \subseteq M$ and $x \notin M$. Thus $x \notin H$. Hence $H \subseteq J$.

5.35. (For those readers acquainted with elementary point-set topology.) Definitions: A *clopen* set of a topological space is a set which is both closed and open. A topological space X is *totally disconnected* if and only if, for any distinct points x and y of X, there exists a clopen set C such that $x \in C$ and $y \notin C$. A topological space which is both compact and totally disconnected is called a *Boolean space.*

(a) Prove that the clopen subsets of a Boolean space X form a field of sets (called the dual algebra $\mathbf{B}_X$).

(b) Let $\mathcal{M}$ be the set of maximal ideals of a Boolean algebra $\mathcal{B}$. For any x in B, let $\Xi(x) = \{M : M \in \mathcal{M} \;\&\; x \notin M\}$. Then if we take arbitrary unions of sets of the form $\Xi(x)$ to be open sets, show that $\mathcal{M}$ becomes a Boolean space (called the *Stone space* of $\mathcal{B}$). Prove also that the sets $\Xi(x)$ are the clopen subsets of $\mathcal{M}$, and that the dual algebra $\mathbf{B}_{\mathcal{M}}$ is isomorphic with the original Boolean algebra $\mathcal{B}$.

(c) If X is a Boolean space, prove that the Stone space $\mathcal{M}$ of the dual algebra $\mathbf{B}_X$ is homeomorphic with the original space X.

Solution:

(a) The complement of an open space is closed and vice versa. Hence the complement of a clopen set is clopen. In addition, the union and intersection of a finite number of closed (open) sets are also closed (open).

(b) Any maximal ideal is a proper ideal and therefore belongs to $\Xi(x)$ for some x. Now assume that M_1 and M_2 are distinct maximal ideals. Then there must be some element $x \in M_1 \sim M_2$. Hence $M_1 \in \Xi(x')$ and $M_2 \in \Xi(x)$. Since $\Xi(x') = \overline{\Xi(x)}$, $\mathcal{M}$ is a totally disconnected space. To prove compactness, assume $\mathcal{M}$ is covered by some collection $\{0_\alpha\}_{\alpha \in A}$ of open sets, i.e. $\mathcal{M} = \bigcup_{\alpha \in A} 0_\alpha$. Let us assume $\mathcal{M}$ is not covered by any finite subset of the collection $\{0_\alpha\}_{\alpha \in A}$, and let us show that this leads to a contradiction. Replace each 0_α by the sets $\Xi(x)$ contained in it. Hence we obtain a covering of $\mathcal{M}$ by a collection U of sets of the form $\Xi(x)$, where x ranges over some set $C \subseteq B$. It follows that no finite subset of U covers $\mathcal{M}$. (Otherwise, replacing each $\Xi(x)$ by a corresponding 0_α containing it, we would obtain a finite covering of $\mathcal{M}$ by 0_α's.) Hence $\Xi(x_1) \vee \cdots \vee \Xi(x_k) \neq \mathcal{M}$ for any $x_1, \ldots, x_k$ in C. But $\Xi(x_1) \vee \cdots \vee \Xi(x_k) = \Xi(x_1 \vee \cdots \vee x_k)$ (by the proof of Theorem 5.30). Hence $\Xi(x_1 \vee \cdots \vee x_k) \neq \mathcal{M}$ for any $x_1, \ldots, x_k$ in C. Therefore $x_1 \vee \cdots \vee x_k \neq 1$ for any $x_1, \ldots, x_k$ in C. By Theorem 5.16 the ideal Gen (C) generated by C is a proper ideal, and therefore by Theorem 5.26 there is a maximal ideal M containing C. Hence for every x in C, $M \notin \Xi(x)$. This contradicts the fact that the set $\mathcal{M}$ of all maximal ideals is covered by the collection U of open sets. Hence the space $\mathcal{M}$ is compact.

By definition, each $\Xi(x)$ is open, and, since $\Xi(x) = \overline{\Xi(x')}$, each $\Xi(x)$ is also closed. Conversely, assume that $\mathcal{Y}$ is a clopen subset of $\mathcal{M}$. Since $\mathcal{Y}$ is closed and $\mathcal{M}$ is compact, $\mathcal{Y}$ is itself compact. Since $\mathcal{Y}$ is open, $\mathcal{Y}$ is a union of sets of the form $\Xi(x)$ and therefore, by compactness, $\mathcal{Y}$ is a union of a finite number of such sets: $\Xi(x_1), \ldots, \Xi(x_m)$. But

$$\Xi(x_1) \vee \cdots \vee \Xi(x_m) \;=\; \Xi(x_1 \vee \cdots \vee x_m)$$

i.e. $\mathcal{Y}$ is of the form $\Xi(y)$.

The isomorphism between the dual algebra $\mathbf{B}_{\mathcal{M}}$ and $\mathcal{B}$ already has been established in the proof of Theorem 5.30.

(c) Given a Boolean space X. For each x in X, let $G(x)$ be the set of all clopen sets A in the dual algebra $\mathbf{B}_X$ such that $x \notin A$. Let us show that $G(x)$ is a maximal ideal in $\mathbf{B}_X$. If A_1 and A_2 are in $G(x)$, then $x \notin A_1$ and $x \notin A_2$, and therefore $x \notin A_1 \cup A_2$, i.e. $A_1 \cup A_2 \in G(x)$. If, in addition, $A_3 \in \mathbf{B}_X$, then $x \notin A_1 \cap A_3$, i.e. $A_1 \cap A_3 \in G(x)$. Thus $G(x)$ is an ideal. Clearly, for any clopen set A, either $x \notin A$ or $x \notin X \sim A$. Hence $G(x)$ is maximal. Thus G is a function from X into the Stone space $\mathcal{M}$ of the dual algebra $\mathbf{B}_X$. To see that G is one-one, observe that if x and y are distinct points of X, then since X is totally disconnected, there is a clopen set containing x but not y, and therefore $G(x) \neq G(y)$. To see that the range of G is all of $\mathcal{M}$, assume M is any maximal ideal in the field of clopen sets and assume for the sake of contradiction that $M \neq G(x)$ for all $x \in X$.

Case 1. For each x in X, there is a clopen set A in $M \sim G(x)$. Hence $x \in A$. Thus the sets of M form a covering of X, and by compactness there must be finitely many sets of M whose union is X. But the union of a finite number of sets in an ideal must again be in the ideal. Therefore the unit element X of the field of clopen sets would have to be in M, and M would not be a proper ideal, contradicting the definition of maximal ideal.

Case 2. There is some element x in X such that there is no clopen set in $M \sim G(x)$ but there is a clopen set A in $G(x) \sim M$. So $A \notin M$, and therefore the clopen set $X \sim A \in M$, since M is maximal. Since there is no clopen set in $M \sim G(x)$, $X \sim A \in G(x)$. Hence both A and $X \sim A$ are in $G(x)$, which is impossible.

It remains to show that G is continuous. (That G^{-1} is also continuous then follows from the fact that $\mathcal{M}$ and X are compact Hausdorff spaces.) Let $x \in X$. Since the open sets of $\mathcal{M}$ are unions of clopen sets of the form $\Xi(A)$, where A is some clopen set of X, it suffices to consider any clopen set $\Xi(A)$ having $G(x)$ as a member. We must show that there is some open set Y containing x as a member such that $G[Y] \subseteq \Xi(A)$. Since $\Xi(A)$ is the set of all maximal ideals of the dual algebra not containing the clopen set A, it follows from the fact that $G(x) \in \Xi(A)$ that $x \in A$. Then A is an open set such that $x \in A$ and, for any y in A, $G(y) \in \Xi(A)$. Hence G is continuous.

INFINITE MEETS AND JOINS

5.36. Prove that a Boolean algebra $\mathcal{B}$ is isomorphic to a Boolean algebra $\mathcal{P}(A)$ of all subsets of some non-empty set A if and only if $\mathcal{B}$ is complete and atomic.

Solution:

We already know that any Boolean algebra $\mathcal{P}(A)$ is atomic and complete. Conversely, assume that $\mathcal{B}$ is atomic and complete, and let A be the set of atoms of $\mathcal{B}$. For any element x in B, let $\Psi(x) = \{b : b \in A \;\&\; b \leqq x\}$. By Theorem 5.7, Ψ is an isomorphism of $\mathcal{B}$ into $\mathcal{P}(A)$. Let $C \in \mathcal{P}(A)$, i.e. $C \subseteq A$. By the completeness of $\mathcal{B}$, C has a lub x. Hence by Problem 5.12, $C = \Psi(x)$. Thus Ψ is an isomorphism of $\mathcal{B}$ onto $\mathcal{P}(A)$.

5.37. Associativity of Meets and Joins. If for each w in a set W, X_w is a set of elements of a given Boolean algebra $\mathcal{B}$, and $X = \bigcup_{w \in W} X_w$, prove:

(a) $\bigvee_{w \in W} \left(\bigvee_{u \in X_w} u \right) = \bigvee_{u \in X} u$

(b) $\bigwedge_{w \in W} \left(\bigwedge_{u \in X_w} u \right) = \bigwedge_{u \in X} u$

in the sense that, if the left-hand sides exist, then so do the right-hand sides and they are equal.

Solution:

Let $z = \bigvee_{w \in W} \left(\bigvee_{u \in X_w} u \right)$. Assume $x \in X$. Then $x \in X_v$ for some $v \in W$. Hence $x \leq \bigvee_{u \in X_v} u \leq z$. Thus z is an upper bound of X. Assume now that y is any upper bound of X. For each v in W, $X_v \subseteq X$. Hence y is an upper bound of X_v, and so $\bigvee_{u \in X_v} u \leq y$. Since this holds for each v in W, $z \leq y$. Therefore $z = \bigvee_{u \in X} u$. This proves (a).

Equation (b) follows from (a) by duality.

5.38. Prove that the following identities hold in any Boolean algebra.

(a) $x \vee \bigvee_{u \in A} u = \bigvee_{u \in A} (x \vee u)$

(b) $x \wedge \bigwedge_{u \in A} u = \bigwedge_{u \in A} (x \wedge u)$

(in the sense that, if the left-hand sides exist, so do the right-hand sides, and they are equal).

Solution:

Observe first that if a set X of elements of a Boolean algebra contains as a member an upper bound z of X, then z is the lub of X. To prove (a), assume $v \in A$. Then $v \leq \bigvee_{u \in A} u$, and so $x \vee v \leq x \vee \bigvee_{u \in A} u$. Thus $x \vee \bigvee_{u \in A} u$ is an upper bound of $\{x \vee u : u \in A\}$. Assume now that y is any upper bound of $\{x \vee u : u \in A\}$. Then $x \vee u \leq y$ for all $u \in A$. We must show that $x \vee \bigvee_{u \in A} u \leq y$, which is equivalent to $x \vee \bigvee_{u \in A} u \vee y = y$. But the latter equation follows by Problem 5.37(a) and the observation at the beginning of this proof.

Equation (b) follows from (a) by duality.

5.39. If $\mathcal{A}$ is a Boolean subalgebra of $\mathcal{B}$, and Y is a set of elements of $\mathcal{A}$ such that $\bigvee^{\mathcal{B}}_{y \in Y} y$ exists and belongs to $\mathcal{A}$, show that $\bigvee^{\mathcal{A}}_{y \in Y} y$ exists and $\bigvee^{\mathcal{A}}_{y \in Y} y = \bigvee^{\mathcal{B}}_{y \in Y} y$ (and similarly, by duality, for meets).

Solution:

This is an obvious consequence of the fact that the partial order $\leq_{\mathcal{A}}$ is the restriction to $\mathcal{A}$ of the partial order $\leq_{\mathcal{B}}$ on $\mathcal{B}$. (For,

$$x \leq_{\mathcal{A}} w \leftrightarrow x \wedge_{\mathcal{A}} w = w, \qquad x \leq_{\mathcal{B}} w \leftrightarrow x \wedge_{\mathcal{B}} w = w$$

and $x \wedge_{\mathcal{A}} w = x \wedge_{\mathcal{B}} w$.)

5.40. If Y is a set of elements of a field of sets $\mathcal{F}$ and if $\bigcup_{y \in Y} y \in \mathcal{F}$, show that $\bigcup_{y \in Y} y = \bigvee^{\mathcal{F}}_{y \in Y} y$ (and, similarly, by duality, for intersections and meets).

Solution:

Clearly, $\bigcup_{y \in Y} y$ is an upper bound in $\mathcal{F}$ of Y. Assume z is an upper bound in $\mathcal{F}$ of Y. Then $y \subseteq z$ for all y in Y. Hence $z \supseteq \bigcup_{y \in Y} y$. Thus $\bigcup_{y \in Y} y$ is the lub in $\mathcal{F}$ of Y.

5.41. Let $\mathcal{F}$ be a field of subsets of a set W such that, for every $w \in W$, $\{w\} \in \mathcal{F}$. Prove that joins (meets) coincide with unions (intersections), i.e. if Y is a collection of sets in $\mathcal{F}$, then $\bigvee^{\mathcal{F}}_{y \in Y} y$ exists if and only if $\bigcup_{y \in Y} y \in \mathcal{F}$ (and therefore by Problem 5.40, $\bigvee^{\mathcal{F}}_{y \in Y} y = \bigcup_{y \in Y} y$).

Solution:

In one direction, if $\bigcup_{y \in Y} y \in \mathcal{F}$, then by Problem 5.40, $\bigcup_{y \in Y} y$ is the lub in $\mathcal{F}$ of Y. Conversely, assume that $\bigvee^{\mathcal{F}}_{y \in Y} y$ exists. Then $y \subseteq \bigvee^{\mathcal{F}}_{y \in Y} y$ for all y in Y. Hence $\bigcup_{y \in Y} y \subseteq \bigvee^{\mathcal{F}}_{y \in Y} y$. Let us assume that equality does not hold and derive a contradiction. Then there is some u in W such that $u \in \bigvee^{\mathcal{F}}_{y \in Y} y$ and $u \notin \bigcup_{y \in Y} y$. Since $\{u\}$ belongs to the field $\mathcal{F}$, $z = \left(\bigvee^{\mathcal{F}}_{y \in Y} y\right) \sim \{u\}$ also belongs to $\mathcal{F}$, and $z < \bigvee^{\mathcal{F}}_{y \in Y} y$. But z is an upper bound of Y, contradicting the fact that $\bigvee^{\mathcal{F}}_{y \in Y} y$ is the lub of Y.

5.42. Is a complete field of subsets of a set X necessarily the field of all subsets of X?

Solution:

If X contains more than one element, then $\{\emptyset, X\}$ is a complete field of subsets. More generally, if A is any non-empty subset of X containing at least two elements, then the collection $\mathcal{F}$ of all subsets $Y \subseteq X$ such that $Y \cap A = \emptyset$ or $A \subseteq Y$ is a complete field of subsets of X not containing any of the non-empty proper subsets of A.

5.43. Prove that any complete field of sets $\mathcal{F}$ is atomic.

Solution:

Let A be any non-empty set belonging to $\mathcal{F}$, and let x_0 be some element of A. Then the intersection H of all sets in $\mathcal{F}$ which contain x_0 is, by the completeness of $\mathcal{F}$, also in $\mathcal{F}$, and it is an atom included in A. To see that H is an atom, assume $W \subseteq H$ and $W \in \mathcal{F}$.

Case 1. $x_0 \in W$. Then $H \subseteq W$ and therefore $H = W$.

Case 2. $x_0 \notin W$. Then $x_0 \in H \sim W \in \mathcal{F}$. Hence $H \subseteq H \sim W$ and therefore $W = \emptyset$.

DUALITY

5.44. (*a*) If $\mathcal{B} = \langle B, \wedge, \vee, ', 0, 1\rangle$ is a Boolean algebra, show that $\mathcal{B}^d = \langle B, \vee, \wedge, ', 1, 0\rangle$ is also a Boolean algebra.

(*b*) Prove that the function f such that $f(x) = x'$ is an isomorphism of $\mathcal{B}$ onto $\mathcal{B}^d$.

Solution:

(*a*) Verification of Axioms (1)-(9) for $\mathcal{B}^d$ is straightforward. Remember that $0_{\mathcal{B}^d} = 1_{\mathcal{B}}$, $1_{\mathcal{B}^d} = 0_{\mathcal{B}}$, $\wedge_{\mathcal{B}^d} = \vee_{\mathcal{B}}$, and $\vee_{\mathcal{B}^d} = \wedge_{\mathcal{B}}$.

(*b*) f is one-one, since $x' = y' \rightarrow x = y$. The range of f is B, since $x = (x')'$. Also, $f(x \wedge y) = (x \wedge y)' = x' \vee y' = f(x) \vee f(y)$, and $f(x') = x'' = f(x)'$.

INFINITE DISTRIBUTIVITY

5.45. Let A be the set of atoms of a Boolean algebra $\mathcal{B}$. Prove that $\mathcal{B}$ is atomic if and only if $\bigvee_{x \in A} x = 1$.

Solution:

By Problem 5.13(*a*), if $\mathcal{B}$ is atomic, then $\bigvee_{x \in A} x = 1$. Conversely, assume $\bigvee_{x \in A} x = 1$. Given any nonzero y in B, we must show that there is an atom $b \leq y$. Let us assume not and derive a contradiction. Then for every atom x, $x \wedge y = 0$. Hence $\bigvee_{x \in A} (x \wedge y) = 0$. But $\bigvee_{x \in A} (x \wedge y) = y \wedge \bigvee_{x \in A} x$ by Theorem 5.35(*a*). Since $\bigvee_{x \in A} x = 1$, we have $0 = \bigvee_{x \in A} (x \wedge y) = y$, contradicting the fact that $y \neq 0$.

5.46. Prove that a Boolean algebra $\mathcal{B}$ is isomorphic to a field $\mathcal{P}(K)$ of all subsets of a non-empty set K if and only if $\mathcal{B}$ is complete and completely distributive.

Solution:

We already know that $\mathcal{P}(K)$ is complete and completely distributive. Conversely, assume $\mathcal{B}$ is complete and completely distributive. Let

$$W = B, \qquad S = \{1, -1\} \qquad \text{and} \qquad x_{w,s} = \begin{cases} w, & \text{if } s \text{ is } 1 \\ w', & \text{if } s \text{ is } -1 \end{cases}$$

By complete distributivity,

$$\bigwedge_{w \in B} (w \vee w') = \bigvee_{f \in S^B} \left(\bigwedge_{w \in B} x_{w,f(w)} \right)$$

Hence

$$1 = \bigvee_{f \in S^B} \left(\bigwedge_{w \in B} x_{w,f(w)} \right)$$

Now by Theorem 5.35(*a*), for any nonzero $u \in B$,

$$u = u \wedge 1 = u \wedge \bigvee_{f \in S^B} \left(\bigwedge_{w \in B} x_{w,f(w)} \right) = \bigvee_{f \in S^B} \left(u \wedge \bigwedge_{w \in B} x_{w,f(w)} \right)$$

Since $u \neq 0$, there must be some $f \in S^B$ such that $u \wedge \bigwedge_{w \in B} x_{w,f(w)} \neq 0$, and therefore

$$\bigwedge_{w \in B} x_{w,f(w)} \neq 0$$

Now observe that if $z = \bigwedge_{w \in B} x_{w,f(w)} \neq 0$, then z is an atom. (To see this, assume $0 \leq v < z$. We must prove that $v = 0$. But $v < z \leq x_{v,f(v)}$. Since $x_{v,f(v)}$ is v or v' and $v < v$ is impossible, it follows that $x_{v,f(v)} = v'$. Hence $v < v'$, which implies that $v = 0$.) Thus for any nonzero $u \in B$ there is an atom z such that $z \leq u$, i.e. $\mathcal{B}$ is atomic. But we already have proved (cf. Problem 5.36) that a complete atomic Boolean algebra is isomorphic to some $\mathcal{P}(K)$ where K is non-empty.

5.47. Regular Open Sets (For those readers acquainted with elementary point-set topology). Let $\mathcal{W}$ be a non-empty topological space. For any $Y \subseteq \mathcal{W}$, we use the notation Y^c for the closure of Y. Recall that $\bar{Y}$ is the complement of Y.

Definitions: $Y^e = \overline{Y^c}$ = the complement of the closure of Y. Y is *regular* if and only if $Y = Y^{ee}$.

Prove the following assertions.

(1) Y^e is open.

(2) $X \subseteq Y \rightarrow Y^e \subseteq X^e$. (Hence $X \subseteq Y \rightarrow X^{ee} \subseteq Y^{ee}$.)

(3) If Y is open, then $Y \subseteq Y^{ee}$.

(4) If Y is regular, Y is open.

(5) Y is open if and only if $Y = Z^e$ for some Z.

(6) If Y is open, then Y^e is regular (i.e. if Y is open, $Y^e = Y^{eee}$).

(7) $\emptyset$ and $\mathcal{W}$ are regular.

(8) If X and Y are open, then $(X \cap Y)^{ee} = X^{ee} \cap Y^{ee}$.

(9) Let B be the set of regular sets of a topological space $\mathcal{W}$. For any sets X and Y in B, define

$$X \wedge Y = X \cap Y, \quad X \vee Y = (X \cup Y)^{ee}, \quad X' = X^e$$

Then $\mathcal{B} = \langle B, \wedge, \vee, ', \emptyset, \mathcal{W}\rangle$ is a complete Boolean algebra (called the *regular open algebra* of $\mathcal{W}$).

Solution:

(1) The complement of a closed set is open.

(2) If $X \subseteq Y$, then $X^c \subseteq Y^c$ and therefore $\overline{Y^c} \subseteq \overline{X^c}$.

(3) Since $Y \subseteq Y^c$, $Y^e = \overline{Y^c} \subseteq \bar{Y}$. Taking closures, we obtain $Y^{ec} \subseteq (\bar{Y})^c = \bar{Y}$, since $\bar{Y}$ is closed. Hence $Y = \bar{\bar{Y}} \subseteq \overline{Y^{ec}} = Y^{ee}$.

(4) This follows immediately from (1).

(5) If $Y = Z^e$, then Y is open by (1). Conversely, if Y is open, and if we let $Z = \bar{Y}$, then Z is closed. Hence $Z^e = \overline{Z^c} = \bar{Z} = Y$.

(6) Assume Y open. Then by (3), $Y \subseteq Y^{ee}$. Hence by (2), $Y^{eee} \subseteq Y^e$. On the other hand, since Y^e is open by virtue of (1), it follows by (3) that $Y^e \subseteq Y^{eee}$.

(7) $\emptyset^e = \overline{\emptyset^c} = \bar{\emptyset} = \mathcal{W}$. Also, $\mathcal{W}^e = \overline{\mathcal{W}^c} = \bar{\mathcal{W}} = \emptyset$. Hence $\emptyset^{ee} = \mathcal{W}^e = \emptyset$, and $\mathcal{W}^{ee} = \emptyset^e = \mathcal{W}$.

(8) Assume X and Y open. First, let us prove

$$X \cap Y^{ee} \subseteq (X \cap Y)^{ee} \qquad (a)$$

To see this, observe that $X \cap Y^c \subseteq (X \cap Y)^c$. (For, let x be any point of $X \cap Y^c$ and let N be any open set containing the point x. We must show that N intersects $X \cap Y$. But $N \cap X$ is an open set containing x, and therefore $N \cap X$ must intersect Y.) Taking complements, we obtain $(X \cap Y)^e \subseteq \bar{X} \cup Y^e$. Taking closures, we have $(X \cap Y)^{ec} \subseteq \bar{X}^c \cup Y^{ec} = \bar{X} \cup Y^{ec}$, and, taking complements again, we obtain the inclusion (a). Now substituting X^{ee} for X in (a), we have $X^{ee} \cap Y^{ee} \subseteq (X^{ee} \cap Y)^{ec}$. But exchanging X and Y in (a), we also have $(Y \cap X^{ee}) \subseteq (Y \cap X)^{ee}$, and therefore by (2), $(Y \cap X^{ee})^{ee} \subseteq (Y \cap X)^{eeee}$. But the last term, by (6), is $(Y \cap X)^{ee}$. Hence $X^{ee} \cap Y^{ee} \subseteq (X \cap Y)^{ee}$. Conversely, since $X \cap Y \subseteq X$ and $X \cap Y \subseteq Y$, two applications of (2) yield $(X \cap Y)^{ee} \subseteq X^{ee} \cap Y^{ee}$.

(9) By (7), $\emptyset \in B$ and $\mathcal{W} \in B$. The operation $\wedge$ is closed in B, for, by (8), $(X \cap Y)^{ee} = X^{ee} \cap Y^{ee} = X \cap Y$. The operation $\vee$ is closed in B, since, by (6),

$$(X \vee Y)^{ee} = (X \cup Y)^{eeee} = (X \cup Y)^{ee} = X \vee Y$$

Similarly, the operation $'$ is closed in B, since $(X')^{ee} = X^{eee} = X^e = X'$. Now we must show that all the axioms for Boolean algebras are satisfied. Axioms (1), (2), (6) and (9) are obvious.

Let us consider Axiom (3):

$$\begin{aligned} X \wedge (Y \vee Z) &= X \cap (Y \cup Z)^{ee} \\ &= X^{ee} \cap (Y \cup Z)^{ee} \\ &= (X \cap (Y \cup Z))^{ee} \quad \text{(by (8))} \\ &= ((X \cap Y) \cup (X \cap Z))^{ee} \\ &= (X \wedge Y) \vee (X \wedge Z) \end{aligned}$$

For Axiom (4):

$$\begin{aligned} X \vee (Y \wedge Z) = (X \cup (Y \cap Z))^{ee} &= ((X \cup Y) \cap (X \cup Z))^{ee} \\ &= (X \cup Y)^{ee} \cap (X \cup Z)^{ee} \quad \text{(by (8))} \\ &= (X \vee Y) \wedge (X \vee Z) \end{aligned}$$

Axiom (5) is easy: $X \vee 0 = (X \cup \emptyset)^{ee} = X^{ee} = X$. Axiom (8) is also easy: $X \subseteq X^c$. Hence $X^e = \overline{X^c} \subseteq \bar{X}$. Therefore $X \cap X^e = \emptyset$. It remains to prove Axiom (7). First, let us show that $(X \cup X^e)^e = \emptyset$. Assume, to the contrary, that some point u lies in $(X \cup X^e)^e = (\overline{X \cup X^e})^c$. Thus $u \notin (X \cup X^e)^c$, which implies that there is an open set N containing u and disjoint from $X \cup X^e = X \cup \overline{X^c}$. Since $N \cap X = \emptyset$, $u \notin X^c$, i.e. $u \in \overline{X^c}$, contradicting the fact that N is disjoint from $X \cup \overline{X^c}$. Taking closures and complements, we obtain $X \vee X' = (X \cup X^e)^{ee} = \mathcal{W} = 1_{\mathcal{B}}$, which is Axiom (7). Thus we have shown that $\mathcal{B}$ is a Boolean algebra.

We still have to prove completeness. Note first that

$$X \leq_{\mathcal{B}} Y \;\leftrightarrow\; X \wedge Y = X \;\leftrightarrow\; X \cap Y = X \;\leftrightarrow\; X \subseteq Y$$

Now let $\mathbf{X}$ be any collection of regular sets. Let us show that $\left(\bigcap_{Y \in \mathbf{X}} Y\right)^{ee}$ is the glb of $\mathbf{X}$. First, for any $Y \in \mathbf{X}$, $\bigcap_{Y \in \mathbf{X}} Y \subseteq Y$ and therefore $\left(\bigcap_{Y \in \mathbf{X}} Y\right)^{ee} \subseteq Y^{ee} = Y$. Thus $\left(\bigcap_{Y \in \mathbf{X}} Y\right)^{ee}$ is a lower bound of $\mathbf{X}$. Assume that Z is a regular set which is a lower bound of $\mathbf{X}$. Then $Z \subseteq Y$ for all $Y \in \mathbf{X}$. Hence $Z \subseteq \bigcap_{Y \in \mathbf{X}} Y$, and therefore

$$Z = Z^{ee} \subseteq \left(\bigcap_{Y \in \mathbf{X}} Y\right)^{ee}$$

This proves the completeness of $\mathcal{B}$. (It may easily be checked that the lub of a collection $\mathbf{X}$ is $\left(\bigcup_{Y \in \mathbf{X}} Y\right)^{ee}$.)

5.48. Show that a complete Boolean subalgebra $\mathcal{B}$ of an algebra of the form $\mathcal{P}(K)$ need not be a complete field of sets (i.e. infinite joins and meets need not coincide with unions and intersections, respectively).

Solution:

Consider the regular open algebra $\mathcal{B}$ of the real line (cf. Problem 5.47). It is easy to verify that every finite open interval is regular. Since every regular set must contain a finite open interval, it follows that the algebra $\mathcal{B}$ is atomless. Hence by Problem 5.43, $\mathcal{B}$ cannot be isomorphic to a complete field of sets.

5.49. Give an example of a complete but not completely distributive Boolean algebra.

Solution:

The regular open algebra of the real line is complete and atomless (cf. Problems 5.47-5.48). Hence by Problems 5.46 and 5.36, the algebra cannot be completely distributive.

$\mathfrak{m}$-COMPLETENESS. σ-ALGEBRAS

5.50. Given σ-algebras $\mathcal{A}$ and $\mathcal{B}$. By a *σ-subalgebra* of $\mathcal{A}$ we mean a subalgebra determined by a subset closed under denumerable joins and meets. By a *σ-homomorphism* of $\mathcal{A}$ into $\mathcal{B}$ we mean a homomorphism g of $\mathcal{A}$ into $\mathcal{B}$ preserving denumerable joins and meets (i.e. such that $g(\bigwedge_i x_i) = \bigwedge_i g(x_i)$; the corresponding equality for joins follows by De Morgan's Laws). By a *σ-ideal* of $\mathcal{A}$ we mean an ideal of $\mathcal{A}$ closed under denumerable joins.

(*a*) If g is a σ-homomorphism of $\mathcal{A}$ into $\mathcal{B}$, then the range $g[\mathcal{A}]$ is a σ-subalgebra of $\mathcal{B}$, the kernel K_g is a σ-ideal of $\mathcal{A}$, and $\mathcal{A}/K_g$ is σ-isomorphic with $g[\mathcal{A}]$.

(*b*) If J is a σ-ideal of $\mathcal{A}$, then $\mathcal{A}/J$ is a σ-algebra $\left(\text{where } \bigvee_i (x_i + J) = \left(\bigvee_i x_i\right) + J\right)$, and $\mathcal{A}/J$ is a σ-homomorphic image of $\mathcal{A}$ under the natural mapping.

Solution:

(*a*) These are just obvious extensions of the results in Problem 5.32.

(*b*) We must show that $\mathcal{A}/J$ is closed under denumerable unions. To see this, we shall show that $\left[\bigvee_i x_i\right] = \bigvee_i [x_i]$. Clearly, $[x_i] \leq \left[\bigvee_i x_i\right]$ and therefore $\left[\bigvee_i x_i\right]$ is an upper bound. Assume now that $[z]$ is an upper bound of the $[x_i]$'s. Note that, in general, $[u] \leq [v]$ if and only if $u \wedge v' \equiv_J 0$. (For, $[u] \leq [v] \leftrightarrow [u \wedge v] = [v] \leftrightarrow u \wedge v \equiv_J v \leftrightarrow (u \wedge v) + v \equiv_J 0 \leftrightarrow (u \wedge v) + (1 \wedge v) \equiv_J 0 \leftrightarrow u \wedge (1 + v) \equiv_J 0 \leftrightarrow u \wedge v' \equiv_J 0$.) Since $[x_i] \leq [z]$, $x_i \wedge z' \equiv_J 0$. Since J is a σ-ideal, $\bigvee_i (x_i \wedge z') \equiv_J 0$, i.e. $z' \wedge \bigvee_i x_i \equiv_J 0$. Hence $\left[\bigvee_i x_i\right] \leq [z]$. That the natural mapping $\phi(x) = x + J$ is a σ-homomorphism of $\mathcal{A}$ onto $\mathcal{A}/J$ is an easy consequence of the fact that $\left[\bigvee_i x_i\right] = \bigvee_i [x_i]$.

5.51. For any subset $C \subseteq \mathcal{P}(K)$, the intersection of all σ-subfields of $\mathcal{P}(K)$ containing C is itself a σ-subfield containing C.

Solution:

The intersection H clearly is closed under denumerable joins and complements. The σ-subfield H is called the *σ-subfield generated by C*.

5.52. Given a subset C of a σ-algebra $\mathcal{A}$. The intersection D of all σ-ideals containing C is itself a σ-ideal containing C (called the *σ-ideal generated by C*). The elements of D are all those $x \leq \bigvee_i c_i$ for elements c_i in C $(1 \leq i < \infty)$.

Solution:

That D is a σ-ideal containing C is obvious. Let E be the set of all $x \leq \bigvee_i c_i$ for some $c_i \in C$. First, if $x_j \leq \bigvee_i c_{ij}$ for $c_{ij} \in C$, then $\bigvee_j x_j \leq \bigvee_{i,j} c_{ij}$. Thus E is closed under denumerable joins. Also, if $x \in E$ and $y \leq x$, then $x \leq \bigvee_i c_i$ for $c_i \in C$, and therefore $y \leq \bigvee_i c_i$, i.e. $y \in E$. Thus E is a σ-ideal containing C. Hence $D \subseteq E$. On the other hand, for any σ-ideal J containing C, if $x \leq \bigvee_i c_i$ for $c_i \in C$, then $\bigvee_i c_i \in J$ and therefore $x \in J$. Hence $E \subseteq D$.

5.53. The Loomis Representation Theorem: Any σ-ideal $\mathcal{A}$ is a σ-homomorphic image of a σ-field of sets, i.e. $\mathcal{A}$ is σ-isomorphic with the quotient algebra of a σ-field of sets by a σ-ideal.

Solution:

Call $Y \in \mathcal{P}(A)$ a *selection* if and only if, for any $a \in A$, Y contains exactly one of a and a'. Let S be the set of selections. Define a function τ from $\mathcal{A}$ into $\mathcal{P}(S)$ by setting $\tau(a)$ equal to the set of all selections Y such that $a \in Y$. Let T be the range of τ. Clearly, T is closed under complements: $\overline{(\tau(a))} = \tau(a')$. Let F be the σ-field of subsets of S generated by T. Let N be the subset of F consisting of countable intersections $\bigcap_i \tau(a_i)$ such that $\bigwedge_i a_i = 0$. Let J be the σ-ideal generated by N. Consider the mapping $\psi(a) = \tau(a) + J \in F/J$. We now show that ψ is a σ-isomorphism of $\mathcal{A}$ onto F/J.

Let $z = \bigvee_i z_i$ in $\mathcal{A}$. Then, $z' \wedge z_i = 0$. Hence $\tau(z') \cap \tau(z_i) \in N$. Therefore

$$\overline{\tau(z)} \cap \bigcup_i \tau(z_i) \quad = \quad \tau(z') \cap \left(\bigcup_i \tau(z_i) \right) \quad = \quad \bigcup_i (\tau(z') \cap \tau(z_i)) \in J$$

Also, $$z \wedge \bigwedge_i (z_i') \quad = \quad z \wedge \left(\bigvee_i z_i \right)' \quad = \quad z \wedge z' \quad = \quad 0$$

Hence $\tau(z) \cap \bigcap_i \tau(z_i') \in N$ and

$$\overline{\tau(z)} \cap \left(\bigcup_i \tau(z_i) \right)' \quad = \quad \tau(z) \cap \bigcap_i \tau(z_i)' \quad = \quad \tau(z) \cap \bigcap_i \tau(z_i') \in J$$

Hence $$\tau(z) + \bigcup_i \tau(z_i) \quad = \quad \left(\overline{\tau(z)} \cap \bigcup_i \tau(z_i) \right) \cup \left(\tau(z) \cap \left(\bigcup_i \tau(z_i) \right)' \right) \in J$$

It follows that $\tau(z) + J = \bigcup_i \tau(z_i) + J$. Therefore

$$\psi(z) \quad = \quad \tau(z) + J \quad = \quad \bigcup_i \tau(z_i) + J \quad = \quad \bigcup_i (\tau(z_i) + J)$$

In addition, $$\psi(z') \quad = \quad \tau(z') + J \quad = \quad (\tau(z) + J)' \quad = \quad (\psi(z))'$$

Hence ψ is a σ-homomorphism. It is readily seen that the range of ψ is F/J. (For, $\bigcup_{a \in \mathcal{A}} \psi(a)$ is a σ-subfield of F containing T and is therefore equal to all of F. Hence $\psi[\mathcal{A}] = F/J$.) It remains to show that ψ is one-one. To do this, we shall show that the kernel K_ψ is $\{0\}$. Assume $a \in K_\psi$. So, $\psi(a) = 0_{F/J}$. Therefore $\tau(a) \in J$. Note that J consists of all β in F such that $\beta \subseteq \bigcup_i \nu_i$, where $\nu_i \in N$. Here $\nu_i = \bigcap_j \tau(a_{ij})$ where $\bigwedge_j a_{ij} = 0$. Therefore $\tau(a) \subseteq \bigcup_i \left(\bigcap_j \tau(a_{ij}) \right)$ for some a_{ij} such that $\bigwedge_j a_{ij} = 0$. Hence

$$\tau(a) \subseteq \bigcup_i \tau(a_{i,f(i)}) \tag{1}$$

where f is any function such that $a_{i,f(i)}$ is defined for all i. Since we wish to prove that $a = 0$, let us assume the contrary, i.e. $a \neq 0$. Hence $1 > a' = a' \vee 0 = a' \vee \left(\bigwedge_j a_{1j} \right) = \bigwedge_j (a' \vee a_{1j})$. Therefore some $a' \vee a_{1j} \neq 1$; say, $a' \vee a_{1f(1)} \neq 1$. Then

$$1 \;>\; a' \vee a_{1f(1)} \;=\; a' \vee a_{1f(1)} \vee 0 \;=\; a' \vee a_{1f(1)} \vee \bigwedge_j a_{2j} \;=\; \bigwedge_j (a' \vee a_{1f(1)} \vee a_{2j})$$

Hence $1 > a' \vee a_{1f(1)} \vee a_{2f(2)}$ for some $j = f(2)$; etc. We obtain a sequence $a_{1f(1)}, a_{2f(2)}, \ldots$ such that

$$1 \;>\; a' \vee a_{1f(1)} \vee a_{2f(2)} \vee \cdots \vee a_{kf(k)} \qquad \text{for each } k$$

Therefore among $a_{1f(1)}, a_{2f(2)}, \ldots$ neither a nor a complement of any $a_{if(i)}$ occurs. Therefore there exists a selection Y containing a and all $a'_{if(i)}$. Thus $Y \in \tau(a)$, but $Y \notin \tau(a_{if(i)})$, contradicting *(1)*.

Remark: This result of Loomis fails to hold for non-denumerable cardinalities [125].

Supplementary Problems

LATTICES

5.54. Which of the partially ordered sets given by the following diagrams are lattices? Among the lattices, which (i) have a zero element, (ii) have a unit element, (iii) are complemented, (iv) are modular, (v) are distributive?

(a)

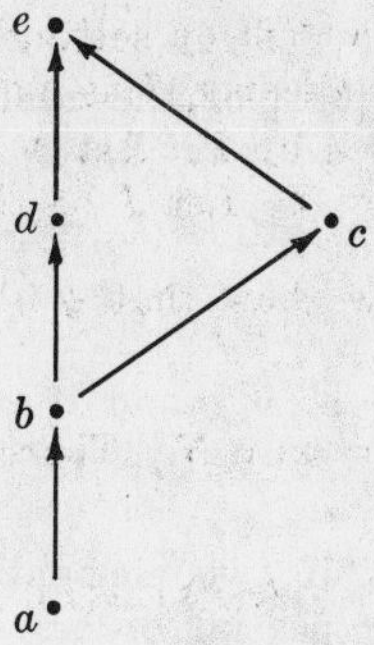

(b)

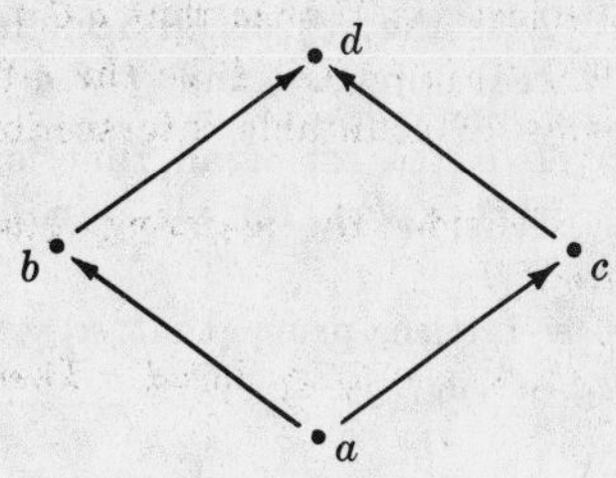

(c)

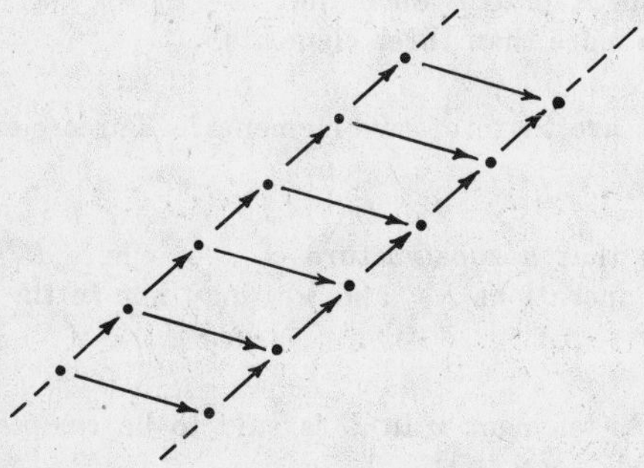

(d)

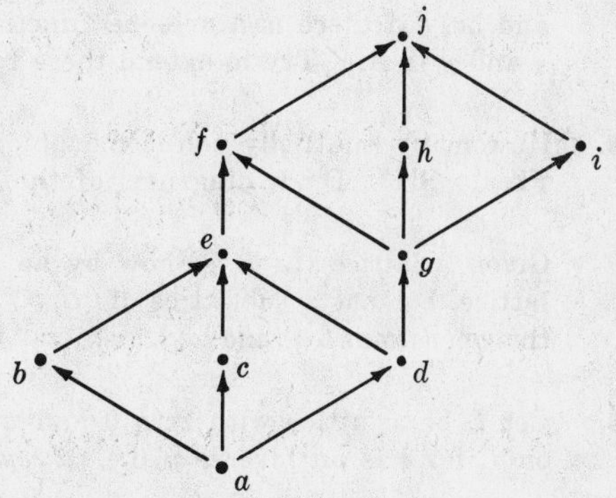

5.55. Which of the following structures $\langle L, \leq \rangle$ are partially ordered sets, totally ordered sets, lattices, distributive lattices, lattices with a zero element, lattices with a unit element, complemented lattices? For those which are lattices, describe the operations $\wedge$ and $\vee$.

(a) L is the set of all finite subsets of a set A and $\leq$ is the inclusion relation $\subseteq$.

(b) Same as (a), except that A itself is also a member of L.

(c) L is the set of complex numbers and $a + bi \leq c + di \leftrightarrow a \leq c$.

(d) L is the set of all complex numbers and $a + bi \leq c + di \leftrightarrow (a < c) \vee (a = c \;\&\; b \leq d)$.

(e) L is the set of all complex numbers and $a + bi \leq c + di \leftrightarrow a \leq c \;\&\; b \leq d$.

(f) L is the set of all subalgebras of a given Boolean algebra, and $\leq$ is the inclusion relation $\subseteq$.

(g) L is the set of all sublattices of a given lattice and $\leq$ is the inclusion relation $\subseteq$.

(h) L is the set of all polynomials with real coefficients and $f \leq g$ means that f divides g.

(i) Same as (h), except that the coefficients of the polynomials must be integers.

(j) L is the set of all subsets of a set A, and $\leq$ is $\subseteq$.

(k) L is the set of positive integers and $x \leq y$ if and only if y is an integral multiple of x (i.e. x divides y).

(l) Assume $\langle A, \leq \rangle$ is a given partially ordered set. Let C be a fixed set. Let L be the set of all functions from C into A. For any f and g in L, let $f \leq g$ if and only if $f(x) \leq g(x)$ for all x in C.

(m) Assume $\langle A, \leq \rangle$ is a given partially ordered set. Let $L = A$, and $x \leq y \leftrightarrow y \leq x$.

(n) L is the set of all infinite subsets of an infinite set A, and $\leq$ is $\subseteq$.

(o) L is the set of all subsets of a set A containing a fixed subset C, i.e. $L = \{Y : C \subseteq Y \subseteq A\}$, and $\leq$ is $\subseteq$.

(*p*) L consists of the empty set $\emptyset$ and all points, lines and planes of three-dimensional Euclidean space, and $\leqq$ is $\subseteq$.

(*q*) L is the set of all subgroups of a group G, and $\leqq$ is $\subseteq$. (This exercise is for those readers acquainted with elementary group theory.)

(*r*) L is the set of all convex planar sets and $\leqq$ is $\subseteq$. (By a *convex* set we mean a set such that, for any two points in the set, all the points on the line connecting the two given points are also in the set.)

(*s*) L is the set of all functions from the unit interval $[0,1]$ of the real line into the set of all real numbers, and $f \leqq g$ means that $f(x) \leqq g(x)$ for all x in $[0,1]$.

(*t*) L is the set of all functions from a fixed set A into a lattice $\langle L_1, \leqq \rangle$, and $f \leqq g$ means that $f(x) \leqq g(x)$ for all x in A.

5.56. In a lattice, prove that $u_1 \vee \cdots \vee u_k$ is the lub of $\{u_1, \ldots, u_k\}$ and $u_1 \wedge \cdots \wedge u_k$ is the glb of $\{u_1, \ldots, u_k\}$.

5.57. How many partial orders can be defined on a fixed set of two elements? Of three elements? (What is the largest number of mutually non-isomorphic partial orders in each case? We say that a partially ordered structure $\langle A, \leqq_A \rangle$ is isomorphic to a partially ordered structure $\langle B, \leqq_B \rangle$ if and only if there is a one-one function f from A onto B such that $x \leqq_A y \leftrightarrow f(x) \leqq_B f(y)$ for all x and y in A.) Try to extend these results to more than three elements.

5.58. How many (mutually non-isomorphic) lattices are there of two elements? Three elements? Four? Five? Six? Draw diagrams of the lattices.

5.59. Given a lattice $\langle L, \leqq \rangle$. Show by an example that a substructure $\langle L_1, \leqq \rangle$, where $L_1 \subseteq L$, may be a lattice, but not a sublattice of $\langle L, \leqq \rangle$ (i.e. the operations $\wedge_{L_1}$ and $\vee_{L_1}$ may not be the restrictions of the operations $\wedge_L$ and $\vee_L$).

5.60. Let L be a lattice with zero 0 and unit 1. An element x in L is said to be *complemented* if and only if x has an inverse y (i.e. $x \wedge y = 0$ and $x \vee y = 1$).

(*a*) If L is distributive, prove that the set of complemented elements is a Boolean algebra (under the operations $\wedge$ and $\vee$ of L).

(*b*) If L is modular (but not distributive), give an example containing six elements to show that the set of complemented elements need not form a sublattice.

(*c*) Show by an example that, if L is not distributive, an element can have more than one complement.

5.61. Show that a lattice $\langle L, \leqq \rangle$ is complemented if and only if L contains a zero element 0 and there is a singulary operation $x \to x'$ on L such that:

$$\text{(i) } x \wedge x' = 0, \quad \text{(ii) } x'' = x, \quad \text{(iii) } (x \vee y)' = x' \wedge y'$$

5.62. Finish the proof of Theorem 5.2, i.e. prove that (L6) implies (L5).

5.63. Using Problem 5.4, state and prove a Duality Theorem for lattices.

5.64. Show that the following properties of a distributive lattice

(*a*) $x \wedge x = x$

(*b*) $x \wedge (y \vee z) = (x \wedge y) \vee (x \wedge z)$

(*c*) $(y \vee z) \wedge x = (y \wedge x) \vee (z \wedge x)$

(*d*) $x \vee 1 = 1 \vee x = 1$

(*e*) $x \wedge 1 = 1 \wedge x = x$

serve to characterize distributive lattices with a unit element 1 in the sense that if the structure $\langle A, \wedge, \vee, 1 \rangle$ satisfies these laws then $\langle A, \wedge, \vee \rangle$ is a distributive lattice with unit element 1, and vice versa [99].

5.65. Prove that the following laws hold in any lattice.

(*a*) $x \wedge y = x \vee y \rightarrow x = y$

(*b*) $x \wedge y \wedge z = x \vee y \vee z \rightarrow x = y = z$

5.66. Prove that the following inequalities hold in any lattice.

(*a*) $(x \wedge y) \vee (u \wedge v) \leqq (x \vee u) \wedge (y \vee v)$

(*b*) $(x \wedge y) \vee (y \wedge z) \vee (z \wedge x) \leqq (x \vee y) \wedge (y \vee z) \wedge (z \vee x)$

5.67. Prove that each of the following conditions is equivalent to distributivity of a lattice.

(*a*) $(x \wedge y \leqq z \ \& \ x \leqq y \vee z) \rightarrow x \leqq z$

(*b*) $(x \wedge z \leqq y \wedge z \ \& \ x \vee z \leqq y \vee z) \rightarrow x \leqq y$ (*Hint*: Use (*a*).)

(*c*) $(x \wedge z = y \wedge z \ \& \ x \vee z = y \vee z) \rightarrow x = y$ (*Hint*: In one direction, use Problem 5.11(*c*).)

(*d*) $(x \vee y) \wedge (y \vee z) \wedge (x \vee z) \leqq (x \wedge y) \vee (y \wedge z) \vee (x \wedge z)$

5.68. Complete the proof of Problem 5.11(*c*).

5.69. Give an example of a distributive lattice $\langle L, \leqq \rangle$ lacking both zero and unit elements such that $\leqq$ is not a total order on L.

5.70[D]**.** If $\langle L, \leqq \rangle$ is a lattice, prove that $\leqq$ totally orders L if and only if all subsets of L are sublattices (i.e. are closed under $\wedge$ and $\vee$).

5.71. Prove that any distributive (modular) lattice can be extended to a distributive (modular) lattice with zero and unit elements simply by adjoining such elements if they are not already present.

5.72. Prove that a lattice is modular if and only if it satisfies the law

$$(x \wedge z = x \wedge y \ \& \ x \vee z = y \vee z \ \& \ x \leqq y) \rightarrow x = y$$

(*Hint*: In one direction, use Problem 5.10.)

5.73. A lattice $\langle L, \leqq \rangle$ is said to be *complete* if and only if every subset of L has a lub and a glb. Prove that in order to verify completeness it suffices to show that every subset has a lub or that every subset has a glb.

5.74. Given lattices $\langle L_1, \leqq_1 \rangle$ and $\langle L_2, \leqq_2 \rangle$ and a function f from L_1 into L_2.

Definitions.

f is an *order-homomorphism* if and only if $x \leqq_1 y \rightarrow f(x) \leqq_2 f(y)$.

f is a *meet-homomorphism* if and only if $f(x \wedge_1 y) = f(x) \wedge_2 f(y)$.

f is a *join-homomorphism* if and only if $f(x \vee_1 y) = f(x) \vee_2 f(y)$.

f is a *lattice-homomorphism* if and only if f is both a meet-homomorphism and a join-homomorphism.

Prove:

(*a*) Every meet-homomorphism is an order-homomorphism.

(*b*) Every join-homomorphism is an order-homomorphism.

(*c*) Every lattice homomorphism is a meet-homomorphism.

(*d*) Every lattice homomorphism is a join-homomorphism.

(*e*) The converses of (*a*)-(*d*) do not hold. (Counterexamples may be found using lattices of at most four elements.)

(*f*) For one-one functions f from L_1 onto L_2, the notions of order-, meet-, join-, and lattice-homomorphism are equivalent.

(*g*) Any order-isomorphism from a Boolean algebra $\mathcal{A}$ onto a Boolean algebra $\mathcal{B}$ is a Boolean isomorphism, i.e. not only is it a lattice-homomorphism, but it also preserves complements: $f(x') = (f(x))'$.

5.75. A lattice $\langle L, \leqq \rangle$ is said to be *relatively pseudo-complemented* if and only if, for any x, y in L, the set $\{z : y \wedge z \leqq x\}$ has a lub. (Such a lub is denoted $y \Rightarrow x$.) By a *pseudo-Boolean algebra*, we mean a relatively pseudo-complemented lattice possessing a zero element 0. Prove ([142]):

(*a*) In a relatively pseudo-complemented lattice:

(i) $z \leqq y \Rightarrow x$ if and only if $y \wedge z \leqq x$.

(ii) The distributive laws hold.

(iii) For any x, $x \Rightarrow x$ is a unit element 1.

(iv) $y \Rightarrow x = 1$ if and only if $y \leqq x$.

(v) $x = y$ if and only if $x \Rightarrow y = y \Rightarrow x = 1$.

(vi) $y \Rightarrow 1 = 1$

(vii) $1 \Rightarrow x = x$

(viii) $y \wedge (y \Rightarrow x) \leqq x$

(ix) If $x \leqq y$, then $y \Rightarrow z \leqq x \Rightarrow z$ and $z \Rightarrow x \leqq z \Rightarrow y$.

(x) $x \leqq y \Rightarrow x$

(xi) $y \wedge (y \Rightarrow x) = y \wedge x$

(xii) $(y \Rightarrow x) \wedge x = x$

(xiii) $(x \Rightarrow y) \wedge (x \Rightarrow z) = x \Rightarrow (y \wedge z)$

(xiv) $(x \Rightarrow z) \wedge (y \Rightarrow z) = (y \vee x) \Rightarrow z$

(xv) $x \Rightarrow (y \Rightarrow z) = (x \wedge y) \Rightarrow z = y \Rightarrow (x \Rightarrow z)$

(xvi) $(x \Rightarrow y) \wedge (y \Rightarrow z) \leqq x \Rightarrow z$

(xvii) $(x \Rightarrow y) \leqq (y \Rightarrow z) \Rightarrow (x \Rightarrow z)$

(xviii) $x \leqq y \Rightarrow (x \wedge y)$.

(*b*) In a pseudo-Boolean algebra, we define: $-x = x \Rightarrow 0$. Then:

(i) $-0 = 1$ and $-1 = 0$

(ii) $x \wedge (-x) = 0$

(iii) $x \leqq y \rightarrow -y \leqq -x$

(iv) $x \leqq --x$

(v) $---x = -x$

(vi) $-(x \vee y) = -x \wedge -y$

(vii) $-(x \wedge y) \geqq -x \vee -y$

(viii) $(-x) \vee y \leqq x \Rightarrow y$

(ix) $x \Rightarrow y \leqq (-y) \Rightarrow (-x)$

(x) $x \Rightarrow (-y) = -(x \wedge y) = y \Rightarrow (-x)$

(xi) $0 \Rightarrow x = 1$

(xii) A subset F is a filter if and only if $1 \in F$ and

$$(x \in F \;\&\; x \Rightarrow y \in F) \rightarrow y \in F$$

(*c*) In a Boolean algebra, show that $y \Rightarrow x = y' \vee x$.

(*d*) (For those readers acquainted with elementary point-set topology.) Show that the lattice of all open sets of a topological space is a pseudo-Boolean algebra, where $A \Rightarrow B$ is the interior of $\bar{A} \cup B$. By taking the special case of the real line, show that the assertions $x \vee (-x) = 1$ and $--x = x$ do not hold for all pseudo-Boolean algebras.

(*e*) By a *Brouwerian lattice* we mean a lattice with a unit element such that, for any x, y in the lattice, the glb of the set $\{z : y \leqq x \vee z\}$ exists (and is denoted $y \dot{-} x$). Show that the notion of Brouwerian lattice is dual to the notion of pseudo-Boolean algebra in the sense that a lattice $\langle B, \leqq \rangle$ is a pseudo-Boolean algebra if and only if the lattice $\langle L, \geqq \rangle$ is a Brouwerian lattice. (For a study of Brouwerian lattices, cf. [133].)

(*f*) In the Boolean algebra of statement bundles (cf. Example 3.5), for any statement forms **A** and **B**, what is the interpretation of $[\mathbf{A}] \Rightarrow [\mathbf{B}]$?

ATOMS

5.76. Prove that two finite Boolean algebras are isomorphic if and only if they have the same number of atoms.

5.77. Show that there is no Boolean algebra containing 28 elements.

5.78. In the Boolean algebra of all divisors of n, where n is a square-free integer > 1 (cf. Problem 3.3), what are the atoms? Note that $x \leq y \leftrightarrow x$ divides y.

5.79. (*a*) How many subalgebras are there of the Boolean algebra of all subsets of a four-element set?

(*b*) Given a Boolean algebra $\mathcal{B}$ with 2^k elements, show that the number of subalgebras of $\mathcal{B}$ is equal to the number of partitions of a set with k elements (where a *partition* is a division of the set into one or more disjoint non-empty sets).

5.80. If $\mathcal{A}$ and $\mathcal{B}$ are Boolean algebras with the same finite number 2^k of elements, how many isomorphisms are there from $\mathcal{A}$ onto $\mathcal{B}$?

5.81. Consider the Boolean algebra given by the field of sets consisting of all finite unions of left-open intervals of real numbers (cf. Problem 2.68). What are the atoms of this algebra?

5.82^D. Does the set of atoms of a Boolean algebra always have a supremum?

5.83. Show that every atomic, uniquely complemented lattice with zero and unit elements is isomorphic to a field of sets and is therefore a Boolean algebra. (*Hint*: Use the proof of Theorem 5.7.)

SYMMETRIC DIFFERENCE. BOOLEAN RINGS

5.84. In an arbitrary Boolean algebra:

(*a*) Does the distributive relation $x + (y \wedge z) = (x + y) \wedge (x + z)$ hold?

(*b*) Is $(x \leq y \;\&\; u \leq v) \rightarrow x + u \leq y + v$ valid?

5.85. In the Boolean algebra of all divisors of n, where n is a square-free integer > 1 (cf. Problem 3.3), find an arithmetic formula for the symmetric difference $x + y$.

5.86. In any Boolean algebra, prove:

(*a*) $x \vee (x + y) = x \vee y$, (*b*) the "dual" of $x + y$ is $(x + y)'$.

5.87. In a ring, $x - y$ is defined to be $x + (-y)$. In a Boolean ring, what is $(-y)$? Is there any difference between $x - y$ in the ring-theoretic sense and $x \sim y$ as defined in Problem 3.1?

5.88. Prove that the uniqueness of $(-x)$ in Axiom (4) for rings need not be assumed (i.e. it can be proved from the other axioms).

5.89. (*a*) Give an example of a Boolean ring without a unit element. (*b*) Show that every Boolean ring without a unit element can be extended (by addition of new elements and extension of the ring operations to the enlarged set) to a Boolean ring with unit element. Prove that the original ring is a maximal ideal of the extension.

5.90. In the axioms for Boolean rings, show that $x + y = y + x$ is not independent.

5.91. Let $\mathcal{R} = \langle R, +, \times, 0\rangle$ be a commutative ring with unit element 1. We say that an element x in R is *idempotent* if and only if $x^2 = x$. Let R^* be the set of idempotent elements of R. For any x and y in R^*, define $x \oplus y = x + y - 2xy$. Prove that $\langle R^*, \oplus, \times, 0\rangle$ is a Boolean ring with unit element 1. Express the Boolean operations $\wedge, \vee, '$ in terms of the original ring operations $+, \times$.

5.92. For any Boolean expression $\tau(u)$, prove that

$$\tau(u) = b_0 + (b_1 \wedge u)$$

where b_1 and b_0 are fixed elements of the Boolean algebra.

5.93. Solve the following equations.

(*a*) $u \wedge w = 0$

(*b*) $u \wedge z = w$

(*c*) $u + w = \tau$ (for u and w in terms of τ)

(*d*) $(\rho \wedge u) \vee (\sigma \wedge w) = \tau$ (for u and w in terms of ρ, σ, τ)

5.94. Solve the simultaneous equations

$$x \vee b = c$$
$$x \wedge b = 0$$

for x in terms of b and c. What further conclusion follows if $b \leq c$?

5.95. Solve the following system of equations.

$$u \leq v \vee w$$
$$v = z \wedge w'$$
$$u' \wedge z \leq w'$$

5.96. If $\tau(u)$ is a Boolean expression and b is an element of a Boolean algebra $\mathcal{B}$, and if $\tau(0) \wedge \tau(1) \leq b \leq \tau(0) \vee \tau(1)$, prove:

(*a*) $\tau(u) = b$ has a solution. Find all solutions.

(*b*) If $\tau(u) = b$ has a unique solution, then, for any c, $\tau(u) = c$ has a unique solution (namely, $\tau(c)$).

(*c*) If $\mathcal{B}$ is finite and there are k atoms which are $\leq \tau(0)' + \tau(1)$, then $\tau(u) = b$ has 2^k solutions.

AXIOMATIZATIONS

5.97^D. In our axiom system for Boolean algebras, determine whether Axiom (5) can be proved from Axioms (1)-(4), (7)-(9). (See Problem 5.22.)

5.98. Determine whether or not each of the axioms (B1)-(B5) for Byrne algebras (cf. Section 5.4) is independent.

5.99. Show that the following variation of the axioms for Byrne algebras (cf. Section 5.4) also may serve as an axiom system for Boolean algebras. Consider structures $\langle B, \wedge, ' \rangle$ satisfying (B1), (B2), (B5), and

(C) $x \wedge y' = z \wedge z' \leftrightarrow x \wedge y = x$

(*Hint*: Prove that $z \wedge z' = w \wedge w'$ for all z and w, and introduce 0 by definition as being equal to this common value of all $z \wedge z'$.)

5.100. (*a*) Prove that the following is a system of axioms for Boolean algebras.

(D1) $x \wedge y = y \wedge x$

(D2) $x \wedge (y \wedge z) = (x \wedge y) \wedge z$

(D3) $(x' \wedge y')' \wedge (x' \wedge y)' = x$

(*b*) Investigate the independence of Axioms (D1)-(D3).

5.101. Prove the independence of Axioms (*a*)-(*f*) for lattices in Problem 5.4.

5.102. [121]. Let L be a complemented lattice. Show that L is a Boolean algebra if and only if, for any x and y in L and for any complement z of y, $x \wedge y = 0 \leftrightarrow x \leq z$.

5.103. [101]. Given a structure $\mathcal{B} = \langle B, \wedge, ' \rangle$. Prove that $\mathcal{B}$ determines a Boolean algebra if and only if the following two laws are satisfied.

(a) $x = x \wedge y \leftrightarrow x \wedge y' = z \wedge z'$

(b) $(x \wedge y) \wedge z = (y \wedge z) \wedge x$

5.104. [122]. Prove that a structure $\mathcal{B} = \langle B, \vee, ' \rangle$ determines a Boolean algebra if and only if the following three laws are satisfied.

(a) $x \vee y = y \vee x$

(b) $(x \vee y) \vee z = x \vee (y \vee z)$

(c) $(x' \vee y')' \vee (x' \vee y)' = x$

IDEALS

5.105. If A is an infinite set, prove that the ideal of all finite subsets of A is not a principal ideal in the field of sets $\mathcal{P}(A)$.

5.106. Prove that a non-empty subset J of a Boolean algebra is an ideal if and only if the condition $x \vee y \in J \leftrightarrow (x \in J \ \&\ y \in J)$ is satisfied.

5.107. What is the ideal generated by the empty subset $\emptyset$ of a Boolean algebra?

5.108. Given a Boolean algebra $\mathcal{B}$. For any ideals J and I of $\mathcal{B}$, let

$$J \vee I = \{x \vee y : x \in J \ \&\ y \in I\}$$

(a) Prove that $J \vee I$ is an ideal.

(b) Show that the set of ideals of $\mathcal{B}$ forms a distributive lattice under the operations of $\cap$ and $\vee$. Are there zero and unit elements? Is the lattice complemented?

5.109. A *non-trivial finitely-additive measure* on a Boolean algebra $\mathcal{B}$ is a function μ from B into the set of non-negative real numbers such that

(1) $\mu(x \vee y) = \mu(x) + \mu(y)$ if $x \wedge y = 0$.

(2) μ is not a constant function.

(a) If μ is a non-trivial finitely-additive measure on a Boolean algebra $\mathcal{B}$, prove:

(i) $\mu(0) = 0$

(ii) $\mu(x_1 \vee \cdots \vee x_k) = \mu(x_1) + \cdots + \mu(x_k)$ if $x_i \wedge x_j = 0$ for $i \neq j$.

(iii) $x \leq y \rightarrow \mu(x) \leq \mu(y)$

(iv) $\mu(x_1 \vee \cdots \vee x_k) \leq \mu(x_1) + \cdots + \mu(x_k)$

(v) $J_\mu = \{x : \mu(x) = 0\}$ is a proper ideal of $\mathcal{B}$.

(vi) If μ is 2-valued, i.e. if the range of μ consists of two numbers (one of which, by (i), must be 0), then J_μ is a maximal ideal.

(vii) If μ is bounded, i.e. the range of μ is bounded above, and if we define $\nu(x) = \mu(x)/\mu(1)$, then ν is a non-trivial finitely-additive measure such that $0 \leq \nu(x) \leq 1$ and $\nu(1) = 1$.

(b) If M is a maximal ideal of $\mathcal{B}$, and if we define

$$\mu(x) = \begin{cases} 0 & \text{if } x \in M \\ 1 & \text{if } x \notin M \end{cases}$$

for any x in B, show that μ is a non-trivial 2-valued finitely-additive measure on $\mathcal{B}$.

(c) If $\mathcal{B}$ is the Boolean algebra $\mathcal{P}(K)$, where K is a finite set, and if we define $\mu(A)$ = the number of elements in A, prove that μ is a non-trivial finitely-additive measure on $\mathcal{B}$.

(d) Prove that every Boolean algebra admits a non-trivial 2-valued finitely-additive measure.

5.110. Given a subset D of a Boolean algebra $\mathcal{B}$, recall (cf. Problem 3.5) that the intersection of all subalgebras containing D is a subalgebra G_D, called the subalgebra *generated* by D.

(*a*) Prove that the elements of G_D are all elements of the form $c_1 \vee \cdots \vee c_k$, where each $c_i = d_{j_1} \wedge \cdots \wedge d_{j_{m_i}}$ and either $d_{j_h} \in D$ or $d'_{j_h} \in D$, i.e. the elements of G_D are finite joins of finite meets of elements of D and complements of elements of D.

(*b*) Show that the subalgebra G_D of $\mathcal{P}(K)$ generated by the set D of all singletons consists of all finite and cofinite sets. Prove that G_D is complete if and only if K is finite.

(*c*) We say that a set D is a set of *generators* of the Boolean algebra $\mathcal{B}$ if and only if the subalgebra G_D generated by D is the whole set B. We say that $\mathcal{B}$ is *finitely generated* if and only if there is a finite set of generators of $\mathcal{B}$. Prove that every finitely generated Boolean algebra is finite.

(*d*) We say that a set D of generators of a Boolean algebra $\mathcal{B}$ is a *free* set of generators of $\mathcal{B}$ if and only if, for every function h from D into a Boolean algebra $\mathcal{C}$, there is an extension g of h which is a Boolean homomorphism of $\mathcal{B}$ into $\mathcal{C}$. The Boolean algebra $\mathcal{B}$ is said to be *free* if and only if there is a free set of generators of $\mathcal{B}$.

(i) For any non-negative integer k, if $\mathcal{B}$ is a Boolean algebra having a free set of k generators, prove that $\mathcal{B}$ has $2^{(2^k)}$ elements.

(ii) Is every subalgebra of a free Boolean algebra also free?

(iii) Show that the cardinal number of any infinite free Boolean algebra $\mathcal{B}$ is equal to the cardinal number of any free set of generators of $\mathcal{B}$.

(iv) Show that if D_1 and D_2 are free sets of generators of the same Boolean algebra $\mathcal{B}$, then D_1 and D_2 have the same cardinal number.

(v) If h is a function from a free set D of generators of a Boolean algebra $\mathcal{B}$ into a Boolean algebra $\mathcal{C}$, show that there is a *unique* homomorphism g from $\mathcal{B}$ into $\mathcal{C}$ such that g is an extension of h.

(vi) If D_1 is a free set of generators of $\mathcal{B}_1$ and D_2 is a free set of generators of $\mathcal{B}_2$ and D_1 and D_2 have the same cardinal number, then $\mathcal{B}_1$ and $\mathcal{B}_2$ are isomorphic.

(vii) Let D be a set of generators of a Boolean algebra $\mathcal{B}$. Show that D is a free set of generators of $\mathcal{B}$ if and only if, for any $u_1, \ldots, u_n$ in B, if $u_i \in D$ or $u'_i \in D$ for each u_i, then $u_1 \wedge \cdots \wedge u_n \neq 0$.

(viii) For any cardinal number $\mathfrak{m}$, prove that there is a Boolean algebra having a free set of generators of cardinality $\mathfrak{m}$. (*Hint*: Generalize the Boolean algebra of statement bundles (Example 3.5) by using a set of statement letters of cardinality $\mathfrak{m}$ instead of a denumerable set of statement letters.)

(ix)^D When is $\mathcal{P}(K)$ a free Boolean algebra?

QUOTIENT ALGEBRAS

5.111. Let J be a proper ideal of a Boolean algebra $\mathcal{B}$ and let K be a proper ideal of the quotient algebra $\mathcal{B}^{\#} = \mathcal{B}/J$. Let $J^{\square} = \{x : x \in B \ \& \ x + J \in K\}$. Prove that $J^{\square}$ is a proper ideal in $\mathcal{B}$ and that the function ψ from $\mathcal{B}/J^{\square}$ onto $\mathcal{B}^{\#}/K$ defined by

$$\psi(x + J^{\square}) \;=\; (x + J) + K$$

is an isomorphism.

5.112. In a Boolean algebra $\mathcal{B}$, B is an ideal. What is $\mathcal{B}/B$? Why isn't it a Boolean algebra?

5.113. (For those readers acquainted with elementary point-set topology.) Let $\mathcal{X}$ be a topological space.

(*a*) Prove that the set $\boldsymbol{F}$ of subsets of $\mathcal{X}$ having nowhere dense boundary is a field of sets.

(*b*) Prove that the set $\boldsymbol{N}$ of nowhere dense sets of $\mathcal{X}$ is an ideal of $\boldsymbol{F}$.

(*c*) Show that the quotient algebra $\boldsymbol{F}/\boldsymbol{N}$ is isomorphic to the algebra of regular open sets of $\mathcal{X}$ (cf. Problem 5.47). (*Hint:* If $A \in \boldsymbol{F}$, show that there is a unique regular set A_1 such that $A + A_1$ is nowhere dense.)

BOOLEAN REPRESENTATION THEOREM

5.114. Prove that every ideal of a Boolean algebra $\mathcal{B}$ is principal if and only if $\mathcal{B}$ is finite.

5.115. For each s in a set S assume that there is an associated Boolean algebra $\mathcal{B}_s$. By the Cartesian product $\prod_{s \in S} \mathcal{B}_s$ we mean the set of all functions f defined on S such that, for each $s \in S$, $f(s) \in B_s$. We define Boolean operations on the Cartesian product in componentwise fashion, e.g. if f and g are in the Cartesian product, let $f \wedge g$ be the function defined on S such that $(f \wedge g)(s) = f(s) \wedge_{\mathcal{B}_s} g(s)$ for each $s \in S$.

(a) Prove that the Cartesian product of Boolean algebras is a Boolean algebra.

(b) Prove that a finite Boolean algebra of cardinality 2^n is isomorphic to a Cartesian product of n copies of the Boolean algebra $\{0,1\}$.

(c) If the cardinal number of a set A is $\mathfrak{m}$, prove that the Boolean algebra $\mathcal{P}(A)$ is isomorphic to a Cartesian product of $\mathfrak{m}$ copies of $\{0,1\}$.

(d) Show that every Boolean algebra is isomorphic to a subalgebra of a Cartesian product of copies of $\{0,1\}$.

5.116. (a) Prove that if x and y are distinct elements of a distributive lattice, then there is a prime ideal containing one of x and y but not the other. (The notion of prime ideal, although originally defined for Boolean algebras, also makes sense for lattices. Hint for the proof: Since $x \neq y$, we may assume $y \nleq x$. Let Z be the set of all ideals containing x but not y, and apply Zorn's Lemma.)

(b) Prove that any distributive lattice is lattice-isomorphic to a lattice of sets. (*Hint*: To each element x of the lattice associate the set of proper prime ideals not containing x.)

(c) Prove the converse of (a), i.e. a lattice is distributive if, for any two distinct elements of the lattice, there is a prime ideal containing one of the elements but not the other.

5.117. Although every Boolean algebra is, by Theorem 5.30, isomorphic to a field of sets, show that there are Boolean algebras $\mathcal{B}$ for which there is no isomorphism to a field of sets preserving infinite joins and meets. (*Hint*: Consider the regular open algebra of the real line (Problem 5.47).)

5.118^D. [137]. Prove that a distributive lattice is a Boolean algebra if and only if every proper prime ideal is maximal.

INFINITE MEETS AND JOINS

5.119. Give an example of a Boolean subalgebra $\mathcal{C}$ of a Boolean algebra $\mathcal{B}$ such that some subset E of C has a lub in $\mathcal{C}$ but not in $\mathcal{B}$.

5.120. In any lattice, prove (assuming that all the indicated lub's and glb's exist):

(a) If $x_w \leq y_w$ for all $w \in W$, then $\bigwedge_{w \in W} x_w \leq \bigwedge_{w \in W} y_w$ and $\bigvee_{w \in W} x_w \leq \bigvee_{w \in W} y_w$.

(b) $\bigvee_{s \in S} \bigwedge_{w \in W} a_{s,w} \leq \bigwedge_{w \in W} \bigvee_{s \in S} a_{s,t}$.

(c) $\bigwedge_{w \in W} x_w \vee \bigwedge_{w \in W} y_w \leq \bigwedge_{w \in W} (x_w \vee y_w)$.

(d) $\bigvee_{w \in W} (x_w \wedge y_w) \leq \bigvee_{w \in W} x_w \wedge \bigvee_{w \in W} y_w$.

(e) None of the inequalities in (a)-(d) can be changed into equalities valid for all lattices.

5.121. An ideal J of a Boolean algebra is said to be *complete* if and only if J is closed under arbitrary joins of its elements.

(a) Prove that every complete ideal is a principal ideal.

(b) If J is a complete ideal of a complete Boolean algebra $\mathcal{B}$, prove that the quotient algebra $\mathcal{B}/J$ is complete.

5.122. (*a*) Prove that the set of sublattices of a lattice is a complete lattice with respect to the inclusion relation $\subseteq$. What are the operations of join and meet?

(*b*) Prove that the set of subalgebras of a Boolean algebra is a complete lattice with respect to the inclusion relation $\subseteq$. Describe the join and meet operations.

5.123. (*a*) Assume that g is a *closure operation* on a set L partially ordered by $\leq$, i.e. g is a function from L into L such that:

(1) $x \leq y \rightarrow g(x) \leq g(y)$

(2) $x \leq g(x)$

(3) $g(g(x)) = g(x)$

An element x in L is said to be *g-closed* if and only if $g(x) = x$. Prove:

(i) x is g-closed if and only if $x = g(y)$ for some y.

(ii) If 1 is a maximum element of L, then 1 is g-closed.

(iii) A glb of a set of g-closed elements is also g-closed. In particular, if x and y are g-closed, so is $x \wedge y$ (if it exists).

(iv) If $\langle L, \leq\rangle$ is a complete lattice, then the set C of g-closed elements is a complete lattice with respect to the original ordering $\leq$ on L, and, for any subset Y of L,

$$\bigwedge^C_{y \in Y} y = \bigwedge^L_{y \in Y} y \quad \text{and} \quad \bigvee^C_{y \in Y} y = g\left(\bigvee^L_{y \in Y} y\right)$$

(*b*) Let $\leq$ be a partial order on L. For any $Y \subseteq L$, let Y^b = the set of upper bounds of Y and Y^s = the set of lower bounds of Y. We say that Y is a *cut* if and only if $Y = Y^{bs}$. Prove:

(i) $Y \subseteq Y^{bs}$

(ii) $Y \subseteq Y^{sb}$

(iii) $X \subseteq Y \rightarrow (Y^b \subseteq X^b \;\&\; Y^s \subseteq X^s) \rightarrow (X^{bs} \subseteq Y^{bs} \;\&\; X^{sb} \subseteq Y^{sb})$

(iv) $X^b = X^{bsb} \;\&\; X^s = X^{sbs}$

(v) $X^{bs} = X^{bsbs} \;\&\; X^{sb} = X^{sbsb}$

(vi) Every set X^{bs} is a cut.

(vii) The function $g(X) = X^{bs}$ is a closure operation on $\langle \mathcal{P}(L), \subseteq\rangle$. The g-closed elements are the cuts.

(viii) The set C of all cuts is a complete lattice with respect to $\subseteq$. Meets are intersections and $\bigvee_{Y \in X} Y = \left(\bigcup_{Y \in X} Y\right)^{bs}$. The function f, such that $f(x) = \{x\}^{bs}$ for any x in L, is an order-isomorphism of $\langle L, \leq\rangle$ into the complete lattice $\langle C, \subseteq\rangle$; f preserves all meets and joins already existing in $\langle L, \leq\rangle$, i.e. if $z = \bigvee_{\alpha \in A} x_\alpha$ in L, then $f(z) = \bigvee_{\alpha \in A} f(x_\alpha)$ in C, and similarly for meets.

(ix) Let $\mathcal{L} = \langle L, \leq\rangle$ determine a Boolean algebra, i.e. it is a distributive, complemented lattice. Define, for any $X \subseteq L$, $X^* = \{y : y \wedge x = 0\}$. Then

(1) $X \cap X^* = \{0\} = 0_C$

(2) $(X \cup X^*)^b = \{1\}$

(3) $(X \cup X^*)^{bs} = L = 1_C$

(4) $X^{**} = X^{bs}$

(5) $\{x'\}^{bs} = \{x\}^{bs*}$

(6) X^{**} is an ideal of L.

(7) The function F such that $F(J) = J^{**}$ for any ideal J of $\mathcal{L}$ is a lattice-homomorphism from the distributive lattice of all ideals of $\mathcal{L}$ onto the lattice of cuts.

(8) Hence C forms a complete Boolean algebra, and the lattice-isomorphism f of (viii) is a Boolean isomorphism. Thus every Boolean algebra is embeddable in a complete Boolean algebra in such a way that unions and meets are preserved [131].

5.124. In Example 5.19, prove that the collection of all singletons $\{x\}$, where $x \in K$, has no lub in $\mathcal{F}$ (in addition to not having its union in $\mathcal{F}$).

$\mathfrak{m}$-COMPLETENESS. σ-ALGEBRAS

5.125. Let $\mathcal{F}$ be the collection of all subsets A of a given set K such that the cardinality of A or the cardinality of $\bar{A}$ is $\leq$ a given infinite cardinal number $\mathfrak{m}$. Prove that $\mathcal{F}$ is an $\mathfrak{m}$-complete field of sets. What happens if we change $\leq$ to $<$? If K has cardinal number $\mathfrak{n} > \mathfrak{m}$ and $\mathfrak{m} < \mathfrak{p} \leq \mathfrak{n}$, is $\mathcal{F}$ a $\mathfrak{p}$-complete field of sets?

5.126. If $\mathcal{F}$ is an $\mathfrak{m}$-complete field of subsets of a set K and $A \in \mathcal{F}$, prove that $\mathcal{F} \mid A = \{Y : Y \in \mathcal{F} \ \& \ Y \subseteq A\}$ is also an $\mathfrak{m}$-complete field of sets.

5.127. Prove that an infinite σ-algebra must have at least $2^{\aleph_0}$ elements. (*Hint*: Problem 5.14.)

5.128. By a *free σ-algebra with $\mathfrak{m}$ generators* we mean a σ-algebra $\mathcal{B}$ having a subset D of cardinality $\mathfrak{m}$ such that any function from D into a σ-algebra $\mathcal{C}$ can be extended to a σ-homomorphism h from $\mathcal{B}$ into $\mathcal{C}$.

(*a*)^D For every cardinal number $\mathfrak{m}$, show that there is a free σ-algebra with $\mathfrak{m}$ generators.

(*b*) Prove the analogues of Problem 5.110(*d*)(iv)-(vi) for free σ-algebras.

(*c*) Prove that any free σ-algebra with $\mathfrak{m}$ generators is isomorphic to a σ-field of sets. (*Hint*: Use Problem 5.53.)

5.129. We say that a Boolean algebra $\mathcal{B}$ satisfies the *countable chain condition* (CCC) if and only if every pairwise-disjoint set of nonzero elements of B is countable. (A set Y is said to be *pairwise-disjoint* if and only if, for any distinct elements y and z in Y, $y \wedge z = 0$.)

(*a*) Prove that a Boolean algebra $\mathcal{B}$ satisfies (CCC) if and only if every subset $Z \subseteq B$ has a countable subset Y such that Y and Z have the same set of upper bounds.

(*b*) Prove that any σ-algebra satisfying (CCC) is complete.

(*c*) Show that the regular open algebra of a topological space with a countable base satisfies (CCC). (Cf. Problem 5.47.)

Appendix A

Elimination of Parentheses

More extensive conventions for eliminating parentheses than those given in Section 1.4, page 5, will be presented here.

(I) We assign a *rank* to the connectives as follows:

$\leftrightarrow$	5
$\rightarrow$	4
$\vee$	3
&	2
$\neg$	1

The *rank* of a statement form **A** will be

(*a*) 0, if **A** is a statement letter;

(*b*) the rank of the principal connective of **A**, otherwise.

We shall describe our procedure for eliminating parentheses by induction on the number of occurrences of connectives in the statement form. (The description will appear complicated, but the simplicity of the procedure will become apparent after a few examples.)

Clearly, if **A** has no connectives, it is a statement letter and has no parentheses. Assume that we have described the procedure for all statement forms having fewer than k occurrences of connectives, and assume that **A** has k occurrences of connectives.

Case (i): **A** is a denial ($\neg$**B**). If **B** itself is of the form ($\neg$**C**), then we apply our procedure to ($\neg$**C**) and omit the outer parentheses (if any) of the resulting expression, obtaining some expression **D**. The final result is taken to be ($\neg$**D**). If **B** is not of the form ($\neg$**C**), and the application of our procedure to **B** yields **E**, then the final result is ($\neg$**E**).

Case (ii): **A** is (**B** α **C**), where α is $\leftrightarrow$, $\rightarrow$, $\vee$ or &. We apply the procedure for eliminating parentheses to **B** and **C**, obtaining **B*** and **C***. At this stage we have (**B*** α **C***). We drop the outermost pair of parentheses (if any) from **B*** if the rank of α is greater than or equal to the rank of **B**. We drop the outermost pair of parentheses (if any) from **C*** if the rank of α is greater than the rank of **C**.

(II) After completion of (I), we omit the outermost pair of parentheses (if any).

Examples.

A.1. $(\neg(\neg A))$.

Applying (I), Case (i), twice, we obtain $(\neg\neg A)$. Then by (II) we have $\neg\neg A$. In general, no parentheses are needed to separate successive negations.

A.2. $((A \vee B) \rightarrow C)$.

The principal connective is $\rightarrow$, of rank 4. Since $(A \vee B)$ has rank 3, we obtain, by (I), Case (ii), $(A \vee B \rightarrow C)$, and finally by (II), $A \vee B \rightarrow C$.

A.3. $(A \vee (B \rightarrow C))$.

The principal connective is $\vee$, of rank 3. Since $(B \rightarrow C)$ has rank 4, we do not drop the parentheses from $(B \rightarrow C)$. Thus in this example (I) allows no elimination of parentheses, and by (II) we obtain $A \vee (B \rightarrow C)$.

A.4. $(((B \vee A) \leftrightarrow B) \rightarrow ((\neg B) \& C))$.

$\rightarrow$ is the principal connective. Application of (I) to $((B \vee A) \leftrightarrow B)$ yields $(B \vee A \leftrightarrow B)$, and application of (I) to $((\neg B) \& C)$ yields $(\neg B \& C)$. Since the rank of $\leftrightarrow$ is greater than that of $\rightarrow$, we leave the outermost pair of parentheses of $(B \vee A \leftrightarrow B)$. However, since the rank of & is less than that of $\rightarrow$, we omit the outermost pair of parentheses from $(\neg B \& C)$. The final result is $(B \vee A \leftrightarrow B) \rightarrow \neg B \& C$.

A.5. $((A \vee B) \vee C)$.

The right-most $\vee$ is the principal connective. Since $(A \vee B)$ has rank equal to that of $\vee$, we may drop the parentheses, obtaining $(A \vee B \vee C)$, and finally by (II), $A \vee B \vee C$.

A.6. $(A \vee (B \vee C))$.

The left-most $\vee$ is the principal connective. Since $(B \vee C)$ has rank equal to that of $\vee$, we cannot drop the parentheses. Thus (I) yields no elimination of parentheses, and by (II) we obtain $A \vee (B \vee C)$.

The results given in Examples A.5 and A.6 illustrate the principle of association to the left. Thus $A \vee B \vee C$ stands for $((A \vee B) \vee C)$. Likewise $A \rightarrow B \rightarrow C$ stands for $((A \rightarrow B) \rightarrow C)$, and the same holds for the other binary connectives. Association to the left is due to our agreement that in $(\mathbf{B}\,\alpha\,\mathbf{C})$, we omit outer parentheses from $\mathbf{B}$ if rank $(\mathbf{B}) \leq$ rank (α), but from $\mathbf{C}$ if rank $(\mathbf{C}) <$ rank (α).

A.7. $((A \vee (B \vee C)) \vee (B \& (\neg(\neg C))))$.

Application of (I) to $(A \vee (B \vee C))$ yields no elimination of parentheses, while $(B \& (\neg(\neg C)))$ becomes $(B \& \neg\neg C)$. Then by (I) we obtain $(A \vee (B \vee C) \vee B \& \neg\neg C)$, and finally $A \vee (B \vee C) \vee B \& \neg\neg C$.

The rough idea of convention (I) is that connectives of higher rank are to have greater scope than those of lower rank. Thus if elimination of parentheses yields $A \vee B \rightarrow C$, this stands for $((A \vee B) \rightarrow C)$. The connective $\rightarrow$, being of higher rank than $\vee$, must "connect" the longest possible statement forms to the left and right. Thus $\rightarrow$ has $(A \vee B)$ as its antecedent rather than just B. Similarly, in $A \rightarrow B \& B \leftrightarrow D \vee B$, $\leftrightarrow$ is the connective of highest rank. Hence the left side of $\leftrightarrow$ should be $(A \rightarrow B \& B)$ and the right side should be $D \vee B$. Thus we have $(A \rightarrow B \& B) \leftrightarrow (D \vee B)$. Within $(A \rightarrow B \& B)$, $\rightarrow$ is stronger, and we obtain $((A \rightarrow (B \& B)) \leftrightarrow (D \vee B))$.

In ordinary arithmetic, without realizing it we already have been taught an analogous ranking of arithmetic operations. Addition is strongest, then comes multiplication, and finally exponentiation. For example, $4 + 2 \cdot 5$ stands for $4 + (2 \cdot 5)$, and not for $(4 + 2) \cdot 5$, while $5 \cdot 2^3$ stands for $5 \cdot (2^3)$, not for $(5 \cdot 2)^3$.

Sometimes, especially in long statement forms, for the sake of clarity we can keep some parentheses which could be omitted according to our conventions. For example, in A.7 above we might write $A \vee (B \vee C) \vee (B \& \neg\neg C)$ instead of $A \vee (B \vee C) \vee B \& \neg\neg C$.

Solved Problems

A.1. Describe an algorithm (i.e. effective procedure) for determining whether a given expression is a statement form, and for determining the principal connective when the expression is a statement form.

Solution:

The description is given by induction on the number of occurrences of connectives. If there are no connectives, then the expression is a statement form if and only if it is a statement letter. Assume now that an expression $\mathbf{A}$ has k connectives, where $k > 0$, and that our algorithm already has been defined for expressions with fewer than k occurrences of connectives. If $\mathbf{A}$ does not begin with a left parenthesis and end with a right parenthesis, then $\mathbf{A}$ is not a statement form. If $\mathbf{A}$ does have the appropriate initial left and terminal right parentheses, omit them, obtaining an expression $\mathbf{B}$.

Case 1. $\mathbf{B}$ has the form $\neg\mathbf{C}$. If $\mathbf{C}$ is a statement form, then $\mathbf{A}$ is a statement form and its principal connective is $\neg$. If $\mathbf{C}$ is not a statement form, then neither is $\mathbf{A}$.

Case 2. $\mathbf{B}$ is not of the form $\neg\mathbf{C}$. For each of the binary connectives $\square$ in $\mathbf{B}$, if $\mathbf{B}$ is $\mathbf{C} \square \mathbf{D}$ and if $\mathbf{C}$ and $\mathbf{D}$ are statement forms, then $\mathbf{A}$ is also a statement form with principal connective $\square$. On the other hand, if the indicated condition does not hold for any of the binary connectives $\square$ in $\mathbf{B}$, then $\mathbf{A}$ is not a statement form.

A.2. Eliminate as many parentheses as possible from:

(*a*) $(((A \vee B) \to (\neg C)) \vee ((\neg B) \& C))$

(*b*) $((A \& (\neg(\neg B))) \leftrightarrow (B \leftrightarrow (C \vee B)))$

(*c*) $((B \leftrightarrow (C \vee B)) \leftrightarrow (A \& (\neg(\neg B))))$

Solution:

(*a*) The second $\vee$ is the principal connective. Since $\vee$ has higher rank than &, we can omit the outer parentheses from $((\neg B) \& C)$. Since $\to$ has higher rank than $\vee$, we cannot drop the outer parentheses from $((A \vee B) \to (\neg C))$. But within $((A \vee B) \to (\neg C))$ we can drop the outer parentheses from $(A \vee B)$, since $\to$ has higher rank than $\vee$:

$$(A \vee B \to (\neg C)) \ \vee \ (\neg B) \& C$$

Finally, we can drop the parentheses around the denials:

$$(A \vee B \to \neg C) \ \vee \ \neg B \& C$$

(*b*) $\leftrightarrow$ is the principal connective. We can omit the outer parentheses of $(A \& (\neg(\neg B)))$, since $\leftrightarrow$ has higher rank than &:

$$A \& (\neg(\neg B)) \ \leftrightarrow \ (B \leftrightarrow (C \vee B))$$

However, we cannot omit the outer parentheses of $(B \leftrightarrow (C \vee B))$, since the latter has the same rank as $\leftrightarrow$ and appears on the right hand side of the biconditional. Within $(B \leftrightarrow (C \vee B))$ we can drop the parentheses of $(C \vee B)$, and, on the other side, we can drop the parentheses around the denials:

$$A \& \neg\neg B \ \leftrightarrow \ (B \leftrightarrow C \vee B)$$

(*c*) This is the same as (*b*) except that the two sides of the biconditional have been reversed. Since $(B \leftrightarrow (C \vee B))$ now appears on the left hand side of the biconditional, we can drop the outer parentheses:

$$B \leftrightarrow (C \vee B) \ \leftrightarrow \ A \& (\neg(\neg B))$$

As before, we then obtain

$$B \leftrightarrow C \vee B \ \leftrightarrow \ A \& \neg\neg B$$

In practice, we would never eliminate all the parentheses in part (*c*), since their complete elimination will not facilitate the interpretation of the original statement form.

A.3. Describe an algorithm for determining whether a given expression **A** has been obtained from a statement form as a result of applying our conventions for eliminating parentheses (and, if it has, find the statement form).

Solution:

We shall describe, by induction on the number of symbols of **A**, a procedure which either will find the statement form abbreviated by **A** or will eventually tell us that there is no such statement form. Clearly, if **A** has one symbol, then **A** abbreviates a statement form (**A** itself) if and only if **A** is a statement letter. Now assume that **A** has k symbols (where $k > 1$) and that our algorithm has been defined for all expressions with fewer than k symbols.

Case 1. **A** either does not begin with a left parenthesis or does not end with a right parenthesis. (Hence if **A** does abbreviate a statement form **F**, then the outer parentheses of **F** were omitted.)

Case 1a. **A** is of the form $\neg\neg\ldots\neg(\mathbf{B})$. If **B** abbreviates a statement form **G** (and hence **B** does not have the outer parentheses of **G**), then **A** abbreviates $(\neg(\neg(\ldots(\neg\mathbf{G})\ldots)))$. Otherwise, **A** is not an abbreviation of a statement form.

Case 1b. **A** is of the form $\neg\neg\ldots\neg\mathbf{B}$, where **B** is a statement letter. Then **A** abbreviates $(\neg(\neg(\ldots(\neg\mathbf{B})\ldots)))$.

Case 1c. **A** is not of the form $\neg\neg\ldots\neg(\mathbf{B})$, and not of the form $\neg\neg\ldots\neg\mathbf{B}$ (where **B** is a statement letter). For each occurrence of a binary connective # in **A**, represent **A** as **C** # **D**. If

(i) **C** abbreviates a statement form **H**,

(ii) **D** abbreviates a statement form **J**,

(iii) **C** contains the outer parentheses of **H** (if any) if the rank of **H** is greater than or equal to that of #,

(iv) **D** contains the outer parentheses of **J** (if any) if the rank of **J** is greater than the rank of #, then **A** abbreviates (**H** # **J**).

If no occurrence of a binary connective in **A** satisfies (i)-(iv), then **A** does not abbreviate a statement form.

Case 2. **A** is of the form (**B**).

Case 2a. Omit the initial left and terminal right parentheses, and then apply Case 1 to **B**. If **B** abbreviates a statement form **C**, then **A** abbreviates **C** also.

Case 2b. If Case 2a does not show **A** to be an abbreviation of a statement form, apply the procedure of Case 1c. If we obtain a statement form **D**, then **A** abbreviates **D**.

Examples.

(*a*) $B \leftrightarrow C \vee B$. Case 1c applies. First, we try $\leftrightarrow$ as principal connective. The left side is B, and the right side $C \vee B$, which abbreviates $(C \vee B)$. Since $(C \vee B)$ has rank less than that of $\leftrightarrow$, the outer parentheses around $C \vee B$ have been legitimately omitted. Thus we obtain $(B \leftrightarrow (C \vee B))$.

(*b*) $A \,\&\, \neg B \,\&\, A$. Case 1c applies. First, we consider the left-most &. The left side is A and the right side is $\neg B \,\&\, A$. Hence we must consider $\neg B \,\&\, A$. Again, Case 1c applies and we consider &. The left side is $\neg B$ (which is an abbreviation of $(\neg B)$) and the right side is A. Thus $\neg B \,\&\, A$ abbreviates $((\neg B) \,\&\, A)$, but, since $((\neg B) \,\&\, A)$ has rank equal to that of &, clause (iv) has been violated. Next, we try the second &. The left side is $A \,\&\, \neg B$, which is easily seen to abbreviate $(A \,\&\, (\neg B))$. Hence the original statement form is $((A \,\&\, (\neg B)) \,\&\, A)$.

(*c*) $\neg\neg A \vee B$. Case 1c applies. The left side of $\vee$ is $\neg\neg A$, which by Case 1b abbreviates $(\neg(\neg A))$. Hence we obtain $((\neg(\neg A)) \vee B)$.

(*d*) $A \vee \neg(A \vee C)$. Case 1c applies and we look at the first $\vee$. The right side is $\neg(A \vee C)$. To the latter, Case 1b applies and we see immediately that $A \vee C$ is a statement form. Hence the original statement form is $(A \vee (\neg(A \vee C)))$.

Supplementary Problems

A.4. Find the ranks of the following statement forms.

(*a*) A; (*b*) $(\neg B)$; (*c*) $(\neg(B \vee C))$; (*d*) $((A \to (\neg B)) \& A)$; (*e*) $(A \to ((\neg B) \& A))$.

A.5. Eliminate as many parentheses as possible from the following.

(*a*) $(((\neg A) \vee (B \& C)) \to (A \leftrightarrow (\neg B)))$

(*b*) $((A \to (\neg(\neg B))) \vee (\neg(A \& C)))$

(*c*) $(((A \to (\neg A)) \to B) \to (A \to B))$

(*d*) $((\neg A) \to (A \to B))$

A.6. Determine whether each of the following expressions is an abbreviation of a statement form, and, if so, construct the statement form.

(*a*) $A \vee \neg B \to (B \leftrightarrow C) \& A$

(*b*) $A \to B \to \neg\neg A \vee B$

(*c*) $\neg(A \vee B) \vee A \& B$

(*d*) $A \vee B \vee (C \to D)$

A.7. Show that if **A** is a statement form, then there is at most one connective □ such that **A** is of the form (**B** □ **C**), where **B** and **C** are statement forms.

A.8. Show that the algorithm of Problem A.3 is correct, i.e. when **B** is an abbreviation of a statement form **A**, then the algorithm applied to **B** yields **A** as its only answer, and when **B** is not an abbreviation of a statement form, then the algorithm says so.

Appendix B

Parenthesis-free Notation

We may avoid the use of parentheses if we redefine the notion of statement form as follows: (i) every statement letter is a statement form; (ii) if **A** and **B** are statement forms, so are $\neg \mathbf{A}$, $\& \mathbf{AB}$, $\vee \mathbf{AB}$, $\rightarrow \mathbf{AB}$, $\leftrightarrow \mathbf{AB}$.

Examples.

The statement form $((\neg A) \& (B \vee A))$ would be rewritten as $\& \neg A \vee BA$. The statement form $(A \rightarrow ((\neg B) \leftrightarrow (A \vee C)))$ would be rewritten as $\rightarrow A \leftrightarrow \neg B \vee AC$.

This way of writing statement forms is sometimes called *Polish notation* (in honor of its inventor J. Lukasiewicz).

Examples.

$\rightarrow \vee AB \vee \neg C \& \neg BC$ is Polish notation for the statement form $((A \vee B) \rightarrow ((\neg C) \vee ((\neg B) \& C)))$. Similarly, $\leftrightarrow \& A \neg\neg B \leftrightarrow B \vee CB$ is Polish notation for $((A \& (\neg(\neg B))) \leftrightarrow (B \leftrightarrow (C \vee B)))$.

Solved Problems

B.1. Write the following statement forms in Polish notation.

(*a*) $((((\neg A) \vee (B \& C)) \rightarrow (A \leftrightarrow (\neg B)))$

(*b*) $((A \rightarrow (\neg(\neg B))) \vee (\neg(A \& C)))$

(*c*) $(((A \rightarrow (\neg A)) \rightarrow B) \rightarrow (A \rightarrow B))$

Solution:

(*a*) $\rightarrow \vee \neg A \& BC \leftrightarrow A \neg B$

(*b*) $\vee \rightarrow A \neg\neg B \neg \& AC$

(*c*) $\rightarrow \rightarrow \rightarrow A \neg AB \rightarrow AB$

B.2. Find the statement forms whose transcriptions into Polish notation are

(*a*) $\rightarrow \vee A \neg B \& \leftrightarrow BCA$

(*b*) $\rightarrow \rightarrow AB \vee \neg\neg AB$

(*c*) $\vee \neg \vee AB \& AB$

(*d*) $\vee \vee AB \rightarrow CD$

Solution:

(*a*) $((A \vee (\neg B)) \rightarrow ((B \leftrightarrow C) \& A))$

(*b*) $((A \rightarrow B) \rightarrow ((\neg(\neg A)) \vee B))$

(*c*) $((\neg(A \vee B)) \vee (A \& B))$

(*d*) $((A \vee B) \vee (C \rightarrow D))$

B.3. Describe an algorithm for determining whether a given expression is a statement form in Polish notation and for constructing the corresponding statement form in its original notation.

Solution:

Assign the integer -1 to statement letters, 0 to $\neg$, and $+1$ to the binary connectives $\&, \vee, \rightarrow, \leftrightarrow$. Then we have

***Theorem*:** An expression **A** is a statement form in Polish notation if and only if

(I) the sum $\#(\mathbf{A})$ of the integers assigned to all the occurrences of symbols in **A** is -1;

(II) the sum $\#(\mathbf{B})$ of the integers assigned to all the occurrences of symbols in every proper initial segment **B**[†] of **A** is non-negative.

Example. $\vee \rightarrow \vee\, A\, B \neg B\, \& \neg B\, C$

1
2
3
2
1
1
0
1
1
0
-1

Proof of the Theorem. First, we shall prove that any statement form in Polish notation satisfies conditions (I) and (II). This is shown by induction on the number k of occurrences of connectives in **A**. If there are no connectives, **A** is a statement letter and conditions (I)-(II) are obvious. (In this case there are no proper initial segments of **A**.) Now assume that the result has been established for all statement forms having fewer than k occurrences of connectives ($k \geqq 1$). By our new definition of statement form, **A** is of one of the forms $\neg\mathbf{B}$, $\&\mathbf{BC}$, $\vee\mathbf{BC}$, $\rightarrow\mathbf{BC}$, $\leftrightarrow\mathbf{BC}$, where **B** and **C** are statement forms (having fewer than k occurrences of connectives). Use of the inductive hypothesis now yields (I)-(II). (For instance, if **A** is $\&\mathbf{BC}$, then $\#(\mathbf{A}) = 1 + \#(\mathbf{B}) + \#(\mathbf{C}) = 1 + (-1) + (-1) = -1$, yielding (I). If **D** is a proper initial segment of **A**, then either **D** is & and $\#(\mathbf{D})$ is $+1 \geqq 0$; or **D** is $\&\mathbf{D}_1$ (where $\mathbf{D}_1$ is a proper initial segment of **B**) and $\#(\mathbf{D}) = 1 + \#(\mathbf{D}_1) \geqq 1 + 0 = 1 > 0$; or **D** is $\&\mathbf{B}$ and $\#(\mathbf{D}) = 1 + \#(\mathbf{B}) = 1 + (-1) = 0$; or **D** is $\&\mathbf{BC}_1$ (where $\mathbf{C}_1$ is a proper initial segment of **C**) and $\#(\mathbf{D}) = 1 + \#(\mathbf{B}) + \#(\mathbf{C}_1) = 1 + (-1) + \#(\mathbf{C}_1) = 0 + \#(\mathbf{C}_1) = \#(\mathbf{C}_1) \geqq 0$.)

Conversely, let us assume now that an expression **A** satisfies (I)-(II). We prove that **A** is a statement form in Polish notation by induction on the number k of symbols in **A**. $k = 1$: then by (I), **A** is a statement letter. Induction step: assume that $k > 1$ and that the result holds for all expressions having fewer than k symbols. Case 1: **A** is of the form $\neg\mathbf{B}$. It is then easy to show that the truth of (I)-(II) for **A** implies the truth of (I)-(II) for **B**, and hence, by inductive hypothesis, **B** is a statement form. Therefore $\neg\mathbf{B}$ is a statement form. Case 2: **A** is of the form **BC**, where **B** is a statement letter. This contradicts (II), since **B** is a proper initial segment of **A**. Case 3: **A** is of the form $\square\mathbf{C}$, where $\square$ is one of the binary connectives $\&, \vee, \rightarrow, \leftrightarrow$. There must be a shortest proper initial segment **B** of **C** such that $\#(\mathbf{B}) = -1$. For, as we move from left to right in **A**, the sum of the symbols begins at $+1$ (the integer for $\square$) and ends with -1 ($= \#(\mathbf{A})$), and moving from one symbol to the next either leaves the sum unchanged or changes it by $+1$ or -1. Hence we must finally arrive at the first proper initial segment of **A** whose sum is 0. This proper initial segment is of the form $\square\mathbf{B}$, where **B** is the shortest proper initial segment of **C** such that $\#(\mathbf{B}) = -1$. Then **B** satisfies (I). As for (II), consider any proper initial segment **D** of **B**. Then

[†] **B** is a *proper initial segment* of **A** if and only if **A** is of the form **BC**, where **C** contains at least one symbol.

since $\square$ **D** is a shorter proper initial segment of **A** than $\square$ **B**, #($\square$ **D**) is > 0, and #(**D**) ≥ 0. Hence (II) holds for **B**, and, by inductive hypothesis, **B** is a statement form in Polish notation. Let **C** be **BE**. Thus **A** is $\square$ **BE**. Since $-1 = \#(\mathbf{A}) = 1 + \#(\mathbf{B}) + \#(\mathbf{E}) = 1 + (-1) + \#(\mathbf{E}) = \#(\mathbf{E})$, **E** satisfies (I). As for (II), let **F** be any proper initial segment of **E**. Then $\square$ **BF** is a proper initial segment of **A**. By (II) for **A**, $0 \leq \#(\square\,\mathbf{BF}) = 1 + \#(\mathbf{B}) + \#(\mathbf{F}) = 1 + (-1) + \#(\mathbf{F}) = \#(\mathbf{F})$. Thus (II) holds for **E**, and, by inductive hypothesis, **E** is a statement form in Polish notation, and therefore so is $\square$ **BE**.

Notice that the second part of the proof of the theorem gives a method for constructing the corresponding statement form in our original notation (since it locates the statement forms out of which our given statement form is constructed).

Example. $\leftrightarrow \& A \neg\neg B \leftrightarrow B \vee CB$

The first proper initial segment whose sum is 0 is $\leftrightarrow \& A \neg\neg B$. Thus we have $(\& A \neg\neg B) \leftrightarrow (\leftrightarrow B \vee CB)$. In $\& A \neg\neg B$, the first proper initial segment whose sum is 0 is $\& A$. Hence we obtain $A \,\&\, \neg\neg B$. In $\leftrightarrow B \vee CB$, the first proper initial segment whose sum is 0 is $\leftrightarrow B$, and we obtain $B \leftrightarrow (\vee\, CB)$. Thus, so far, $(A \,\&\, \neg\neg B) \leftrightarrow (B \leftrightarrow (\vee\, CB))$. Finally, $\vee\, CB$ corresponds to $C \vee B$, and the statement form in our original notation is $(A \,\&\, \neg\neg B) \leftrightarrow (B \leftrightarrow (C \vee B))$.

Supplementary Problems

B.4. Write the following statement forms in Polish notation.

(a) $B \,\&\, \neg(A \vee B)$

(b) $[(A \to B) \to \neg(B \to A)] \leftrightarrow (A \leftrightarrow B)$

(c) $[(A \to B) \to (C \to D)] \to [E \to \{(D \to A) \to (C \to A)\}]$

B.5. Determine whether each of the following expressions is a statement form in Polish notation, and, if it is, find the corresponding statement form in our original notation.

(a) $\to \neg \vee AB \leftrightarrow C \,\&\, \neg AB$

(b) $\neg \,\&\, A \to B \to A$

(c) $\vee \vee \to \to AB \neg AA \,\&\, BA$

(d) $\leftrightarrow \to A \to BC \to \& ABC$

Appendix C

The Axiom of Choice Implies Zorn's Lemma

By the axiom of choice we mean the assertion that, for any set x, there is a function f (called a *choice function* for x), defined on $\mathcal{P}(x) \sim \{\emptyset\}$, such that, for any non-empty subset u of x, $f(u) \in u$.

We shall say that a collection $\boldsymbol{A}$ of sets is *well-ordered by inclusion* if and only if $\boldsymbol{A}$ is an $\subset$-chain and every non-empty subset $\boldsymbol{C}$ of $\boldsymbol{A}$ has a least element b (i.e. if $u \in \boldsymbol{C}$, then $b \subseteq u$). Given a collection $\boldsymbol{A}$ well-ordered by inclusion, and given any set y in $\boldsymbol{A}$, the *segment* determined by y (denoted $\mathrm{Seg}(\boldsymbol{A}, y)$) is defined to be the set of all x in $\boldsymbol{A}$ such that $x \subset y$. Notice that, if $\boldsymbol{S}$ is a *section* of $\boldsymbol{A}$ (i.e. if $\boldsymbol{S}$ is a subset of $\boldsymbol{A}$ such that $(y \in \boldsymbol{S} \;\&\; x \subseteq y) \rightarrow x \in \boldsymbol{S})$) and if $\boldsymbol{S} \neq \boldsymbol{A}$, then $\boldsymbol{S} = \mathrm{Seg}(\boldsymbol{A}, u)$, where u is the least element of $\boldsymbol{A} \sim \boldsymbol{S}$.

Theorem. The axiom of choice implies Zorn's Lemma.

Proof. Assume that a set $\boldsymbol{Z}$ of sets has the property that, for every $\subset$-chain $\boldsymbol{C}$ in $\boldsymbol{Z}$, the union $\bigcup_{A \in \boldsymbol{C}} A$ is also in $\boldsymbol{Z}$. Let $\boldsymbol{F}$ be a choice function for $\boldsymbol{Z}$. Thus if $\emptyset \neq \boldsymbol{D} \subseteq \boldsymbol{Z}$, then $F(\boldsymbol{D}) \in \boldsymbol{D}$.

Let us assume that $\boldsymbol{Z}$ has no $\subset$-maximal element. We shall derive a contradiction from this assumption.

For any y in $\boldsymbol{Z}$, the set $\boldsymbol{Y}$ of all elements x of $\boldsymbol{Z}$ such that $y \subset x$ is non-empty (since there are no $\subset$-maximal elements). Let $f(y) = F(\boldsymbol{Y})$. Thus for any y in $\boldsymbol{Z}$, $f(y)$ is an element of $\boldsymbol{Z}$ such that $y \subset f(y)$.

By a *ladder* we mean any subset $\boldsymbol{L}$ of $\boldsymbol{Z}$ such that $\boldsymbol{L}$ is well-ordered by inclusion, and, for any $x \in \boldsymbol{L}$, $f\big(\bigcup_{u \in \mathrm{Seg}(\boldsymbol{L},x)} u\big) = x$. (By hypothesis, we know that $\bigcup_{u \in \mathrm{Seg}(\boldsymbol{L},x)} u \in \boldsymbol{Z}$, since $\mathrm{Seg}(\boldsymbol{L}, x)$ is an $\subset$-chain.)

Let L be the set of all ladders.

(1) Given two different ladders $\boldsymbol{L}_1$ and $\boldsymbol{L}_2$, we shall show that one of them is a segment of the other. Let $\boldsymbol{K}$ be the set of all $u \in \boldsymbol{L}_1 \cap \boldsymbol{L}_2$ such that $\mathrm{Seg}(\boldsymbol{L}_1, u) = \mathrm{Seg}(\boldsymbol{L}_2, u)$. Clearly, $\boldsymbol{K}$ is a section of both $\boldsymbol{L}_1$ and $\boldsymbol{L}_2$. Hence if $\boldsymbol{K} = \boldsymbol{L}_1$ or $\boldsymbol{K} = \boldsymbol{L}_2$, then one of $\boldsymbol{L}_1$ or $\boldsymbol{L}_2$ is a segment of the other. Thus we must show that $\boldsymbol{K} \subset \boldsymbol{L}_1$ and $\boldsymbol{K} \subset \boldsymbol{L}_2$ do not simultaneously hold. To this end, assume $\boldsymbol{K} \subset \boldsymbol{L}_1$ and $\boldsymbol{K} \subset \boldsymbol{L}_2$. Let u_1 be the least element of $\boldsymbol{L}_1 \sim \boldsymbol{K}$, and let u_2 be the least element of $\boldsymbol{L}_2 \sim \boldsymbol{K}$. Then $\mathrm{Seg}(\boldsymbol{L}_1, u_1) = \boldsymbol{K}$ and $\mathrm{Seg}(\boldsymbol{L}_2, u_2) = \boldsymbol{K}$. By definition of *ladder*,

$$u_1 = f\left(\bigcup_{u \in \mathrm{Seg}(\boldsymbol{L}_1, u_1)} u\right) = f\left(\bigcup_{u \in \boldsymbol{K}} u\right)$$

and

$$u_2 = f\left(\bigcup_{u \in \mathrm{Seg}(\boldsymbol{L}_2, u_2)} u\right) = f\left(\bigcup_{u \in \boldsymbol{K}} u\right)$$

Hence $u_1 = u_2$ and $\text{Seg}\,(\boldsymbol{L}_1, u_1) = \boldsymbol{K} = \text{Seg}\,(\boldsymbol{L}_2, u_2)$. Thus $u_1 \in \boldsymbol{K}$, which contradicts the fact that $u_1 \in \boldsymbol{L}_1 \sim \boldsymbol{K}$.

(2) The union of all ladders $\boldsymbol{H} = \bigcup_{\boldsymbol{L}_1 \in L} \boldsymbol{L}_1$ is again a ladder. For, by (1) it is clear that $\boldsymbol{H}$ is an $\subset$-chain. $\boldsymbol{H}$ is well-ordered by inclusion. (In fact, if $\emptyset \neq \boldsymbol{W} \subseteq \boldsymbol{H}$ and $u \in \boldsymbol{W}$, then $u \in \boldsymbol{L}_1$ for some $\boldsymbol{L}_1 \in L$ and the least element of $\boldsymbol{W} \cap \boldsymbol{L}_1$ must be the least element of $\boldsymbol{W}$.) Finally, for any $x \in \boldsymbol{H}$, $x \in \boldsymbol{L}_1$ for some $\boldsymbol{L}_1 \in L$, and, by (1), $\text{Seg}\,(\boldsymbol{H}, x) = \text{Seg}\,(\boldsymbol{L}_1, x)$. Hence $x = f\Big(\bigcup_{u \in \text{Seg}(\boldsymbol{L}_1, x)} u\Big) = f\Big(\bigcup_{u \in \text{Seg}(\boldsymbol{H}, x)} u\Big)$.

Since $\boldsymbol{H}$ is an $\subset$-chain, the union $v = \bigcup_{u \in \boldsymbol{H}} u \in \boldsymbol{Z}$, by hypothesis. Hence $\boldsymbol{H} \cup \{f(v)\}$ is a ladder, and therefore $f(v) \in \boldsymbol{H}$.

It follows that $f(v) \subseteq v$, contradicting $v \subset f(v)$. ▸

Appendix D

A Lattice-Theoretic Proof of the Schröder-Bernstein Theorem

Lemma. Let $\langle L, \leq \rangle$ be a complete lattice and let ϕ be a function from L into L such that ϕ is order-preserving, i.e. $x \leq y \rightarrow \phi(x) \leq \phi(y)$. Then ϕ has a fixed point b in L, i.e. $\phi(b) = b$.

Proof. Let $W = \{x : x \in L \,\&\, x \leq \phi(x)\}$, and let $b = \bigvee_{x \in W} x$. We shall show that $\phi(b) = b$. First, $\phi(b)$ is an upper bound of W. For, if $x \in W$, $x \leq b$, and therefore $\phi(x) \leq \phi(b)$. But since $x \in W$, $x \leq \phi(x)$. Hence $x \leq \phi(b)$. Thus $\phi(b)$ is an upper bound of W, and so $b = \bigvee_{x \in W} x \leq \phi(b)$. On the other hand, since $b \leq \phi(b)$, it follows that $\phi(b) \leq \phi(\phi(b))$, i.e. $\phi(b) \in W$. Hence $\phi(b) \leq \bigvee_{x \in W} x = b$. From $b \leq \phi(b)$ & $\phi(b) \leq b$ we obtain $\phi(b) = b$. ▶

Schröder-Bernstein Theorem. If there is a one-one correspondence f between X and a subset of Y and a one-one correspondence g between Y and a subset of X, then there is a one-one correspondence between X and Y. (In terms of cardinal numbers $\mathfrak{m}$ and $\mathfrak{n}$, if $\mathfrak{m} \leq \mathfrak{n}$ and $\mathfrak{n} \leq \mathfrak{m}$, then $\mathfrak{m} = \mathfrak{n}$.)

Proof. For every subset $Z \subseteq X$, let $\phi(Z) = X \sim g[Y \sim f[Z]]$ (Fig. D-1). (Recall that, for any function h, $h[C] = \{h(u) : u \in C\}$.)

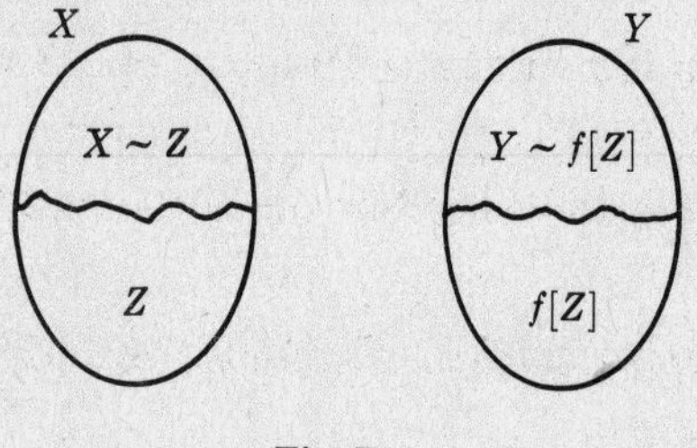

Fig. D-1

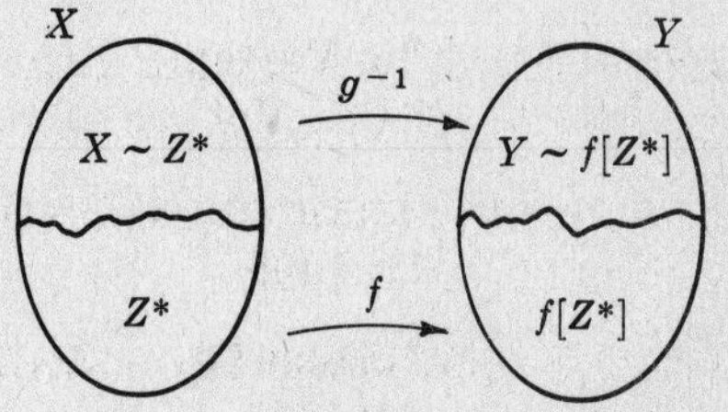

Fig. D-2

Now,

$$\begin{aligned} Z_1 \subseteq Z_2 &\rightarrow f[Z_1] \subseteq f[Z_2] \rightarrow Y \sim f[Z_2] \subseteq Y \sim f[Z_1] \\ &\rightarrow g[Y \sim f[Z_2]] \subseteq g[Y \sim f[Z_1]] \\ &\rightarrow \underbrace{X \sim g[Y \sim f[Z_1]]}_{\phi(Z_1)} \subseteq \underbrace{X \sim g[Y \sim f[Z_2]]}_{\phi(Z_2)} \end{aligned}$$

Hence ϕ is an order-preserving function from the complete lattice $\langle \mathcal{P}(X), \subseteq \rangle$ into itself. Hence by the lemma above, ϕ has a fixed point Z^*, i.e.

$$Z^* = \phi(Z^*) = X \sim g[Y \sim f[Z^*]]$$

Therefore

$$g[Y \sim f[Z^*]] = X \sim Z^*$$

It is easy now to verify, using Fig. D-2 as a guide, that the following function h,

$$h(x) = \begin{cases} f(x) & \text{if } x \in Z^* \\ g^{-1}(x) & \text{if } x \in X \sim Z^* \end{cases}$$

is a one-one correspondence between X and Y. ▶

Bibliography

The Bibliography is divided into two parts: the first on switching circuits, logic circuits and minimization, the second on Boolean algebras and related areas.

I. SWITCHING CIRCUITS, LOGIC CIRCUITS AND MINIMIZATION

1. Ádám, A.: *Truth Functions and the Problem of their Realization by Two-terminal Graphs,* Akadémiai Kiadó, Budapest, 1968.
2. Adefio, S. A., and C. F. Nolan: *Principles and Applications of Boolean Algebras for Electronic Engineers,* Iliffe Books, London, 1966.
3. Akers, S. B., Jr.: A Truth Table Method for the Synthesis of Combinational Logic Circuits, IRETEC,† EC-10, 604-615, 1961.
4. Ashenhurst, R. L.: The Theory of Abstract Two-terminal Switching Networks, Harvard Comp. Lab. BL-5, VII, 1-15, 1954.
5. Ashenhurst, R. L.: The Decomposition of Switching Functions, Proc. Int. Symp. Theory of Switching, 1957, Annals, Harvard Comp. Lab., 1959.
6. Beatson, T. J.: Minimization of Components in Electronic Switching Circuits, AIEE, I, Comm. and Electr., 77, 283-291, 1958.
7. Booth, T. M.: The Vertex-Frame Method for Obtaining Minimal Propositional-Letter Formulas, IRETEC, Vol. EC-11, 144-154, 1962.
8. Calabi, L.: A Solution of the Minimization Problem for Boolean Formulas, Parke Math. Lab. Rept. 7-3471, 1960.
9. Calabi, L.: Relations between Switching Networks and Boolean Formulas, Parke Math. Lab. Rept. 8-2471, 1961.
10. Calabi, L., and J. A. Riley: Inessentiality in Minimal Networks and Formulas, IRETEC, EC-11, 5, 711-713, 1962.
11. Calabi, L., and J. A. Riley: The Algebra of Boolean Formulas: Some Criteria for Minimality, Proc. Third Ann. Symp. Switching Circuit Theory & Logical Design, S-141, 33-47, 1962.
12. Calabi, L., and E. W. Samson: On The Theory of Boolean Formulas: Minimal Including Sums, I, *J. Soc. Industr. Appl. Math.,* 11, 212-234, 1963.
13. Caldwell, S.: *Switching Circuits and Logical Design,* Wiley, New York, 1958.
14. Chang, D. M. Y., and T. H. Mott, Jr.: Computing Irredundant Normal Forms from Abbreviated Presence Functions, IEEETEC, EC-14, 335-342, 1965.
15. Choudhury, A. K., and M. S. Basu: A Mechanized Chart for Simplification of Switching Functions, IRETEC, EC-11, 713-714, 1962.

---

†IRE stands for Institute of Radio Engineers, recently changed to Institute of Electrical and Electronics Engineers (IEEE). TEC stands for Transactions on Electronic Computers.

16. Chu, J. T.: A Generalization of a Theorem of Quine for Simplifying Truth Functions, IRETEC, EC-10, 165-168, 1961.

17. Cobham, A. R., R. Fridshal and J. North: Switching Circuit Theory & Logical Design, Proc. 2nd Ann. Symp. Switching Theory & Logical Design, Detroit, AIEE, New York, 1961.

18. Curtis, H. A.: *A New Approach to the Design of Switching Circuits*, Van Nostrand, Princeton, 1962.

19. Das, S. R., and A. K. Choudhury: Maxterm Type Expressions of Switching Functions and their Prime Implication, IEEETEC, EC-14, 920-923, 1965.

20. Dietmeyer, D. L., and J. R. Duley: Generating Prime Implicants via Ternary Encoding and Decimal Arithmetic, Comm. Assoc. Comp. Mach., 11, 520-523, 1968.

21. Dunham, B., and R. Fridshal: The Problem of Simplifying Logical Expressions, *J. Symb. Logic*, 24, 17-19, 1959.

22. Flegg, H. G.: *Boolean Algebra and its Applications*, Wiley, New York, 1964.

23. Flores, I.: *Logic of Computer Arithmetic*, Prentice-Hall, Englewood Cliffs, N. J., 1963.

24. Fridshal, R.: The Quine Algorithm, Summaries of Talks, Summer Inst. Symb. Logic, Cornell, 211-212, 1957.

25. Ford, L. R., and D. R. Fulkerson: *Flows in Networks*, Princeton Univ., Princeton, 1962.

26. Ghazala, M. J.: Irredundant Disjunctive and Conjunctive Normal Forms of a Boolean Function, *IBM J. Res. and Dev.*, 1, 171-176, 1957.

27. Gimpel, J. F.: A Reduction Technique for Prime Implicant Tables, Proc. Fifth Ann. Symp. Switching Circuit Theory & Logical Design, S-164, IEEE, 174-182, 1964.

28. Gimpel, J. F.: A Method of Producing a Boolean Function Having an Arbitrarily Prescribed Prime Implicant Table, IEEETEC, EC-14, 3, 485-488, 1965.

29. Gimpel, J. F.: A Reduction Technique for Prime Implicant Tables, IEEETEC, EC-14, 4, 535-541, 1965.

30. Hall, F. B.: Boolean Prime Implicants by the Binary Sieve Method, AIEE Trans. I, Comm. and Electr., 80, 709-713, 1962.

31. Hammer, P., and S. Rudeanu: *Boolean Methods in Operations Research*, Springer Verlag, New York, 1968.

32. Harris, B.: An Algorithm for Determining Minimal Representations of a Logic Function, IRETEC, EC-6, 2, 103-108, 1957.

33. Harrison, M. A.: *Introduction to Switching and Automata Theory*, McGraw-Hill, New York, 1965.

34. Hill, F. J., and G. R. Peterson: *Introduction to Switching Theory and Logical Design*, Wiley, New York, 1968.

35. Hockney, R.: An Intersection Algorithm Giving All Irredundant Forms from a Prime Implicant List, IEEETEC, EC-11, 2, 289-290, 1962.

36. Hoernes, G., and M. Heilweil: *Introduction to Boolean Algebra and Logic Design*, McGraw-Hill, New York, 1964.

37. Hohn, F. E.: *Applied Boolean Algebra*, Macmillan, New York, 1960.

38. Hohn, F. E., and L. R. Schissler: Boolean Matrices in the Design of Combinational Switching Circuits, *Bell System Tech. J.*, 34, 177-202, 1955.

39. Holst, P. A.: Bibliography on Switching Circuits & Logical Algebra, IRETEC, EC-10, 638-661, 1961.

40. House, R. W., and T. Rado: A Generalization of Nelson's Algorithm for Obtaining Prime Implicants, *J. Symb. Logic*, 30, 8-12, 1965.

41. Hu, S. T.: *Mathematical Theory of Switching Circuits & Automata*, Univ. of California Press, Berkeley, 1968.

42. Karnaugh, M.: The Map Method for Synthesis of Combinational Logic Circuits, Trans. AIEE, I, 72, 593-599, 1953.

43. Karp, R. M.: Functional Decomposition & Switching Circuit Design, *J. Soc. Ind. Appl. Math.*, 291-335, 1963.

44. Kautz, W. H.: A Survey & Assessment of Progress in Switching Theory & Logical Design in the Soviet Union, IEEETEC, EC-15, 164-204, 1966.

45. Keister, W. A., A. Ritchie and S. Washburn: *The Design of Switching Circuits*, Van Nostrand, Princeton, 1951.

46. Lawler, E. L.: An Approach to Multi-Level Boolean Minimization, *J. Assoc. Comp. Mach.*, 11, 283-295, 1964.

47. Lawler, E. L., and G. A. Salton: The Use of Parenthesis-Free Notation for the Automatic Design of Switching Circuits, IRETEC, EC-9, 342-352, 1960.

48. Luccio, F.: A Method for the Selection of Prime Implicants, IEEETEC, EC-15, 2, 205-212, 1966.

49. Lupanov, O. B.: On the Synthesis of Contact Networks, Doklady Akad. Nauk, USSR, 119, 1, 23-26, 1958 (Russian).

50. Lupanov, O. B.: On Asymptotic Estimates of Complexities of Formulas which Realize Functions of the Algebra of Logic, Ibid. 128, 3, 464-467, 1959 (Russian).

51. Lupanov, O. B.: On Realization of Functions of the Algebra of Logic Using Formulas of Limited Depth in the Basis of &, $\vee$, and $^{-}$, Ibid., 136, 5, 1041-1042, 1961 (Russian).

52. Maley, G. A.: Simplifying Switching Circuits with Boolean Algebra, *Electrotechnology*, 67, 101-106, 1961.

53. Marcus, M.: *Switching Circuits for Engineers*, Prentice-Hall, Englewood Cliffs, N. J., 1962.

54. McCluskey, E. J.: Minimization of Boolean Functions, *Bell System Tech. J.*, 35, 1417-1444, 1956.

55. McCluskey, E. J.: Minimal Sums for Boolean Functions Having Many Unspecified Fundamental Products, Proc. Second Ann. Symp. Switching Circuit Theory & Logical Design, AIEE, 10-17, 1961.

56. McCluskey, E. J.: *Introduction to the Theory of Switching Circuits*, McGraw-Hill, New York, 1965.

57. McCluskey, E. J., and T. C. Bartee: *A Survey of Switching Circuit Theory*, McGraw-Hill, New York, 1962.

58. Meo, A. R.: On the Determination of the *ps* Maximal Implicants of a Switching Function, IEEETEC, EC-14, 6, 830-840, 1965.

59. Mileto, F., and G. Putzolu: Average Values of Quantities Appearing in Boolean Function Minimization, IEEETEC, EC-13, 2, 87-92, 1964.

60. Mott, T. H., and C. C. Carroll: Numerical Procedures for Boolean Function Minimization, IEEETEC, EC-13, 4, 470, 1964.

61. Mott, T. H.: Determination of the Irredundant Normal Forms of a Truth Function by Iterated Consensus of the Prime Implicants, IRETEC, EC-9, 2, 245-252, 1960.

62. Moortgat, M. J.: Simplification of Boolean Functions, Gen. Electric Co. Tech. Information Series R64CD18, Falls Church, Va., 1964.

63. Muller, D. E.: Application of Boolean Algebra to Switching Circuit Design and to Error Detection, IRETEC, EC-3, 3, 6-12, 1954.

64. Muller, D. E.: Complexity in Electronic Switching Circuits, IRETEC, EC-5, 1, 15-19, 1956.

65. Nelson, R. J.: Weak Simplest Normal Truth Functions, *J. Symbolic Logic,* 20, 3, 232-234, 1955.

66. Nelson, R. J.: Simplest Normal Truth Functions, Ibid., 20, 2, 105-108, 1955.

67. Ninomiya, I.: A Study of the Structures of Boolean Functions and Applications to the Synthesis of Switching Circuits, Mem. Faculty Eng. Nagoya Univ., 13, 2, 149-363, 1961.

68. Peterson, W. W.: *Error-correcting Codes,* Wiley, New York, 1961.

69. Petrick, S. R.: A Direct Determination of the Irredundant Forms of a Boolean Function from the Set of Prime Implicants, Air Force Cambridge Res. Center Tech. Report 56-110, April 1956.

70. Phister, M.: *Logical Design of Digital Computers,* Wiley, New York, 1958.

71. Prather, R. E.: *Introduction to Switching Theory: A Mathematical Approach,* Allyn and Bacon, Boston, 1967.

72. Pyne, I. B., and E. J. McCluskey: An Essay on Prime Implicant Tables, *J. Soc. Industr. Appl. Math.,* 9, 604-631, 1961.

73. Pine, I. B., and E. J. McCluskey: The Reduction of Redundancy in Solving Prime Implicant Tables, IRETEC, EC-11, 4, 473-482, 1962.

74. Quine, W. V.: The Problem of Simplifying Truth Functions, *Amer. Math. Monthly,* 59, 521-531, 1952.

75. Quine, W. V.: A Way to Simplify Truth Functions, Ibid., 62, 627-631, 1955.

76. Quine, W. V.: On Cores and Prime Implicants of Truth Functions, Ibid., 66, 755-760, 1959.

77. Quine, W. V.: Two Theorems about Truth Functions, *Bol. Soc. Math. Mexicana,* 10, 64-70, 1953.

78. Richards, R. K.: *Arithmetic Operations in Digital Computers,* Van Nostrand, Princeton, 1955.

79. Roth, J. P.: Algebraic Topological Methods for the Synthesis of Switching Systems, I, Trans. Amer. Math. Soc., 88, 301-326, 1958.

80. Roth, J. P.: Algebraic Topological Methods in Synthesis, Proc. Int. Symp. on Theory of Switching, Harvard Univ., 1959.

81. Rudeanu, S.: Axiomatization of Certain Problems of Minimization, *Studia Logica,* 20, 37-61, 1967.

82. Samson, E. W., and B. E. Mills: Circuit Minimization: Algebra and Algorithm for New Boolean Canonical Expressions, Air Force Cambridge Res. Center Tech. Report, 54-21, 1954.

83. Samson, E. W., and R. Mueller: Circuit Minimization: Minimal and Irredundant Boolean Sums by Alternative Set Method, Air Force Cambridge Res. Center Tech. Report, 55-109, 1955.

84. Scheinman, A. H.: A Method for Simplifying Boolean Functions, *Bell System Tech. J.*, 41, 4, 1337-1346, 1962.

85. Schubert, E. J.: The Relative Merits of Minimization Techniques for Switching Circuits, IEEETEC, EC-12, 321-322, 1963.

86. Seshu, S., and M. B. Reed: *Linear Graphs and Electrical Networks,* Addison-Wesley, Reading, Mass., 1961.

87. Semon, W.: Synthesis of Series-parallel Network Switching Functions, *Bell System Tech. J.*, 37, 4, 877-898, 1958.

88. Semon, W.: Matrix Methods in the Theory of Switching, Proc. Int. Symp. Switching Theory, Harvard, 1959.

89. Shannon, C. E.: A Symbolic Analysis of Relay and Switching Circuits, Trans. Amer. Inst. Electr. Eng., 57, 713-723, 1938.

90. Shannon, C. E.: The Synthesis of Two-terminal Switching Circuits, *Bell System Tech. J.*, 28, 59-98, 1949.

91. Torng, H. C.: *Introduction to the Logical Design of Switching Systems,* Addison-Wesley, Reading, Mass., 1964.

92. Vasil'ev, Yu. L.: Minimal Contact Networks for Boolean Functions of Four Variables, Doklady Akad. Nauk, USSR, 127, 2, 2420245, 1959 (Russian).

93. Veitch, E. W.: A Chart Method for Simplifying Truth Functions, Proc. Assoc. Comp. Mach., 127-133, 1952.

94. Washburn, S. H.: An Application of Boolean Algebra to the Design of Electronic Switching Circuits, AIEE, I, Comm. & Electr., 72, 380-388, 1953.

95. Yablonskii, S. V.: The Algorithmic Difficulty of Synthesizing Minimal Switching Circuits, Problems of Cybernetics, 2, 401-457, 1961 (Translated from Russian).

96. Zacharov, B.: *Digital Systems Logic and Circuits,* American Elsevier, New York, 1968.

97. Žuravlev, Yu. I.: On Various Concepts of Minimality for Disjunctive Normal Forms, *Siberian Math. J.*, 1, 609-610, 1960 (Russian).

II. BOOLEAN ALGEBRA AND RELATED TOPICS

(Sikorski [1958] is the best and most authoritative text. Halmos [1963] is very stimulating.)

98. Birkhoff, G.: Lattice Theory, American Math Society Colloquium Publications XXV, New York, 1940, Second Edition, 1948, 1961.

99. Birkhoff, G., and G. D. Birkhoff: Distributive Postulates for Systems like Boolean Algebras, Trans. Amer. Math. Soc., 60, 3-11, 1946.

100. Birkhoff, G., and O. Frink: Representation of Lattices by Sets, Trans. Amer. Math. Soc. 64, 299-316, 1948.

101. Byrne, L.: Two Brief Formulations of Boolean Algebra, *Bull. Amer. Math. Soc.*, 52, 269-272, 1946.

102. Byrne, L.: Short Formulations of Boolean Algebra Using Ring Operations, *Canadian J. Math.*, 3, 31-33, 1951.

103. Chang, C. C.: On the Representation of α-complete Boolean Algebras, Trans. Amer. Math. Soc., 85, 208-218, 1957.

104. Chang, C. C.: Algebraic Analysis of Many-valued Logics, Trans. Amer. Math. Soc., 88, 467-490, 1958.

105. Cohn, P. M.: *Universal Algebra,* Harper & Row, New York, 1965.

106. Curry, H. B.: *Foundations of Mathematical Logic,* McGraw-Hill, New York, 1963.

107. Dilworth, R. P.: Lattices with Unique Complements, Trans. Amer. Math. Soc., 57, 123-154, 1954.

108. Dilworth, R. P.: Structure and Decomposition Theory of Lattices, Proc. Symp. Pure Math., Lattice Theory, 3-16, 1961.

109. Dunford, N., and M. H. Stone: On the Representation Theorem for Boolean Algebras, Rev. Ci. Lima, 43, 743-749, 1941.

110. Dwinger, P.: *Introduction to Boolean Algebra,* Würzburg, 1961.

111. Enomoto, S.: Boolean Algebras and Fields of Sets, *Osaka Math. J.*, 5, 99-115, 1953.

112. Frink, O.: Representations of Boolean Algebras, *Bull. Amer. Math. Soc.*, 47, 755-756, 1941.

113. Gardner, M.: *Logic Machines and Diagrams,* McGraw-Hill, New York, 1958.

114. Goodstein, R. L.: *Boolean Algebra,* Pergamon, Oxford, 1963.

115. Halmos, P. R.: *Algebraic Logic*, Chelsea, New York, 1962.

116. Halmos, P. R.: *Lectures on Boolean Algebras*, Van Nostrand, 1963.

117. Henkin, L.: *La Structure Algebrique des Theories Mathematiques*, Gauthier-Villars, Paris, 1956.

118. Henkin, L., and A. Tarski: Cylindric Algebras, Proc. Symp. Pure Math. II, Lattice Theory, 83-133, 1961.

119. Hermes, H.: *Einführung in die Verbandstheorie,* Springer Verlag, Berlin, 1955.

120. Horn, A., and A. Tarski: Measures in Boolean Algebras, Trans. Amer. Math. Soc., 64, 467-497, 1948.

121. Huntington, E. V.: Sets of Independent Postulates for the Algebra of Logic, Trans. Amer. Math. Soc., 5, 288-309 (1904).

122. Huntington, E. V.: New Sets of Independent Postulates for the Algebra of Logic, Trans. Amer. Math. Soc., 35, 274-304, 557-558, 971, 1933.

123. Jonsson, B.: A Boolean Algebra without Proper Automorphisms, Proc. Amer. Math. Soc., 2, 766-770, 1951.

124. Jonsson, B., and A. Tarski: Boolean Algebras with Operators, *Amer. J. of Math.*, (I) 73, 891-939, 1951; (II) 74, 127-162, 1952.

125. Karp, C. R.: A Note on the Representation of α-Complete Boolean Algebras, Proc. Amer. Math. Soc., 14, 705-707, 1963.

126. Kolmogorov, A. N.: *Foundations of the Theory of Probability,* Chelsea, New York, 1950.

127. Kripke, S.: An Extension of a Theorem of Gaifman-Hales-Solovay, Fund. Math., 61, 29-32, 1967.

128. Kuratowski, K., and A. Mostowski: *Set Theory,* North Holland, Amsterdam, 1968.

129. LeVeque, W. G.: *Topics in Number Theory,* Vol. I, Addison-Wesley, Reading, Mass., 1956.

130. Loomis, L. H.: On the Representation of σ-complete Boolean Algebras, *Bull. Amer. Math. Soc.,* 53, 757-760, 1947.

131. MacNeille, H.: Partially Ordered Sets, Trans. Amer. Math. Soc., 42, 416-460, 1937.

132. McKinsey, J. C. C., and A. Tarski: The Algebra of Topology, *Ann. of Math.,* 45, 141-191, 1944.

133. McKinsey, J. C. C., and A. Tarski: On Closed Elements in Closure Algebras, *Ann. of Math.,* 47, 122-162, 1946.

134. McKinsey, J. C. C., and A. Tarski: Some Theorems about the Sentential Calculi of Lewis and Heyting, *J. Symb. Logic,* 13, 1-15, 1948.

135. Mendelson, E.: *Introduction to Mathematical Logic,* Van Nostrand, Princeton, 1964.

136. Mostowski, A.: Abzählbare Boolesche Körper und ihre Anwendungen auf die allgemeine Metamathematik, Fund. Math., 29, 34-53, 1937.

137. Nachbin, L.: Une propriété caractèristique des algebres booleiennes, Portugal Math., 6, 115-118, 1947.

138. Neumann, J., von, and M. H. Stone: The Determination of Representative Elements in the Residual Classes of a Boolean Algebra, Fund. Math., 25, 353-376, 1935.

139. Newman, M. H. A.: A Characterization of Boolean Lattices and Rings, *J. London Math. Soc.,* 16, 256-272, 1941.

140. Pierce, R. S.: Distributivity in Boolean Algebras, *Pacific J. Math.,* 7, 983-992, 1957.

141. Pierce, R. S.: *Introduction to the Theory of Abstract Algebras,* Holt, New York, 1968.

142. Rasiowa, H., and R. Sikorski: The Mathematics of Metamathematics, Polish Academy of Sciences, Monografie Matematyczne, Warsaw, 1963.

143. Rieger, L.: *Algebraic Methods of Mathematical Logic,* Academic Press, New York, 1967.

144. Rosenbloom, P.: *The Elements of Mathematical Logic,* Dover, New York, 1950.

145. Rudeanu, S.: Boolean Equations and their Applications to the Study of Bridge Circuits, I, *Bull. Math. Soc. Math. Phys. R.P.R.,* 3, 447-473, 1959.

146. Scott, D.: The Independence of Certain Distributive Laws in Boolean Algebras, Trans. Amer. Math. Soc., 84, 258-261, 1957.

147. Sikorski, R.: On the Representation of Boolean Algebras as Fields of Sets, Fund. Math., 35, 247-258, 1948.

148. Sikorski, R.: *Boolean Algebras,* Springer Verlag, Berlin, 1958; 2nd Edition, 1964.

149. Smith, E. C., and A. Tarski: Higher Degrees of Distributivity and Completeness in Boolean Algebras, Trans. Amer. Math. Soc., 84, 230-257, 1957.

150. Stabler, E. R.: Boolean Representation Theory, *Amer. Math. Monthly,* 51, 129-132, 1944.

151. Stone, M. H.: The Theory of Representations for Boolean Algebras, Trans. Amer. Math. Soc., 40, 37-111, 1936.

152. Stone, M. H.: Applications of the Theory of Boolean Rings to General Topology, Trans. Amer. Math. Soc., 41, 321-364, 1937.

153. Stone, M. H.: The Representation of Boolean Algebras, *Bull. Amer. Math. Soc.*, 44, 807-816, 1938.

154. Szasz, G.: *Introduction to Lattice Theory,* Academic Press, New York, 1963.

155. Tarski, A.: Zur Grundlegung der Boole'schen Algebra I, Fund. Math. 24, 177-198, 1935.

156. Tarski, A.: Der Aussagenkalkül und die Topologie, Fund. Math. 31, 103-134, 1938.

157. Tarski, A.: *Cardinal Algebras,* Oxford Univ. Press, New York, 1949.

158. Tarski, A.: A Lattice-Theoretical Fixed-Point Theorem and its Applications, *Pacific J. Math.*, 5, 285-310, 1955.

INDEX

Index of Symbols and Abbreviations